Lectins
Analytical Technologies

T0348613

Lectins
Analytical Technologies

Edited by

Carol L. Nilsson

National High Magnetic Field Laboratory
Florida State University
Tallahassee, FL, USA

ELSEVIER

Amsterdam – Boston – Heidelberg – London – New York – Oxford – Paris
San Diego – San Francisco – Singapore – Sydney – Tokyo

Elsevier
Radarweg 29, PO Box 211, 1000 AE Amsterdam, The Netherlands
Linacre House, Jordan Hill, Oxford OX2 8DP, UK

First edition 2007

Library of Congress Cataloging-in-Publication Data
A catalog record for this book is available from the Library of Congress

British Library Cataloguing in Publication Data
A catalogue record for this book is available from the British Library

ISBN: 978-0-444-53077-6

For information on all Elsevier publications
visit our website at books.elsevier.com

Printed and bound in the United Kingdom

Transferred to Digital Print 2010

Working together to grow
libraries in developing countries

www.elsevier.com | www.bookaid.org | www.sabre.org

ELSEVIER   BOOK AID International   Sabre Foundation

Table of Contents

Preface

The intention with the production of this volume is to provide a toolbox of new and classical techniques for lectin and lectin–carbohydrate characterization. New lectins are still being discovered and new ways to exploit the specificity of lectin binding properties are being developed. The number of recent publications in which lectins are either described or used as biochemical tools points to a large and diverse group of scientific activities. Between January 1, 2004 and January 1, 2007, nearly 6,000 research articles that contained the search term "lectin" were posted in the PubMed database (http://www.ncbi.nlm.nih.gov/entrez/query.fcgi). During this short period, new analytical techniques have appeared and existing ones have been refined to meet the demands of higher sensitivity and expanding applications. It is expected that the number of citations will continue to rise along with the pace of future developments.

Because lectins are translators of the sugar code, they appear in important biological settings and can be extremely useful analytical tools in organism-wide studies (i.e. proteomics and glycomics). The trend towards miniaturized assays is also very strong. From structural characterization of lectins to technical aspects of glycoprotein enrichment, this volume offers an in-depth guide to essential modern methodologies. A handbook for the study of lectin structure and protein–carbohydrate energetics, use of lectins as analytical tools, and even the design of new lectins from other proteins is offered. It is designed with both the novice and advanced researcher in mind, who work in fields such as biology (including glycobiology), biomedicine, and analytical biology and analytical chemistry.

Because the study of lectins and their employment in analytical settings is quite diverse, a large number of techniques are presented by experts in the field. The beginning chapters offer up-to-date descriptions of classical techniques employed in lectinology, including X-ray crystallography, NMR spectroscopy, microcalorimetry, and surface plasmon resonance. The following chapters present a wide variety of useful methods, including the use of carbohydrate microarrays, lectin affinity chromatography, proteomic identification of adhesins, development of bioaffinity mass spectrometry probes, and frontal affinity chromatography, to name a few. Many of the chapters include practical, hands-on tips for the experimentalist. The introductory chapter includes references to web-based lectin resources, which have the advantage of changing quickly with the pace of new developments, unlike the printed word. Descriptions of many analytically useful lectins are provided in one appendix. Lastly, because abbreviations have become commonplace in most specialty areas, the reader is provided with an explanatory list at the end of the book for easy reference.

In conclusion, the future of lectin science appears extremely bright. The chapters that follow have been written by fellow enthusiasts in the field and provide invaluable information through their expertise. Without my co-authors' time and dedication, this book could not have been produced; therefore, I offer each contributor my sincere gratitude for joining me in this project.

Carol L. Nilsson
National High Magnetic Field Laboratory,
Florida State University, 1800 E.
Paul Dirac Dr., Tallahassee,
FL 32310, USA

List of Contributors

Julie T. Adamson — Department of Chemistry, University of Michigan, 930 North University Avenue, Ann Arbor, MI 48109-1055, USA

Ivan M. Belyanchikov — Shemyakin and Ovchinnikov Institute of Bioorganic Chemistry RAS, 117997, ul. Miklukho-Maklaya 16/10, Moscow, Russia

Nicolai V. Bovin — Shemyakin and Ovchinnikov Institute of Bioorganic Chemistry RAS, 117997, ul. Miklukho-Maklaya 16/10, Moscow, Russia

C. Fred Brewer — Department of Molecular Pharmacology, Albert Einstein College of Medicine, Bronx, New York 10461, USA

F. Javier Cañada — Department of Protein Science, Centro Investigaciones Biológicas – CSIC, Ramiro de Maeztu 9, 28040 Madrid, Spain

Angeles Canales — Department of Protein Science, Centro Investigaciones Biológicas – CSIC, Ramiro de Maeztu 9, 28040 Madrid, Spain

Elisabet Carlsohn — Proteomics Core Facility at Göteborg University, Box 435, SE 405 30 Göteborg, Sweden National High Magnetic Field Laboratory, Florida State University, 1800 E. Paul Dirac Dr., Tallahassee, FL 32310, USA

M. Isabel Chávez — Department of Protein Science, Centro Investigaciones Biológicas – CSIC, Ramiro de Maeztu 9, 28040 Madrid, Spain

Instituto de Quimica, Universidad Nacional Autónoma de México, Ciudad Universitaria, Ciudad, niversitaria, 04510 Coyoacán (México DF), Mexico

Caroline S. Chu — Chemistry Department, University of California, Davis, One Shields Avenue, Davis, CA 95616, USA

Tarun K. Dam   Department of Molecular Pharmacology, Albert
                           Einstein College of Medicine, Bronx, New York
                           10461, USA

Bengt Danielsson   Pure and Applied Biochemistry, Lund University,
                           SE-221 00 Lund, Sweden

Dolores Díaz   Department of Protein Science,
                           Centro Investigaciones Biológicas – CSIC,
                           Ramiro de Maeztu 9, 28040 Madrid, Spain

Jukka Finne   Department of Medical Biochemistry and
                           Molecular Biology, University of Turku,
                           Kiinamyllynkatu 10, FI-20520 Turku, Finland

Patrick Groves   Department of Protein Science,
                           Centro Investigaciones Biológicas – CSIC,
                           Ramiro de Maeztu 9, 28040 Madrid, Spain

Kristina Håkansson   Department of Chemistry, University of Michigan,
                           930 North University Avenue, Ann Arbor,
                           MI 48109-1055, USA

Jari Helin   Glykos Finland Ltd., Viikinkaari 6, 00790
                           Helsinki, Finland

Jun Hirabayashi   Research Center for Glycoscience, National
                           Institute of Advanced Industrial Science and
                           Technology (AIST), Central 2,1-1-1, Umezono,
                           Tsukuba, Ibaraki 305-8568, Japan

Stephen B. Hooser   Department of Comparative Pathobiology,
                           Purdue University, West Lafayette, IN, USA
                           Animal Disease Diagnostic Laboratory,
                           Purdue University, West Lafayette, IN, USA

Anne Imberty   CERMAV-CNRS, BP 53,
                           38041 Grenoble cedex 9,
                           France

Elina Jakobsson   Department of Medical Biochemistry and
                           Molecular Biology, University of Turku,
                           Kiinamyllynkatu 10, FI-20520 Turku, Finland

Jesús Jiménez-Barbero   Department of Protein Science,
                           Centro Investigaciones Biológicas – CSIC,
                           Ramiro de Maeztu 9, 28040 Madrid, Spain

Anne Jokilammi   Department of Medical Biochemistry and
                           Molecular Biology, University of Turku,
                           Kiinamyllynkatu 10, FI-20520 Turku, Finland

Kwanyoung Jung	Department of Chemistry, Purdue University, West Lafayette, IN, USA
Crystal Kirmiz	Chemistry Department, University of California, Davis, One Shields Avenue, Davis, CA 95616, USA
Junko Kominami	Research Center for Glycoscience, National Institute of Advanced Industrial Science and Technology (AIST), Central 2,1-1-1, Umezono, Tsukuba, Ibaraki 305-8568, Japan
	Fine Chemical & Foods Laboratories, J-Oil Mills, Inc., 11, Kagetoricho, Totsuka-ku, Yokohama 245-0064, Japan
Miikka Korja	Department of Medical Biochemistry and Molecular Biology, University of Turku, Kiinamyllynkatu 10, FI-20520 Turku, Finland
Elena I. Kovalenko	Shemyakin and Ovchinnikov Institute of Bioorganic Chemistry RAS, 117997, ul. Miklukho-Maklaya 16/10, Moscow, Russia
Ute Krengel	Department of Chemistry, University of Oslo, P.O. Box 1033 Blindern, N-0315 Oslo, Norway
Roger A. Laine	Department of Biological Sciences, Louisiana State University, Baton Rouge, LA 70803, USA Department of Chemistry, Louisiana State University and A&M College, Baton Rouge, LA 70803, USA Anomeric, Inc., 755 Delgado Dr., Baton Rouge, LA 70808, USA
Carlito B. Lebrilla	Chemistry Department, University of California, Davis, One Shields Avenue, Davis, CA 95616, USA
Bingcheng Lin	Dalian Institute of Chemical Physics, CAS, Dalian 116023, China
Jennifer W.-C. Lo	Department of Biological Sciences, Louisiana State University, Baton Rouge, LA 70803, USA Anomeric, Inc., 755 Delgado Dr., Baton Rouge, LA 70808, USA
Milan Madera	Department of Chemistry, Indiana University, Bloomington, IN 47405, USA
Xiuli Mao	Dalian Institute of Chemical Physics, CAS, Dalian 116023, China

Jana Masárová   Pure and Applied Biochemistry, Lund University, SE-221 00 Lund, Sweden Institute of Molecular Biology and Biotechnology, FORTH, GR-711 00 Heraklion, Greece

Yehia Mechref   Department of Chemistry, Indiana University, Bloomington, IN 47405, USA

Halina Miller-Podraza Institute of Biomedicine, Department of Medical Chemistry and Cell Biology, Göteborg University, P.O. Box 440, SE 405 30, Göteborg, Sweden

Sachiko Nakamura-Tsuruta Research Center for Glycoscience, National Institute of Advanced Industrial Science and Technology (AIST), Central 2,1-1-1, Umezono, Tsukuba, Ibaraki 305-8568, Japan

Jari Natunen   Glykos Finland Ltd., Viikinkaari 6, 00790 Helsinki, Finland

Ritva Niemelä   Glykos Finland Ltd., Viikinkaari 6, 00790 Helsinki, Finland

Carol L. Nilsson   National High Magnetic Field Laboratory, Florida State University, 1800 E. Paul Dirac Dr., Tallahassee, FL 32310, USA

                         Proteomics Core Facility at Göteborg University, Box 435, SE 405 30 Göteborg, Sweden

Milos V. Novotny   Department of Chemistry, Indiana University, Bloomington, IN 47405, USA

Małgorzata Palczewska Department of Molecular and Cellular Biology, National Center for Biotechnology, CNB-CSIC, Campus Cantoblanco, 28049 Madrid, Spain

Jianhua Qin   Dalian Institute of Chemical Physics, CAS, Dalian 116023, China

Eugenia M. Rapoport   Shemyakin and Ovchinnikov Institute of Bioorganic Chemistry RAS, 117997, ul. Miklukho-Maklaya 16/10, Moscow, Russia

Fred E. Regnier   Department of Chemistry, Purdue University, West Lafayette, IN, USA

Noboru Uchiyama   Research Center for Glycoscience, National Institute of Advanced Industrial Science and Technology (AIST), Central 2,1-1-1, Umezono, Tsukuba, Ibaraki 305-8568, Japan

Denong Wang	Carbohydrate Microarray Laboratory, Department of Genetics, and Neurology and Neurological Sciences, Stanford University School of Medicine, Beckman Center B007, Stanford, CA 94305-5318, USA
Krista Weikkolainen	Department of Biological and Environmental Sciences, Faculty of Biosciences, University of Helsinki, Viikinkaari 5, 00790 Helsinki, Finland
Christina R. Wilson	Department of Comparative Pathobiology, Purdue University, West Lafayette, IN, USA Animal Disease Diagnostic Laboratory, Purdue University, West Lafayette, IN, USA
Fredrik Winquist	Applied Physics, Linköping University, SE-581 83 Linköping, Sweden
Albert M. Wu	Glyco-Immunochemistry Research Laboratory, Institute of Molecular and Cellular Biology, College of Medicine, Chang-Gung University, Kwei-san, Tao-yuan, 333, Taiwan
Betty C.-R. Zhu	Department of Biological Sciences, Louisiana State University, Baton Rouge, LA 70803, USA

Lectins: Analytical Technologies
C.L. Nilsson (Editor)

Chapter 1

Lectins: Analytical Tools from Nature

Carol L. Nilsson

National High Magnetic Field Laboratory, Florida State University, 1800 E. Paul Dirac Dr., Tallahassee, FL 32310, USA

1. Introduction

Lectins are proteins of non-immune origin that recognize and bind to specific carbohydrate structural epitopes without modifying them. This group of carbohydrate-binding proteins function as central mediators of information transfer in biological systems and perform their duties by interacting with glycoproteins, glycolipids and oligosaccharides. Whether extracted from natural sources or expressed in cell cultures, lectins provide models for the study of protein–carbohydrate interactions and exquisite tools for the analysis of carbohydrates, in either free form or bound to lipids or proteins. Also, because of their presence at carbohydrate recognition events, lectins may be therapeutic targets or may be used to deliver drugs to their site of action [1].

1.1. Lectin functions

Because of the pivotal role that lectins play in many life processes, they are ubiquitous. Lectins have been identified in microorganisms, animals and plants. It could be safely postulated that only a small fraction of the total number of lectins that exist in the natural world have yet been identified. Lectins can be identified based on functional assays (for instance, hemagglutination) or by amino acid sequence homology (putative lectins) with known lectin sequences. At the time of writing this chapter, a simple search of the Swiss-Prot and TrEMBL protein databases (http://ca.expasy.org) yielded about 2,200 amino acid sequences; however, the primary structure of all known lectins has not yet been determined and many new ones are yet to be discovered. Only a small fraction of the lectins that have been discovered to date have been carefully characterized with respect to protein structure, binding affinities for extended carbohydrates, binding thermodynamics and other properties.

The most fully characterized group of lectins are those from the plant kingdom because they are frequently hydrophilic and produced in large amounts, such as is the case with the seed lectins. However, lectins are found in many cellular layers of a large number of organisms, and their localization reflects their diversity of function. Intracellular lectins are involved in protein trafficking. Membrane-bound lectins mediate microbial adhesion, lymphocyte homing and cell–cell recognition. Secreted plant lectins may be highly toxic, such as in the case of ricin; in contrast, human galectin-1 has both intra and extracellular functions and is associated with the invasive potential of malignant brain tumors (*glioblastoma multiforme*) [2, 3]. Some cell nuclei are known to stain positively for galectin as well, although the role of intranuclear galectin is not yet fully understood [4]. One plant lectin, the tobacco agglutinin (Nictaba), is synthesized in the cytoplasm and is partly translocated into the nucleus [5]. The lectin has a high affinity for high mannose and complex *N*-glycans, but it is unclear how this binding profile is related to the nuclear compartment.

One well-known function associated with intracellular lectins in animals regards the sorting of *N*-linked glycoproteins (for review, see Yamashita et al.) [6]. During *N*-linked glycoprotein synthesis, $Glc_3Man_9GlcNAc_2$-dolicholpyrophosphate may be covalently attached to asparagine residues within the consensus sequence Asn-X-Ser/Thr (Cys) of proteins in the endoplasmic reticulum (ER). Extensive processing of the glycan follows this step and occurs in both the ER and Golgi apparatus. After synthesis and modification are completed, glycoproteins are sorted into lysosomes, secretory vesicles or the plasma membrane. Lectins associated with *N*-linked glycoprotein sorting include calnexin, calreticulin, ER–Golgi intermediate compartment (ERGIC)-53 and mannose-6-phosphate receptors.

Some lectins have putative defense functions for the parent organism. For instance, plant lectins can possess insecticidal activity [7, 8]. In the animal kingdom, some lectins have the ability to recognize molecules that are "non-self" in origin and are thus a component of innate immunity. One lectin from the sea cucumber (CEL-III) has been demonstrated to both agglutinate and lyse red blood cells [9]. Horseshoe crabs, a phylogenetically ancient family of arthropods, produce tachylectins (TL) that recognize surface saccharides on pathogens [10]. Many crab lectins have specificity toward acetyl groups. TL-1, -3 and -4 recognize sugar moieties on bacterial LPS, whereas TL-2 binds to GlcNAc or GalNAc and recognizes lipoteichoic acids. In mammals, collectins and ficolins fulfill similar functions, binding to oligosaccharide structures on the surface of microorganisms, which in turn trigger complement activation and phagocytosis [11]. Collectins and ficolins are oligomeric structures that possess an N-terminal Cys-containing segment, a rigid middle collagen-like domain and C-terminal carbohydrate-recognizing domains (CRDs). Mannan-binding lectin recognizes several Gram-negative bacteria such as *Staphylococcus aureus* and *Escherichia coli*, fungi such as *Aspergillus fumigatus* and viruses such as human immunodeficiency virus (HIV). L-ficolin binds to surface saccharides of the enteric pathogen *Salmonella typhimurium*.

The TL, collectins and ficolins function as defense proteins against microbial invasion, but pathogens themselves also express a wide array of lectins that recognize host glycoconjugates and aid in cellular adhesion and invasion. Microbial lectins are thus of high interest from the perspective of vaccine and anti-adhesion therapies. The human gastric pathogen *Helicobacter pylori* has several known carbohydrate-binding affinities, and at least two adhesins have been identified in the *H. pylori* genome, the Lewis[b]-binding and sialic acid binding adhesins [12–14]. Hemagglutinin (HA) is a protein on the surface of the flu virus that recognizes and binds to exposed sialic acids on host cells. This viral pathogen is estimated to have caused about 50 million deaths during the global pandemic of 1918. The appearance of bird flu strains that have the ability to infect and kill humans has led to a resurgence of interest in understanding how small changes in the sialic acid binding pocket can shift the tropism of the virus from avian (α2, 3-linked sialic acids) to human (α2, 6-linked sialic acids) tissues, one prerequisite for the emergence of a new flu pandemic. Crystal structures of HA from the 1918 virus have revealed that structural features could be identified in avian HAs, that only two structural changes separate avian from human tropism and that HA from the highly pathogenic flu strain H5N1 closely resembles the HA from the 1918 flu strain [15, 16].

1.2. Known lectins can be discovered in new places

Previously characterized lectins are being assigned new biological functions because of the increasing frequency of studies that employ global differential genomic or proteomic monitoring techniques. This is not surprising in itself, but emphasizes the increasing need for scientists who use functional genomic assays to learn more about lectins. Some recent examples of new roles for galectins and selectins are given below.

Galectins are a family of animal lectins that bind β-galactoside epitopes on extracellular components such as laminin and fibronectin or the exposed carbohydrates of gangliosides. Zebrafish may express a more limited number of galectins than higher animals, but this organism has not yet been fully investigated. Zebrafish have been used as a model for studying the role of galectin in early embryogenesis [4]. It has been established that the protogalectin Drgal1-L2 expression is specific to the notochord of the developing fish embryo. Recently, Drgal1-L2 surfaced in another setting, as a promoter of proliferation of Muller glial stem cells in a cDNA microarray study of regeneration of the zebrafish retina [17]. Human galectin-1 overexpression has been implicated in connection with progression in at least a dozen different types of malignant tumors [2]. In a proteomic investigation of glioblastoma cells treated with wild-type p53 and cytotoxic chemotherapy that was initiated by a neuro-oncologist, a clear association between galectin-1 and p53 expression was demonstrated (Fig. 1) [3].

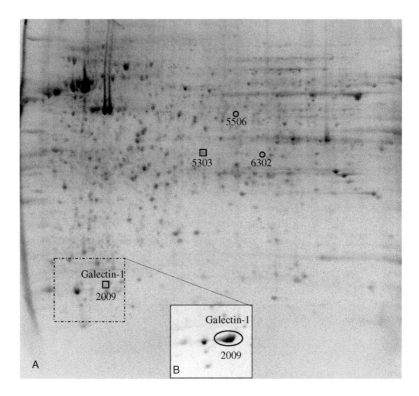

Figure 1. Photograph of a 2D gel of human glioblastoma cells treated with an adenovirus vector carrying wild-type p53 prior to topoisomerase (SN-38) treatment (panel A) compared to a 2D gel from cells treated with an empty adenovirus vector and then SN-38 (panel B). Galectin-1 is greatly up-regulated in the latter cells, which are resistant to apoptosis [3].

One class of lectins, the selectins, are transmembrane glycoproteins that recognize a subset of sialyl-Le^x-containing carbohydrate antigens [18]. L-selectins, which bind lymphocytes to high endothelial venules during inflammation, have been extensively studied in the vascular system. Recently, L-selectin was identified unexpectedly in cytotrophoblast (CTB) cells as a key protein in the establishment of human pregnancy (Fig. 2) [19].

2. A Short History of Lectins

The term "lectin" was first coined in 1954 by William C. Boyd of Boston University in order to describe agglutinins of plant origin that were blood group specific. The word lectin is derived from the past participle of the Latin verb *legere*, meaning to select, gather, or read. Selectivity of lectins and analytical

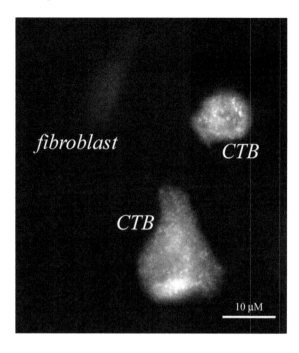

Figure 2. A new place for an "old" lectin: human trophoblast adhesion is mediated by L-selectin [19]. The image shows primary cytotrophoblasts (CTB) stained for L-selectin. The fibroblast, which does not express L-selectin, served as a negative control. Reprinted from Ref. [20] with permission.

technologies are inextricably linked to one another. Following the naming of lectins, it became apparent that proteins from sources other than plants could also bind to carbohydrates, and the term lectin gained more broad usage. It also became increasingly clear that the functions of lectins had been observed many years prior to their identification as a separate group of proteins. In 1974, a paper was published that described the asialoglycoprotein receptor of the liver [21]. The authors claimed at the time that their work described the first lectin of mammalian origin. However, Charcot and Robin had observed inclusions (Charcot–Leyden crystals) in some diseased human tissues as early as 1853 [22]. The crystals were later demonstrated to comprise a nearly pure form of a protein that had the ability to bind carbohydrate, and was named galectin-10 [23]. Thus, in this instance, the lectin protein was discovered prior to the elucidation of its carbohydrate-binding function. Before 1860, S. Weir Mitchell observed that certain snake venoms had the ability to clump together (agglutinate) red blood cells [24]. Later in the same century, Hermann Stillmark presented a medical doctoral dissertation in the country now known as Estonia, which described the agglutination properties of ricin, a highly toxic protein isolated from the seeds of the castor bean, *Ricinus communis* [25].

The identification and characterization of lectins in the latter part of the 1900s accelerated thanks to the early work of scientists in the field such as Nathan Sharon and colleagues, and many others. Sharon and Lis performed early, ground-breaking studies of plant lectins such as soybean agglutinin, peanut agglutinin and the seed lectins of the coral tree (*Erythrina*) [26] and have published a large number of widely read scientific articles and textbooks on the subject. Since those early days, lectins have been discovered and characterized from a large number of plants, animals and microorganisms.

One of the first analytical uses of lectins was their application in the elucidation of the ABH histo-blood group antigens [27]. A large number of blood group binding plant lectins have been identified, but animal lectins are also useful. For instance, a blood group A-specific lectin was discovered in a bivalve (*Saxidomus*), snails (*Helix*), and the hemolymph of horseshoe crabs (*Limulus*). Blood group B-specific lectins can be found in crabs (*Scylla* and *Charybdis*). The lectin of the eel *Anguilla anguilla* played an important role in the demonstration of fucose as a feature of the blood group O antigen [28]. Today, the relative affinities of lectins are often studied by analytical methods such as inhibition of hemagglutination and precipitation. These techniques are still widely used for lectin characterization; however, the methods have been described extensively in other textbooks [29] and thus are not included in this volume.

3. Lectin Structure and Function

Lectins may be classified according to either structural or functional characteristics. A detailed overview of structural determination of lectins is given in the next chapter and includes excellent figures to illustrate some of the different lectin structures. The first lectin structures determined were from the plant kingdom; the first animal lectin structure, a galectin, was published in 1993 [30]. So far, more than a dozen structural families of animal lectins have been discovered. Until recently, only seven plant lectin families were known. However, two new families of lectins could be assigned recently to plants, a mannose-binding lectin in the liverwort *Marchantia polymorpha* that resembles lectins from the fungus *Agaricus bisporus* [31], and a lectin from the bark of the black locust (RobpsCRA, *Robinia pseudoacacia*) that shares high sequence identity with class V chitinases but lacks chitinase activity [32]. It remains clear that the plant kingdom has not been fully investigated with respect to lectins.

There are a number of secondary structures, canonical protein folds, which can be found in plant and animal lectins, recently reviewed by Remy Loris [33]. These can be identified in proteins from closely related or phylogenetically different organisms and may lack amino acid sequence homology; the latter suggests convergent evolution. One example is the legume lectin fold, an antiparallel β-sandwich, which can be identified in galectins and pentraxins. The β-trefoil fold contains three each of four-stranded antiparallel sheets, which together form a globular structure. The β-trefoil fold may be found not only in ricin (plant lectin) but also in the mannose receptor (animal). The hevein domain is a short disulfide-rich amino acid sequence that can be often found in plant lectins, but was also determined in a cardiotoxin from cobra venom [34]. A complete list of lectin structures classified by organism (plant, animal, bacterial, viral and fungal) and fold is provided in Table 1 of Chapter 2.

The affinity of single lectin carbohydrate-binding domains may be low, but lectin avidity can be increased through multivalency. A large number of lectins form multimeric aggregates (dimers, trimers, tetramers, pentamers) and this increases their ability to maintain contact with their target oligosaccharide (avidity). Many lectins require the presence of divalent metal ions in order to maintain their binding ability. The C-type lectins depend on calcium ions to perform their functions and yet others may require manganese.

One family of lectins, the I-type lectins, is defined by the overall structural similarity to immunoglobulins [35]. It is not yet known whether all of the I-type lectins share an in common ancestral gene or if they also display evolutionary convergence. Many of the I-type lectins show affinity for sialic acids, such as the siglecs (sialic acid binding immunoglobulin superfamily lectins) [36], CD83, and cell adhesion molecule L1.

Lectins can also be assigned to functional groups, but members of the same functional group sometimes show little structural homology. Galectins are a conserved protein family that share a consensus sequence of around 130 amino acids and a carbohydrate domain with affinity for *N*-acetyllactosamine [37]. Many galectins are homodimers but some are active also in monomeric form (galectins-5,-7,-10). The selectins are another functional group of lectins that recognize a subset of sialyl-Le^x-containing carbohydrate antigens [18]. The L, E and P selectins bind to carbohydrates on lymph node vessels (L), endothelium (E) or activated blood platelets (P) and perform important intermediate functions in diseases such as inflammation that make them interesting as drug targets.

4. Lectins as Analytical Tools

The discovery of new lectins in biological systems means that reliable techniques for determining lectin properties are needed. Because of the complex nature of protein–carbohydrate interactions, no single technique can provide universal characterization. In this volume, a large number of analytical techniques are presented that can be applied to the characterization of new or existing lectins, or the use of lectins as analytical tools. X-ray crystallography of purified lectins in complex with saccharides can provide high-resolution structural data and a visual tool to probe protein–carbohydrate interactions (Chapter 2). Although NMR of lectin–carbohydrate complexes can be challenging, this technique has been proven to be a useful alternative technique for three-dimensional structure determination (Chapter 3). Further characterization of the specificity and energetics of the binding site in solution can be accomplished elegantly with isothermal calorimetry, as described in Chapter 4. Surface plasmon resonance can be employed as a biosensor of lectin–saccharide interactions (Chapter 5). Several new high-throughput techniques have been described recently that also may help to describe lectin-binding characteristics. Carbohydrate microarrays are a relatively new tool that will shed light on the nature of glyco-epitopes (Chapter 7); their use is rapidly expanding for characterization of lectin-binding affinities. The interactions between lectins and glycans in a chemical library can be measured in a systematic and quantitative manner by frontal affinity chromatography (FAC, Chapter 10). Another promising and relatively simple analytical tool to study lectin–carbohydrate interactions is fluorescence polarization [38]. One newer method that requires a high-resolution mass spectrometer can also provide binding data [39, 40].

In the post-genomic era, there is a need to integrate experimental data from the levels of proteome, glycome and metabolome in order to fully understand biological and pathological processes. At each of these levels, lectins can play crucial analytical roles, due to their carbohydrate specificities. Previously uncharted lectins may be discovered in tissues by the use of glycoprobes (Chapter 17). In proteomics, prefractionation of complex samples is often necessary in order to probe the depth of the proteome. Lectin affinity methods can be employed to profile changes in protein glycosylation (Chapters 8, 9 and 11), and several new methodologies have been developed that increase the sensitivity of analysis.

Lectins can be used to construct bioaffinity probes for oligosaccharides, using sensitive mass spectrometric techniques for detection, as in the study of binding partners of cortical granule lectin (*Xenopus laevis*, described in Chapter 13). Pathogenic microorganisms often express lectins, also known as adhesins, in their outer membranes that can bind to host glycoconjugates. Because adhesins are important virulence factors and putative drug and vaccine targets, methods have been developed to identify the proteins using proteomic techniques (Chapter 12) and use glycolipid libraries to define adhesin-binding characteristics (Chapter 6). In the area of glycomics, lectins play an important role because they recognize the sugar code. One exciting new development is the ability to design new lectins from enzymes (Chapters 15 and 16).

5. Lectin Resources

There are a wide variety of resources of lectin information available, in both electronic (Table 1) and paper formats. One definitive textbook is the 2nd edition of *Lectins* [41] by Nathan Sharon and Halina Lis. This book gives an excellent

Table 1. Currently, a large number of databases with useful information for lectin scientists can be accessed through the Internet.

Database	Web Address
Bacterial carbohydrate structural database	http://www.glyco.ac.ru/bcsdb
Carbohydrate-active enzymes	http://www.cazy.org/CAZY/
Consortium for Functional Glycomics	http://www.functionalglycomics.org/static/consortium
Genomic Resource for Animal Lectins	http://www.imperial.ac.uk/research/animallectins
Glyco Forum	http://www.glycoforum.gr.jp/science/word/lectin/LE_E.html
Glyco 3D	http://www.cermav.cnrs.fr/glyco3d/index.php
Lectins and food	http://www.ansci.cornell.edu/plants/toxicagents/lectins/lectins.html
Pathogen–sugar binding database	http://sugarbinddb.mitre.org
Plant lectin database	http://nscdb.bic.physics.iisc.ernet.in/lectindb/abtlectindb.html
U.K. Glycoarrays Consortium	http://www.glycochips.org.uk

overview of lectin research, where lectins are found, and their properties, biosynthesis and genetics. Definitive reviews of animal lectins were published in 1999 in a special issue of *Biochimica et Biophysica Acta*, volume 1473, edited by Hans-Joachim Gabius. David Kilpatrick's *Handbook of Animal Lectins: Properties and Biomedical Applications* was published the following year [42]. The corresponding reference for plant lectins is *Handbook of Plant Lectins: Properties and Biomedical Applications* [43]. Volumes 362/363 of *Methods in Enzymology*, published in 2003, contain several reviews of lectin analytical techniques.

The recent establishment of online resources (Table 1) is of great benefit to scientists because the information is continuously updated. Glyco Forum, developed by Jun Hirabayashi, is a good reference and introductory tool. It contains descriptions of a broad range of lectin types, including galectins, legume lectins, collectins and TL. Genomic, proteomic and glycomic data can be found as well. The Genomics Resource for Animal Lectins is a separate website developed by

Kurt Drickamer and contains two parts, structures and functions of animal lectins and C-type lectin domains.

A plant lectin database (Lectin DB) contains statistics that were compiled from protein databases for approximately 3,600 lectins regarding structural and functional annotation. Information regarding which lectins are known mitogens, can be purchased as conjugates or have commercial antibodies directed against them can also be found at this website. The most comprehensive structural database is Glyco3D, which contains structures, resolution of measurements and references for more than 550 lectins, many in complex with oligosaccharides. This database, developed by Anne Imberty and colleagues, is discussed further in the next chapter of this book.

For the lectin scientist interested in microbial lectins and adhesins, a pathogen–sugar binding database has been established by Elaine Mullen. Extensive literature searches were performed in order to build this resource. Lists of known carbohydrate sequences to which pathogens adhere can be searched by pathogen/toxin name and/or saccharide sequence. Over 8,000 carbohydrate structures from Gram-negative and many Gram-positive bacteria are stored in a separate database (glyco.ac.ru/bcsdb). The database can be searched by microorganism, glyco-epitope or bibliography. A search interface for NMR data is being developed as well.

In some instances, such as the carbohydrate-binding resource at the Consortium for Functional Glycomics (CFG), lectin data is only a mouse click away from a live literature search. This site, still under development, already provides a well-integrated presentation of animal lectin data from public data. The CFG has received a sizable grant from the National Institute of General Medical Sciences to enhance the availability of existing glycoresources and specialty databases. General information regarding lectins is provided and has been written by well-recognized experts in the field such as Kurt Drickamer. Data provided includes specificities of glycan-binding proteins, cell-type expression and glycans recognized by lectins and lectin structures. In addition to lectin resources, the CFG processes a large number of requests annually for glycan array screening and profiling of *N*- and *O*-linked glycans, which are lectin ligands. A European resource, U.K. Glycoarrays Consortium, is currently under development as well. This resource will provide carbohydrate arrays, carbohydrate libraries, expression of lectins and the analysis of their binding characteristics.

There are a large number of commercial vendors that sell essential analytical reagents for lectin research, such as conjugated and native lectins and oligosaccharides. A partial list of 15 companies and their web addresses is provided in Table 2. The identity of the companies was revealed by a search of the World Wide Web and thus does not include lectin vendors that do not have an internet site.

The future of lectin research appears exceedingly bright, judging by the surge in publications related to lectins and protein–carbohydrate interactions. New techniques continue to be developed in order to study lectins and to utilize their

Table 2. Some commercial vendors of lectin products.

Company	Web Address
Associates of Cape Cod, Incorporated	http://www.acciusa.com
Biomeda	http://biomeda.com/si_l/lectins.htm?s=1142968399-628203483
Biotinylated Lectins	http://www.galab.de/technologies/products/biotinylated_lectins.html
EMD Biosciences	http://www.emdbiosciences.com/g.asp?f=CBC/home.html
Europa Bioproducts	http://www.europa-bioproducts.com/catalog/39
EY Laboratories	http://www.eylabs.com
Glycorex	http://finechem.glycorex.se
GlycoTech	http://www.glycotech.com
Glygen Corporation	http://glygen.com
Invitrogen	http://probes.invitrogen.com/handbook
Prozyme	http://www.prozyme.com
Sigma-Aldrich	http://www.sigmaaldrich.com
U.S. Biologicals	http://www.usbio.net/Category.aspx?catName=Lectins
Vector Laboratories	http://www.vectorlabs.com/frames/FRLectins.htm
Worthington Biochemical Company	http://www.worthington-biochem.com

specificities in functional assays. It is hoped that the methods presented in this volume will provide both new ideas and a toolbox in a single reference work for both the beginner and experienced lectin scientists.

Acknowledgments

The support of the NSF National High-Field FT-ICR Mass Spectrometry Facility (DMR 0084173) is gratefully acknowledged.

References

[1] C. Bies, C.-M. Lehr and J.F. Woodley, Adv. Drug Deliv. Rev., 56 (2004) 425–435.

[2] I. Camby, M. Le Mercier, F. Lefranc and R. Kiss, Glycobiology, 16 (2006) 137R–157R.

[3] M. Puchades, C.L. Nilsson, M.R. Emmett, K.D. Aldape, Y. Ji, F.F. Lang, T.J. Liu and C.A. Conrad, J. Proteome Res., (2006), published on the Web December 21, 2006.

[4] H. Ahmed, S.J. Du, N. O'Leary and G.R. Vasta, Glycobiology, 14 (2004) 219–232.

[5] N. Lannoo, E. Van Pamel, R. Alvarez, W.J. Peumans and E.J.M. Van Damme, FEBS Lett., 580 (2006) 6329–6337.

[6] K. Yamashita, S. Hara-Kuge and T. Okhura, Biochim. Biophys. Acta, 1473 (1999) 147–160.

[7] I.M. Vasconcelos and J.T. Oliveira, Toxicon, 44 (2004) 385–403.

[8] K.D. Cox, D.R. Layne, R. Scorza and G. Schnabel, Planta, 224 (2006) 1373–1383.

[9] T. Hatekeyama, H. Nagatomo and N. Yamasaki, J. Biol. Chem., 270 (1995) 3560–3564.

[10] S. Kawabata and R. Tsuda, Biochim. Biophys. Acta, 1572 (2002) 414–421.

[11] J. Lu, C. Teh, U. Kishore and K.B.M. Reid, Biochim. Biophys. Acta, 1572 (2002) 387–400.

[12] D. Ilver, A. Arnqvist, J. Ogren, I.M. Frick, D. Kersulyte, E.T. Incecik, D.E. Berg, A. Covacci, L. Engstrand and T. Boren, Science, 279 (1998) 373–377.

[13] T. Larsson, J. Bergström, C. Nilsson and K.A. Karlsson, FEBS Lett., 469 (2000) 155–158.

[14] J. Mahdavi, B. Sonden, M. Hurtig, F.O. Olfat, L. Forsberg, N. Roche, J. Angstrom, T. Larsson, S. Teneberg, K.A. Karlsson, S. Altraja, T. Wadstrom, D. Kersulyte, D.E. Berg, A. Dubois, C. Petersson, K.E. Magnusson, T. Norberg, F. Lindh, B.B. Lundskog, A. Arnqvist, L. Hammarstrom and T. Boren, Science, 297 (2002) 573–578.

[15] J. Stevens, O. Blixt, L. Glaser, J.K. Taubenberger, P. Palese, J.C. Paulson and I.A. Wilson, J. Mol. Biol., 355 (2006) 1143–1155.

[16] J. Stevens, O. Blixt, T.M. Tumpey, J.K. Taubenberger, J.C. Paulson and I.A. Wilson, Science, 312 (2006) 404–410.

[17] S.E. Craig, A.A. Calinescu and P. Hitchcock, in Neuroscience 2006, Society for Neuroscience, Atlanta, Georgia, 2006.

[18] J.B. Lowe, Biochem. Soc. Symp., 69 (2002) 33–45.

[19] O.D. Genbacev, A. Prakobphol, R.A. Foulk, A.R. Krtolica, D. Ilic, M.S. Singer, Z.Q. Yang, L.L. Kiessling, S.D. Rosen and S.J. Fisher, Science, 299 (2003) 405–408.

[20] C.L. Nilsson, Anal. Chem., 75 (2003), 348A–353A.

[21] R.J. Stockert, A.G. Morell and I.H. Scheinberg, Science, 186 (1974) 365–366.

[22] J.M. Charcot and C. Robin, C. R. Mem. Soc. Biol., 5 (1853) 44–50.

[23] G.J. Swaminathan, D.D. Leonidas, M.P. Savage, S.J. Ackerman and K.R. Acharya, Biochemistry, 38 (1999) 13837–13843.

[24] D.C. Kilpatrick, Biochim. Biophys. Acta, 1572 (2002) 187–197.

[25] H. Stillmark, Doctoral Dissertation University of Dorpat, Dorpat (Tartu), 1888.

[26] N. Sharon, Protein Sci., 7 (1998) 2042–2048.

[27] W.M. Watkins, Trends Glycosci. Glycotechnol., 11 (1999) 391–411.

[28] W.M. Watkins, Transfus. Med., 11 (2001) 243–265.

[29] I.E. Liener, N. Sharon and I.J. Goldstein, The Lectins, Academic Press, New York, 1986.

[30] Y.D. Lobsanov, M.A. Gitt, H. Leffler, S.H. Barondes and J.M. Rini, J. Biol. Chem., 268 (1993) 27034–27038.

[31] W.J. Peumans, E. Fouquaert, A. Jauneau, P. Rouge, N. Lannoo, H. Hamada, R. Alvarez, B. Devreese and E.J.M. Van Damme, Plant Physiol., (2006) in press.

[32] W.J. Peumans, R. Culerrier, A. Barre, P. Rouge and E.J.M. Van Damme, Plant Physiol., 144 (2007) 1–11.

[33] R. Loris, Biochim. Biophys. Acta, 1572 (2002) 198–208.

[34] H.V. Patel, A.A. Vyas, K.A. Vyas and W. Wu, J. Biol. Chem., 272 (1997) 1484–1492.

[35] T. Angata and E.C.M. Brinkman-Van der Linden, Biochim. Biophys. Acta, 1572 (2002) 294–316.

[36] P.R. Crocker, E.A. Clark, M. Filbin, S. Gordon, Y. Jones, J.H. Kehrl, S. Kelm, N. Le Douarin, L. Powell, J. Roder, R.L. Schnaar, D.C. Sgroi, K. Stamenkovic, R. Schauer, M. Schachner, T.K. Van den Berg, P.A. Van der Merwe, S.M. Watt and A. Varki, Glycobiology, 8 (1998) v.

[37] S.H. Barondes, V. Castronovo, D.N. Cooper, R.D. Cummings, K. Drickamer, T. Feizi, M.A. Gitt, J. Hirabayashi, C. Hughes and K. Kasai, Cell, 76 (1994) 597–598.

[38] P. Sorme, B. Kahl-Knutsson, M. Huflejt, U.J. Nilsson and H. Leffler, Anal. Biochem., 334 (2004) 36–47.

[39] E.N. Kitova, D.R. Bundle and J.S. Klassen, J. Am. Chem. Soc., 124 (2002) 5902–5913.

[40] W. Wang, E.N. Kitova, J. Sun and J.S. Klassen, J. Am. Soc. Mass Spectrom., 16 (2005) 1583–1594.

[41] N. Sharon and H. Lis, Lectins, 2nd edition, Kluwer Academic Publishers, Amsterdam, 2004.

[42] D.C. Kilpatrick, Handbook of animal lectins: Properties and biomedical applications, Wiley, Chichester, England, 2000.

[43] E.J.M. Van Damme, W.J. Peumans, A. Pusztai and S. Bardocz, Handbook of plant lectins: Properties and biomedical applications, Wiley, Chichester, England, 1998.

Lectins: Analytical Technologies
C.L. Nilsson (Editor)
© 2007 Elsevier B.V. All rights reserved.

Chapter 2

Crystallography and Lectin Structure Database

Ute Krengel[a] and Anne Imberty[b]

[a]*Department of Chemistry, University of Oslo, P.O. Box 1033 Blindern, NO-0315 Oslo, Norway*
[b]*CERMAV-CNRS, (affiliated with Université J. Fourier and member of ICMG) BP 53, 38041 Grenoble cedex 9, France*

1. Introduction

Lectins are carbohydrate-binding proteins of non-immune origin. They comprise a very diverse group of proteins, which come in many different sizes and folds. The characteristic binding properties of lectins, i.e. their ligand specificity and affinity, are determined by the molecular architecture of the carbohydrate-binding site, which is often also called the "combining site," since many lectins are multivalent and able to cross-link glycoconjugates, e.g. on the cell surface.

Several methods exist that allow the detailed three-dimensional characterization of protein structures and their ligand complexes, such as NMR, X-ray crystallography, neutron diffraction, and electron microscopy. Each of these methods has their weaknesses and strengths and they all complement each other. Two of the methods stand out, as they are the most widely used. These are X-ray crystallography and NMR (the latter method is discussed in Chapter 3). These two methods are sensitive to totally different physical properties of the molecules: while NMR detects signals related to the magnetic spin of the nuclei, X-ray crystallography relies on the diffraction of X-rays by the electrons of the atoms. X-ray crystallography is particularly powerful in terms of the range of problems that can be studied (from small molecules to huge protein complexes such as the ribosome) and the atomic precision that is obtained. The power of this method is also reflected in the number of protein structures deposited in the Protein Data Bank (PDB, http://www.pdb.org) [1], which far outnumbers those from any of the other methods named.

2. X-Ray Crystallography

The foundation of X-ray crystallography was laid in 1895, when Wilhelm Conrad Röntgen discovered radiation with previously uncharacterized properties while experimenting with cathode rays. He called the new rays "X-rays" and received the first Nobel Prize in physics for his discovery in 1901. A few years later, it was proposed that the wavelength of X-rays were of the same magnitude as inter-atomic distances, and in 1912, a crystal diffraction experiment confirmed this hypothesis. The first crystal structure was solved in the same year. It then took a few decades until this method could be successfully applied to large and fragile macromolecules such as proteins and DNA. Dorothy Crowfoot Hodgkin played a leading role in this work. The year 1962 marked a period of particular recognition of macromolecular X-ray structure analysis, as both the Nobel Prize in physiology and that in chemistry went to landmark X-ray structures: Watson, Crick, and Wilkins received honors for their famous 3D-structural model of DNA and Kendrew and Perutz for the first protein structures solved, those of myoglobin and hemoglobin.

As the name "X-ray crystallography" suggests, protein crystals are a necessary precondition for the study of proteins or protein ligand complexes by this method. Crystals are needed in order to enhance the scattering power of the subjects under investigation. Normally, molecules scatter X-rays only weakly, but if they are regularly arranged in a crystal (Fig. 1), scattering by one molecule is reinforced by all the other molecules in the crystal and an X-ray diffraction pattern can be recorded. Crystallizing a protein is not that difficult in pure technical terms (see Section 5 and the book by Bergfors [2] for practical instructions), but in practice, the path to well-diffracting protein crystals is often long and thorny. In fact, protein crystallization is one of the two main bottlenecks of X-ray crystallography. Practically, a good advice is to start out with ultra-pure protein. This will in many cases significantly increase the chances for success.

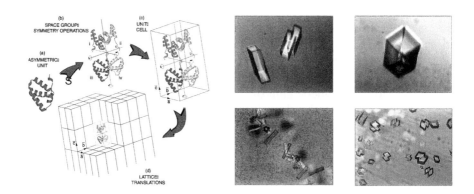

Figure 1. Protein crystals and crystal architecture. Crystals are three-dimensional macroscopic objects that are constructed from smaller units, so-called "unit cells," by translation along the unit cell axes. The unit cells in turn contain even smaller units termed "asymmetric units," which are related to each other by crystallographic symmetry, such as the rotation about a symmetry axis. The crystal pictures were kindly provided by Åsa Holmner Rocklöv and the picture of the crystal architecture was reprinted with permission from [5].

The second major bottleneck in X-ray crystallography is to overcome the so-called *Phase Problem*, the problem that only the amplitudes but not the phases of the diffracted X-rays can be obtained from conventional X-ray diffraction experiments. This problem is all the more serious as the phases contain much more important information than the amplitudes. For macromolecular structure analysis, there are mainly two methods to circumvent the phase problem: Molecular replacement (MR) and heavy-atom phasing methods. The first method is relatively simple. It involves placing a similar molecule in the lattice of the crystal under investigation (in the same orientation and position as the target molecule) and then calculating the phases from this model by means of a Fourier analysis. This method is quick and works rather well if the similarity between target and search model is significant and the crystal symmetry not too high. The method, however, can only be applied if a suitable search model exists (which happens increasingly often as more and more structures are solved and deposited in databases). If no such search model is available, there is usually no way around heavy-atom phasing methods, such as multiple isomorphous replacement (MIR). These techniques require the introduction of heavy atoms (such as mercury and lead) into the native crystals. Heavy atoms scatter X-rays particularly strongly (as they contain many electrons, which are the main scattering particles of the atoms) and leave traces in the X-ray diffraction pattern, from which initial phases can be derived.

Once the relative phases are determined for all reflections ("reflections" being the proper scientific term[1] for the diffraction spots of a crystal diffraction pattern), the information from amplitudes and phases can be combined in a Fourier synthesis to yield a three-dimensional electron density map of the crystal structure, which can in turn be interpreted in terms of atomic coordinates (for an overview of an X-ray crystal structure analysis, see Fig. 2).

The final steps of an X-ray structure analysis are crystallographic refinement and validation. During the refinement of the model, the differences between observed and calculated (model-derived) amplitudes are being minimized, in order to obtain a precise and accurate model of the crystal structure under investigation.

A number of very good textbooks are available that cover protein crystallography in more detail, e.g. the books by Blow [3], Drenth [4], McPherson [5], and Rhodes [6].

2.1. Recent technical advances in the field

During the past two decades, many advances in the field of X-ray crystallography helped to significantly speed up the process from crystal to final structural model. Recombinant techniques [7–9] are now routinely employed to obtain large quantities

---

[1] The term "reflection" refers to Bragg's interpretation of X-ray diffraction in terms of reflection of the X-ray beam at crystal lattice planes. Constructive interference only occurs when Bragg's law is fulfilled, meaning that the X-rays hit a certain set of lattice planes with spacing d at an angle θ, such that $2d\sin\theta = \lambda$.

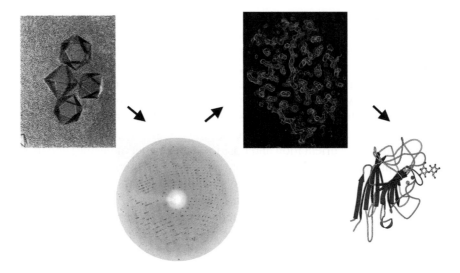

Figure 2. Overview over an X-ray crystallographic analysis (from crystal to refined structural model). Steps required are (1) protein crystallization, (2) X-ray data collection, (3) phasing to obtain (4) a 3D-electron density map (by Fourier Synthesis), into which (5) a structural model can be built that needs to be (6) refined and validated before publication. The picture shows the real-life success story for the *Erythrina crystagalli* lectin [139] (picture kindly provided by Cecilia Cronet).

of homogenous protein preparations, while traditionally, lectins were directly extracted from their natural sources. It should be stated here that there is nothing wrong with using the traditional methods, especially if the natural lectin sources are abundant, but work can benefit from the use of recombinant techniques in several different ways: Apart from the obvious advantage of reproducibly obtaining homogenous protein sample, molecular cloning opens the door to mutagenesis studies for functional investigations. But more than that, recombinant techniques may also be used to introduce unnatural amino acids like seleno-methionine (Se-Met) and seleno-cysteine (Se-Cys) [10–13] into the protein sequence. This allows, among others, the rational introduction of heavy atoms for phasing and thus avoids the often very time-consuming trial-and-error searches for suitable heavy-atom derivatives. In effect, biomodification significantly speeds up the structure determination process.

These new labeling methods, however, can only take effect due to the development of a new phasing method, called multiple anomalous dispersion (MAD) [14]. This method, also requiring the introduction of heavy atoms, is similar to the traditional MIR method, but it makes use of the anomalous scattering of heavy atoms at the absorption edge. Instead of collecting several datasets of different isomorphous heavy-atom derivatives and comparing them with the X-ray data from the native, unmodified protein crystal, datasets are collected from, ideally, one (derivatized) crystal, at several different wavelengths. This method requires far weaker heavy-atom scatterers and has become increasingly popular in recent years [15]. In some cases, the method works without any modification of the protein at all, e.g. if the protein naturally contains metal ions (as in Fe–S clusters); and there have even been reports of phasing based on the anomalous scattering of the native sulfurs in cysteine residues [16–18]. If combined with biomodification, i.e. Se-Met labeling, one may not only rapidly and rationally obtain an experimental electron density map, but in addition take advantage of the known sequence positions when building the model. Phasing based on Se-Met labeling can be done even for relatively large proteins, provided that the proteins contain enough methionine residues (≥ 1 per 100 amino acid residues). If the number of Se-Met residues is very high, traditional Patterson methods fail to identify the large number of peaks. However, the selenium substructures can be located by advanced direct method approaches [19]. These new approaches can even identify other kinds of substructures like sulfur or halide atoms or even *missing* sulfur atoms from radiation-induced decay [20, 21] and hence open the way to completely new phasing methods [22–24].

The MAD method relies on data collection at the absorption edge of the anomalous scatterers and this in turn requires the use of very specific wavelengths. For traditional X-ray sources, the wavelength is determined by the characteristics of the anode material, which is in general incompatible with the wavelength(s) needed for a MAD experiment. In order for the MAD method to become generally applicable, it was therefore essential that a tunable X-ray source was available. These requirements are exquisitely met by synchrotrons. Originally developed for particle physicists, synchrotrons with time became one of the most important tools for protein crystallographers [25–29]. Nowadays, every synchrotron has its dedicated protein crystallography beamline(s). Main advantage, apart from the tunable wavelength, is the high brilliance of the X-ray beam, a nearly parallel beam of very high intensity, which allows data collection even for tiny and poorly diffracting crystals, to the limit of their diffraction power. In recent years, X-ray data for the great majority of reported crystal structures has been collected at synchrotrons. Interestingly, this trend could soon start to reverse, as also the home sources are becoming more and more powerful, especially those equipped with advanced micro-focusing optics. Even table-top synchrotrons are now appearing on the market [30].

A high intensity X-ray beam, however, also comes at a price: the higher the intensity, the more serious is the radiation damage that the crystals have to withstand. As a result, crystals quickly loose diffraction power when exposed to synchrotron radiation. They "die" in the X-ray beam. Their lifetime can, however, be extended, if the experiments are performed at low temperature. Around the beginning of the 1990s, cryotechniques were developed [31–33], which employ low-molecular weight organic compounds (glycerol, ethylene glycol, sugars, etc.) or oils as cryoprotectants to prevent the formation of ice crystals that would destroy the crystal. If transferred quickly ("flash-freezing") into liquid nitrogen or propane, the aqueous solvent instead freezes like a glass, allowing high-quality data to be collected for extended periods of time. At the same time, data collection was speeded up, due to the advent of larger X-ray detectors with high spatial resolution, such as image plates and CCD detectors.

An enormous time-saving factor for X-ray crystallographic analyses further came about by the explosive increase in computing speed (and disk storage). This has in particular significantly reduced the time it takes to refine a structure. In addition, it has left its marks on the development of more computer-intensive software, e.g. for advanced MR searches and automatic phasing [34], model building [35, 36], refinement [37], and the calculation of unbiased simulated annealing composite OMIT maps [38].

Another area that has seen strong development in the past 15 years is structure validation. Whereas in early times, the crystallographic R-factor (see Section 2.2) was basically the only criterion used to verify a crystal structure apart from the electron density itself, today, a whole battery of quality control parameters stand to the side of both the crystallographer and the user of the crystallographic models [39, 40].

The main bottleneck that now remains is to obtain diffraction-quality crystals. Also this step of the crystallographic analysis has seen improvements (e.g. through the development of experience-based crystallization screens [41, 42], the establishment of biomolecular crystallization databases (BMCD, http://www.bmcd.nist.gov:8080/bmcd/bmcd.html [43, 44], and the development of methods for the crystallization of membrane proteins [45–48]); however, success of this step more than any other crystallographic step benefits from long years of experience. Especially telling apart promising early signs of crystallization from bad leads (or even dust particles or glass chips) are true challenges, but often critical for success. Since long-term financing of qualified and experienced personnel is increasingly hard to obtain, the trend is moving towards tackling the problem by increasing the input rather than by optimizing the output. More and more labs, and in particular structural genomics initiatives, now invest in the parallel production (mostly by recombinant techniques) of many homologous proteins from different sources, which are then subjected to robotized crystallization screening in the hope to obtain diffracting crystals [49].

2.2. How reliable are the crystal structures?

Crystallographers usually spend a significant amount of time solving and refining "their" protein structures. They therefore know very well, to which extent the models are reliable and where to find possible shortcomings. The readers of crystallographic publications on the other hand may be stunned by the beautiful graphical representations of the models and unintentionally misled into believing that all they see is equally "true." This section of the chapter shall therefore provide nonspecialists with the knowledge they need to judge the quality of X-ray crystal structures, both overall and with concern to particular regions of the structural model (for good references see also Refs. [39, 40, 50]).

The traditional overall indicator of how well the model matches the experimental data is the so-called R-factor[2] ("reliability" factor or "residual"), which gives the relative error of the calculated structure factor amplitudes. The conventional R-factor (called R_{cryst}), however, is only meaningful if the number of data at least equals and better greatly exceeds the number of parameters that are refined. Such parameters are the coordinates x, y and z for each atom of the protein, possibly the coordinates of tightly bound water molecules, alternative conformations of protein side chains and so-called B- or temperature factors, which give a measure of the mobility or disorder for the parts of the structure they are assigned to. To avoid overfitting, the amount of detail in the model must be balanced against the number of measured observations, which increases with the inverse cube of the resolution. For example, it makes no sense to fit water molecules into an electron density at a resolution of 3 Å or worse. Crystallographic publications usually list the number of independent ("unique") reflections (the experimental data) as well as the number of refined atoms. The data-to-parameter ratio can therefore be checked rather easily. If individual B-factors are assigned to each atom, the number of parameters will generally be four times the number of atoms refined (and, as a rule of thumb, there are roughly ten non-hydrogen atoms per amino acid residue). The data-to-parameter ratio may be improved if exploiting the information available from non-crystallographic symmetry (see Section 2.2.2) or geometric restraints (keeping the stereochemistry of the model as close as possible to ideal values obtained from extremely high-resolution small molecule structures [51].

The R-factor is not only used to judge the quality of the final structural model, but it serves in addition as quality indicator during the process of crystallographic refinement. This is not unproblematic, as the target of refinement[3] is highly correlated with the quality indicator.

The risk of overfitting the data has been considerably reduced when the concept of cross-validation was introduced [52]. This procedure involves setting aside a certain proportion of the data, the test set (ca. 5–10% of the data), which does not enter refinement, but is only taken for quality control. In this way, an independent quality indicator is generated that has a high correlation with the overall phase error and thus with the accuracy of the structural model. The free R-factor (R_{free}) should remain as close as possible to the conventional crystallographic R-factor during refinement. Usually, the two values stay within 2–5% of each other if refinement proceeds well, with R_{cryst} having final values of 18–25% and R_{free} of 22–30%. The lower the value, the better is the match between experimental data and structural model.

---

[2] $R = \Sigma_h \, \|Fobs| - |Fcalc\| / \Sigma_h \, |Fobs|$.
[3] In least squares refinement, the target of refinement is $\Sigma_h w_h(|Fobs| - |Fcalc|)^2$, weighted by w_h for each reflection.

2.2.1. Limitations

The final structural models are deposited in the PDB [1], from which the atomic coordinates and structure factors (that represent the experimental data) can then be retrieved by any interested individual. The databases as well as the scientific publications contain in addition a wealth of information, from which the limits of the structural models can be estimated.

The most important factor determining the overall precision of the structural model is given by the resolution of the diffraction data. To be precise, one should consider the *effective* resolution of the data, which can be judged from the quality and completeness of the diffraction data in the highest resolution shells [40]. For example, if a crystal diffracts to 2.5 Å resolution, but the data are incomplete or very weak beyond 3.0 Å resolution, the effective resolution is only 3.0 Å even though the nominal resolution (given by the diffraction limit) is higher (2.5 Å in this case).

From the effective resolution, one may obtain a good estimate as to which features of the structure can be resolved. To give some examples, already at very low resolution (ca. 9 Å) α-helices appear as tubes of electron density. β-Sheets can be identified at approximately 4 Å resolution, aromatic side chains at intermediate resolution of ca. 3.5 Å, smaller side chains around 3.0 Å resolution, bulbs for main-chain carbonyl groups and ordered water molecules at ca. 2.5 Å resolution, and finally, individual atoms and alternative side-chain conformations can be identified at approximately 1.5 Å resolution.

Even if the resolution is high, however, this does not mean that all parts of the structure can be determined with high precision. Some parts of the structure may be mobile or disordered and cannot be resolved for this reason (see Section 2.2.4). Other parts may be resolvable but involved in crystal contacts and are therefore not particularly reliable when it comes to judging their conformation in a true biological system. It is hence of great importance to keep in mind the structural context, when analyzing the structures and drawing functional conclusions. While crystals of biological macromolecules usually contain extensive water-filled solvent channels (in average 30–70% water) and thus give a formidable picture of the native structure in solution, this is not equally true for those regions involved in contacts with symmetry-related molecules in the crystal.

2.2.2. Symmetry

Many lectins are multimeric proteins that contain more than one copy of a particular domain. These domains are usually arranged in a symmetric fashion, as dimers, trimers, tetramers, pentamers, etc. The biological purpose of such multimeric arrangements is to increase the usually low carbohydrate-binding affinity of the lectins through avidity.[4]

---

[4] "Avidity" is defined as the apparent affinity due to multivalent binding.

In crystal structures, the molecular symmetry is often reflected in the crystal packing. Symmetry axes that relate the individual domains often coincide with crystallographic symmetry axes. Since the deposited structures in the PDB usually consist of the smallest crystallographic unit, the so-called "asymmetric unit," it is often necessary for the user of a deposited structure to generate the whole biologically relevant molecule from the deposited coordinates by applying the information given by the PDB (e.g. by clicking on "Biomolecule" on the left hand menu of the PDB or by applying given transformation matrices to the content of the asymmetric unit).

It is also possible that the asymmetric unit already contains the complete biomolecule or even several copies of the biomolecule. For example, when the molecules form pentamers, the individual protomers are never related by crystallographic symmetry, simply because there is no way to continuously fill space with underlying fivefold symmetry. In these cases, one speaks of "non-crystallographic" symmetry or NCS (in contrast to the crystallographic symmetry discussed earlier). Non-crystallographic symmetry can be exploited in the X-ray analysis in several different ways. For example, as mentioned above, non-crystallographic symmetry can be used to improve the data-to-parameter ratio in crystallographic refinement [53]. At low resolution, it is advisable not to allow any deviation from non-crystallographic symmetry ("constrained" refinement), whereas at higher resolution, the restraints may be loosened, as there are more diffraction data available ("restrained" or "un-restrained" refinement).

Besides the use of non-crystallographic symmetry in refinement, NCS can be very useful when analyzing crystal packing effects. Depending on the crystal environment, some parts of the molecules will be involved in crystal contacts, while others will be solvent-exposed. When comparing the molecules in different crystal environments, regions that are involved in crystal contacts in some molecules will be positioned at water-filled cavities in others, which often makes it possible to discriminate crystal packing artifacts from the natural conformation of the molecules in solution, hence allowing for functional conclusions even in these regions.

2.2.3. Electron density

It is important to keep in mind that the published structures are only models. Resolution and *R*-factors are important quality indicators, but the best criterion remains the electron density, which is directly calculated from the original experimental data. For this reason, crystallographers are now actively encouraged to deposit their experimental data along with the structural coordinates. For those structures where the data have been deposited, it is possible to download electron density maps from the electron density server (EDS, http://eds.bmc.uu.se/eds/) [54] through a link from the PDB. Freeware programs like Swiss-PDBViewer (http://ca.expasy.org/spdbv/) can then be used to display both the structural coordinates and electron density maps for a detailed analysis.

There are several different types of electron density maps [55]. The most standard type of map is a Fourier map of the 2*Fo–Fc* type, where *Fo* and *Fc* stand for *observed* and *calculated* structure factor amplitudes, respectively. This type of map is continuous along the polypeptide chain. The 2m*Fo–*D*Fc* map that can be downloaded from the electron density server falls into this category. Usually, one displays a Fourier map side by side with a difference Fourier map of the *Fo–Fc* type, in order to more easily identify errors in the model. Positive peaks in the difference density map point to features that have not been modeled, whereas negative peaks indicate that an atom has been placed wrongly into this position. If everything was correctly modeled, the difference density map should be flat and only correspond to random noise.

Even though the electron density more than any other quantity reflects the experimental data, it should be kept in mind that it is biased by the phases of the diffracted X-rays and not calculated from the reflection amplitudes alone. For electron density maps, which are calculated from experimentally determined phases, this is less of a problem, but phases obtained from a MR search will certainly leave their marks in the maps and show resemblance to the placed search model, whether or not it was correctly placed in the crystal lattice. To obtain a less biased view of the electron density, it is therefore very helpful to use so-called OMIT maps (best simulated annealed OMIT maps [38]), which are calculated from partial models.

2.2.4. B-factors and real-space correlation coefficients

Not all parts of the structural model are equally well defined by electron density. Surface residues, for example, often have very mobile side chains, which cannot be seen in the electron density maps. The same is true for mobile loops. At low temperature, these parts of the model are no longer flexible, but freeze out in different conformations. One then speaks of disorder. There are several ways for crystallographers to communicate static or dynamic disorder to the end user of crystallographic models. Most popular are the atomic displacement parameters (so-called *B-* or temperature factors). High *B*-factors indicate high disorder and thus low reliability of the concerned atomic positions. If the electron density gets very diffuse, one may no longer be able to model this residue at all. In this case, the depositor may assign an occupancy of zero to the concerned residue (or even clip off the residue or its side chain entirely).

A weak correlation between the structural model and the electron density, due to disorder or other reasons, is also indicated by a low real-space correlation coefficient. As the name suggests, it gives the correlation, per residue, of the structural model and the supporting electron density. Values close to 100% indicate perfect correlation, while values below 50% indicate problem regions.

Similar to real-space correlation coefficients, also *B*-factors can indicate more than just disorder. Especially at low resolution, they can also function as error sinks, absorbing various kinds of errors of the structural model [40]. When modeling is attempted based on deposited coordinates, one should therefore take the *B*-factors into account.

2.3. Glycobiology specifics

The above-stated facts are valid for all X-ray structure analyses, no matter if we are interested in isolated proteins, protein–protein complexes, or protein complexes with small ligands, such as, for example, small oligosaccharides. There are, however, some specifics with respect to glycobiological targets, which shall be discussed separately in this section.

Let us first consider glycoproteins: proteins to which a carbohydrate chain is covalently attached. Glycoproteins in many cases present a considerable challenge to X-ray structure analysis, because they do not easily form crystals. This is because the attached carbohydrate chains are often heterogenous and flexible and interfere with the formation of crystal contacts. A well-known strategy to tackle this problem is to deglycosylate the glycoproteins prior to crystallization (see Section 5). This is usually done with endoglycosidases such as PNGase F (resulting in virtually complete removal of all N-linked glycans) or Endo H, Endo F1, F2, F3 (for partial deglycosylation). Also, exoglycosidases can be used, alone or in combination with endoglycosidases. Alternatively, the glycosylation sites can be mutated (substitution of S/T to Ala or N to Gln or Asp in the conserved NXS/T motif). Expression in lower organisms like *Escherichia coli* or *Pichia pastoris* is also an option, or – if a less radical procedure is required – genetically modified expression systems can be used that do not contain certain glycosyltransferases (e.g. CHO Lec cells). As long as the protein remains functional upon deglycosylation, such treatments should be unproblematic. In case the protein does loose its activity upon deglycosylation, one might nevertheless resort to using the deglycosylated structural model despite of its shortcomings if extensive trials have shown that the protein only crystallizes in this way. In such a case it is advisable to model the carbohydrate chains onto the crystal structure in retrospect, thus combining experimental and in silico-methods to obtain a functional model.

Carbohydrates may also bind non-covalently to proteins, as is the case for lectin–carbohydrate complexes. Structures of these complexes can be obtained either by co-crystallizing the protein with its sugar ligand or by soaking the protein crystals with carbohydrate-containing mother liquor (see Section 5). A positive side effect of such a procedure, if sufficiently high sugar concentrations are used, may be that the crystals at the same time are cryoprotected for low temperature studies. It is even possible to use modified carbohydrates, which contain heavy atoms at certain positions, for MAD phasing. At the present time, only seleno-sugar [56] and bromo-sugar [57] have been used, but the potential of the method is quite large since it allows with only one data collection to phase the structure and characterize the binding site. There are, however, also cases were neither co-crystallization nor soaking give any results, either because the ligand does not bind strongly enough or because ligand binding is not compatible with crystal formation. (Tip: Binding can easily be checked by calculating difference Fourier maps of either *Fo–Fo* or *Fo–Fc* type for the obtained crystals.) If no electron density is visible for the carbohydrate ligand, the option remains to obtain further insight into ligand binding by modeling the compound into the combining site of the protein structure.

Modeling the ligand into the binding site, with or without supporting electron density, requires a 3D structural model of the carbohydrate ligand. This can be obtained either from scratch (e.g. using the modeling tool "Sweet" from the http://www.glycosciences.de web site, which converts carbohydrate sequences into 3D models) or from a structure database (via the Uppsala HIC-Up server at http://xray.bmc.uu.se/hicup/) or directly from the PDB (http://www.pdb.org/) or other appropriate databases such as the Cambridge Structural Database (CSD, http://www.ccdc.cam.ac.uk/) or the Glyco3D server of CERMAV (http://www.cermav.cnrs.fr/glyco3d). The model of the carbohydrate ligand should then be refined and checked as carefully as the protein structure itself. Currently, however, tools for analyzing carbohydrate structures are not as widely known and applied as those for protein or DNA structures [50]. A good tip is to check out the web site at http://www.glycosciences.de. A couple of tools, such as *pdb-care* (to check carbohydrate residues in pdb files for errors [58]) or *carp* (which generates Ramachandran-like plots for carbohydrates [59]) or even *GlyProt* (to identify glycosylation sites in proteins and automatically attach them in silico) make life of structural glycobiologists significantly easier. Another database, currently under development, is EuroCarbDB (http://www.eurocarbdb.org/databases). And questions concerning carbohydrate nomenclature may be resolved by consulting the web site http://www.chem.qmul.ac.uk/iupac/2carb/.

2.4. Electron microscopy and neutron diffraction

Two other methods for macromolecular structure determination exist that are very similar to X-ray crystallography. These are electron microscopy [60] and neutron diffraction [61]. Neutron diffraction is more of an oddity in structural studies. In this method, the obtained 3D-crystals are exposed to a neutron beam instead of X-rays. Neutrons, produced in a nuclear reactor, have no electrical charge, a mass almost the same as a proton, a nuclear spin of 1/2, and a magnetic moment. Thermalized fission neutrons (thermal neutrons) have as in the case of X-rays, wavelengths of the same order of magnitude as the diameter of atoms and are therefore well suited for the study of condensed matter. The interaction between neutrons and the atoms in the crystal lattice differs in several respects from the interaction between atoms and X-rays. The major difference is caused by the fact that X-rays are diffracted by electron clouds, whereas neutrons are scattered by nuclei. The principal problem associated with neutron crystallography of proteins is caused by the low flux of neutrons through the sample. The interaction of neutrons with atoms in the biomaterial is therefore weak and the crystals have to be very large, generally at least $1 \, mm^3$ in volume. In addition, the exposure times become very long, typically several weeks or more. The required large crystal size poses a serious limit to most structural studies of protein crystals with this method. In cases where such large crystals can be obtained, however, as a reward, hydrogen atoms in the structure can be resolved. Knowledge of the exact location of hydrogen atoms in a given macromolecule is crucial for the understanding of many biological problems, and this feature is probably the most important reason for using the neutron diffraction technique [62].

Electron microscopy has traditionally been used to study heavy-atom soaked tissue sections at relatively low resolution [63]. This method can, however, also be used to study the structures of macromolecules. This is done with two variations of the technique: electron crystallography [64] and single-particle cryoelectron microscopy [65–67]. Single-particle studies require relatively large molecule sizes (corresponding to molecular weights of at least 100–200 kDa). The resolution of the obtained structures is not particularly impressive (20–30 Å); nevertheless, single-particle studies are invaluable in cases where no crystals can be obtained. Even if only a rough molecular envelope can be deduced, this can provide starting phases for higher resolution X-ray crystal structures and, more importantly, available atomic-resolution models of macromolecule domains can be modeled into the molecular envelope of a complete protein complex to give a picture of the biological system in action [68–70].

In contrast to single-particle studies, electron crystallography requires crystals. This technique is analogous to X-ray crystallography, only that an accelerated electron beam is used instead of an X-ray beam, and 2D-crystals are used instead of 3D-crystals. Nevertheless, in order to obtain a three-dimensional image, the 2D-crystals need to be placed on a grid support and then tilted in the electron beam. Since usually the highest tilt angles obtainable are ca. 70°, electron diffraction data are usually less complete than X-ray diffraction data. They are also much weaker. Electron crystallography, however, has one huge advantage over X-ray crystallography: it does not have a phase problem. Since electrons can be focused with magnetic lenses, a direct image of the structure can be obtained in addition to the electron diffraction pattern. Nevertheless, the diffraction pattern is valuable due to its superior data quality. As for X-ray diffraction, also electron diffraction gives a discontinuous diffraction pattern featuring spots (the reflections) on lattice planes. Between the reflections, the intensity should be zero, which can be taken as a condition to accurately estimate the background of the diffraction data and hence derive high-quality primary data.

Electron crystallography is a method that is particularly suitable for macromolecules that readily form 2D-crystals. As such, this method is predestined for the study of membrane proteins. Until very recently, the structures obtained by this method were of much lower resolution than X-ray structures and 7–8 Å resolution could be seen referred to as "high" resolution. Since many membrane proteins adopt an α-helical structure, nevertheless the underlying fold could often be revealed. There is, however, no principal reason, why electron crystal structures should not be obtainable at much higher resolution and in fact a very recent structure of human aquaporin AQP0 at 1.9 Å resolution has proven that atomic resolution electron crystal structures are no longer out of reach [71].

3. Lectin Databases

The large amount of biological and structural information that is available on lectins is at present not available on one unique Internet site. A textbook-style compilation of general information regarding the biological function of lectins (and other glyco-topics) can be found on the Glycoforum site (http:// www.glycoforum.gr.jp/science/word/lectin/LE_E.html). The web site developed by K. Drickamer describes the classification of animal lectins (http://www. imperial.ac.uk/research/animallectins/), while the newly developed site LectinDB (http://nscdb.bic.physics.iisc.ernet.in/) provides information about the sequence and predicted folds of plant lectins [72].

The 3D-lectin database (http://www.cermav.cnrs.fr/lectines/) developed at CERMAV contains information for almost all available three-dimensional structures of lectins. The different families have been classified according to their folds. Useful links exist to retrieve bibliographic information and to download the atomic coordinates from the PDB. In addition, related structural databases such as SCOP (http://scop.mrc-lmb.cam.ac.uk/scop/) [73] are accessible for each entry. For every lectin, images of quaternary structure, overall fold and binding sites are provided in order to illustrate the specificities of protein–carbohydrate interactions.

The lectin structure database can be browsed according to the origin of lectins (plant and animal lectins being the most strongly represented) or by fold families. A searchable expert mode further allows for specific requests concerning different fields, such as the biological origin, author name or carbohydrate ligand. For example, the search for complexes involving a fucose residue will yield 32 results that can be tabulated and further analyzed (Fig. 3).

3.1. Statistics

More than 551 structures are now listed in the 3D-lectin database (Table 1). Most of them are X-ray crystal structures, although concanavalin A has also been studied by means of neutron crystallography after soaking the crystals in heavy water,

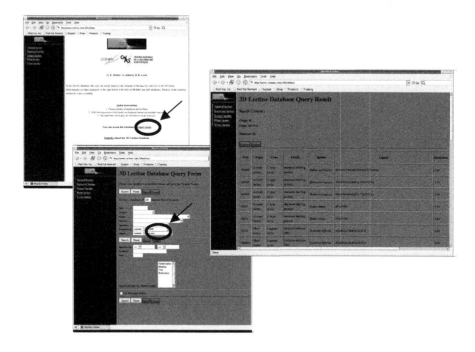

Figure 3. Interface of the 3D-lectine database and an example of its use.

hence yielding the precise location of bound water deuterium atoms [74]. For a few low molecular weight lectins, such as hevein [75], also NMR structures are available.

Structures of plant lectins are the most numerous and represent 42% of the crystal structures in the database; most of these are legume lectins. These proteins are present in very large quantities in many beans or peas and are easily purified by affinity chromatography, which explains why the structure of concanavalin A from horse beans was the first lectin structure solved [76]. Legume lectins share the same monomeric fold but exhibit a variety of oligomeric states. The sequence variations in the ligand-binding loops can be correlated with the large range of observed lectin specificities [77]. Animal lectin structures account for 27% of the database entries. These proteins adopt a large variety of different folds, related to their numerous functions [78]. Best studied of these are the galectins, soluble galactose-binding lectins with a fold very similar to the legume lectins, and the C-lectins, calcium-binding lectins that exhibit the particularity of a calcium ion directly coordinating two hydroxyl groups of the carbohydrate ligand. Structures of lectins from bacteria and viruses currently correspond to 16% and 8% of the database entries, respectively, but their number is growing rapidly. Finally, only 5% of structures are from fungal origin, a family that has only recently attracted significant interest.

Table 1. Number of structures available in the 3D-lectin database, classified by biological origin and fold (the number of carbohydrate complexes is given within parentheses).

Plant lectins		Bacterial lectins	
β-Prism II lectin	14 (9)	AB$_5$ toxin	27 (14)
Knottin (hevein-like)	26 (14)	Bacterial neurotoxin (trefoil)	13 (8)
β-Prism I lectin	32 (24)	Staphylococcal toxin	2 (2)
β-Trefoil lectin	18 (12)	Pili adhesin	12 (9)
Legume lectin		Cyanobacterial lectins	12 (4)
(Con A-like)	152 (95)	2-Ca β-sandwich	15 (13)
		1-Ca β-sandwich	3 (1)
Animal lectins		β-Propeller	3 (3)
C-type lectin	66 (41)	Toxin repetitive domain	2 (1)
R-type lectin (β-trefoil)	5 (3)		
I-type lectin	9 (5)	*Virus lectins*	
Pentraxin	7	Coat protein	5 (4)
P-type lectin	6 (4)	Hemagglutinin	28 (11)
Galectin	35 (26)	Tailspike protein	9
Spider toxin	1	Capsid spike protein	2 (1)
Tachylectin	2 (2)	Fiber knob	3 (2)
Chitin-binding protein	1		
TIM-lectin	8 (6)	*Fungal lectins*	
L-type lectin		Ig-like	1
(ERGIC, VIP)	1	Galectin	10 (8)
Calnexin-calreticulin	2	Actinoporin-like	7 (4)
Fucolectin	1 (1)	β-trefoil pore forming	3 (3)
H-type lectin	2 (1)	6-bladed β-propeller	2 (2)
		7-bladed β-propeller	4 (3)
Total	*551 (336)*		

3.2. Protein–carbohydrate interactions

More than 60% of the structures in the database are present in the form of carbohydrate ligand complexes. The database therefore represents a mine of information for dissecting the molecular bases of protein–carbohydrate interactions. Previous analyses of carbohydrate-binding sites revealed several common principles governing carbohydrate recognition by lectins (Fig. 4) [79, 80]. First of all, due to the large number of hydroxyl groups characteristic for carbohydrates, polar interactions play a very important role. In particular, there are a large number of hydrogen bonds between the many hydroxyl groups of the carbohydrate ligand and the amino acid residues in the binding site. In addition, the same hydroxyl groups can engage in polar interactions with divalent cations present in the binding site of some animal and bacterial lectins. The interactions depend on the orientation of the hydroxyl groups, related to the stereochemistry of the monosaccharide.

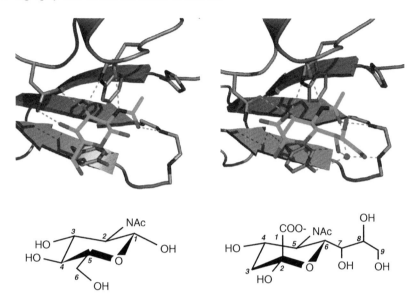

Figure 4. Protein–carbohydrate interactions. This figure shows a comparison of the GlcNAc and NeuAc-binding modes in one of the PVL-binding sites. Below: schematic representation of GlcNAc and Neu5Ac, in the same orientation as above, together with the numbering of the carbon atoms.

For example, affinity for galactose is almost invariably correlated with strong hydrogen bonds to its axial O-4 hydroxyl group.

Also hydrophobic interactions generally contribute to a significant extent to the stabilization of sugar–lectin interactions. Aromatic residues often interact face to face with the sugar rings. This parallel interaction, often referred to as stacking, plays an important role in the specificity since it requires strict steric complementarity. Aromatic stacking has been attributed to weak hydrogen bonds between the non-polar C-H groups of the sugar ring and the π-electron cloud of the aromatic system [81]. Other important hydrophobic interactions involve the methyl moieties of the *N*-acetyl group or the monosaccharide fucose, which are often surrounded by two or three aromatic amino acids, thereby creating a strong hydrophobic cluster.

The role of the solvent is more difficult to evaluate. Whereas the most buried carbohydrate residues tend to satisfy their hydrogen bonding potential with the amino acids of the protein, the more exposed ones establish hydrogen bonds with water molecules that are already bound to the protein or they interact with the bulk solvent. There is, however, great variation in the importance of water-mediated interactions for carbohydrate recognition. In extreme cases, the majority of protein–carbohydrate interactions can be water-mediated [82] and in several cases, a single such indirect interaction has been shown to suffice to enhance binding affinity to levels that generated binding specificity [83–85].

4. Recent Research Highlights

In recent years, the interest in lectin structures has shifted from the plant kingdom – even though some very interesting structures are yet to be analyzed – towards animal lectins, including lower life, and microbial lectins. Mammalian lectins have always been considered of high interest, particularly when they have an important biological function. A recent example is the structural investigation of the cellular receptor DC-SIGN, a C-type lectin present on the surface of dendritic cells, which is implicated in infection processes by several pathogens. The crystal structure of the carbohydrate recognition domain of this protein in complex with high-mannose oligosaccharides present on enveloped viruses deciphered the atomic basis for recognition and further shed light on the physiological function of DC-SIGN, acting both as an adhesin and as a mediator of endocytosis, e.g. of the human immunodeficiency virus [86].

Lectins from lower animals are often involved in innate immunity and have the capacity to agglutinate pathogens. After pioneering work on tachylectin from horseshoe crab [87], the crystal structures of eel lectin (fucolectin) and then snail agglutinin (HPA) have been solved, revealing new folds and new binding modes towards carbohydrates [88, 89]. In both cases, a trimeric arrangement organizes the binding sites in an optimal fashion for binding to the surface of bacteria.

4.1. Fungal lectins

To date, only few structures of fungal lectins have been solved (Fig. 5). So far, the focus has mainly been on lectins that can be extracted from the fruiting bodies or from the mycelium of mushrooms. These proteins play a role in toxicity, defense mechanisms, and mycorrhization. The first crystal structure of a fungal lectin, i.e. the fucose-binding lectin AAL from the orange peel mushroom *Aleuria aurantia*, was found to exhibit a six-bladed β-propeller fold, different from any lectin fold known before [90]. More recently, a seven-bladed β-propeller fold was reported for PVL, a GlcNAc-binding lectin from *Psathyrella velutina* [91]. Together with tachylectin 2 from horseshoe crab, which adopts a five-bladed β-propeller fold [87], three different shapes of β-propeller folds have now been reported for lectins. Their tandem architecture with wheel-type structure distributing the binding sites on a circle appears to be perfectly adapted for multivalency.

AAL and PVL are not the only fungal lectins that adopt folds never before observed for lectins. The same is true for Fve from *Flammulina velutipes* [92], which is structurally similar to human fibronectin, and for the lectins from *Xerocomus chrysenteron* (XCL) and from *Agaricus bisporus* (ABL), which both resemble actinoporins, a family of pore-forming toxins from sea anemones [93, 94]. In contrast, two folds that have previously been widely known for other lectin families have now also been observed in fungal lectins. ACG from *Agrocybe cylindracea* and CGL2 from *Coprineus cinerea* appear to be galectins [95, 96],

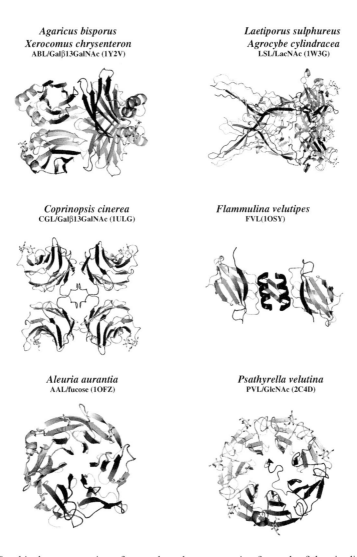

Agaricus bisporus
Xerocomus chrysenteron
ABL/Galβ13GalNAc (1Y2V)

Laetiporus sulphureus
Agrocybe cylindracea
LSL/LacNAc (1W3G)

Coprinopsis cinerea
CGL/Galβ13GalNAc (1ULG)

Flammulina velutipes
FVL(1OSY)

Aleuria aurantia
AAL/fucose (1OFZ)

Psathyrella velutina
PVL/GlcNAc (2C4D)

Figure 5. Graphical representation of one selected representative for each of the six different folds observed for crystal structures of fungal lectins. The protein chains are represented as ribbons and the carbohydrate ligand, when present, as capped sticks.

therefore belonging to a large family of lectins with members in all classes of vertebrates, while LSL, the lectin from *Laetiporus sulphureus* contains a ricin-B domain [97], a β-trefoil fold observed in many lectins and carbohydrate-binding domains and referred to as the $(QXW)_3$ domain. This domain has been identified not only in bacteria, fungi, and plants, but also in sponge, insects, and mammals, while generally conserving its role of targeting a sugar-coated substrate.

4.1.1. Fungal lectins – A case study
PVL, the lectin from *P. velutina*, represents an interesting case study for the molecular basis of protein–carbohydrate interactions, because of its dual specificity. The lectin was first described as a GlcNAc-binding protein [98], but later reported to also bind sialic-acid containing glyconjugates, although with moderate affinity [99]. Three crystal structures have recently been solved for PVL: one concerning the unliganded state (1.5Å resolution) and two complexes with GlcNAc (2.6Å) and Neu5Ac (1.8Å) [91]. Electron density for six GlcNAc residues was clearly visible, indicating that the sugar-binding sites are pockets located in the upper part of the β-propeller, in a space between two consecutive blades, accessible to the solvent. The weaker binding to NeuAc resulted in lower occupancy, with only two to three sites occupied per monomer. Comparison of the two binding modes demonstrated that GlcNAc and NeuAc bind in the same pocket, albeit with different orientation. In both cases, the *N*-acetyl group establishes the same hydrogen bonding network and the same hydrophobic contacts to a histidine residue (Fig. 4).

Such dual specificity of a lectin-binding site for both GlcNAc and Neu5Ac has already been observed for the isolectins of wheat germ agglutinin. Also in this case, crystal structures (of WGA2) are available for both ligand complexes: one with chitobiose [100] and one with Neu5Ac [101]. As observed for PVL, the *N*-acetyl group of both ligands is buried in the same fashion in the protein-binding site, while the orientation of the sugar ring is different. These two cases illustrate the fact that the protein recognizes a scaffold of hydroxyl groups (and other prominent functional groups like *N*-acetyl or methyl groups). Since the stereochemistry of carbohydrates allows for different configuration at each position of the ring, it may happen that different monosaccharides, viewed from different perspectives, present similar arrangements of bioactive groups. Such an observation opens the route for the design of non-carbohydrate mimetics that present the same active groups but carried by different scaffolds, such as peptides.

4.2. Bacterial lectins

Bacteria use several strategies for targeting host sugars. Some bacteria use lectins as virulence factors that enable the bacteria to recognize and bind to the glyco-conjugate receptors (glycolipids or glycoproteins) on the surface of their host cells, in a first step to confer toxicity. Bacterial toxins that contain lectin domains have been crystallized from several pathogenic organisms such as *Vibrio cholera,* enterotoxigenic *E. coli*, and *Bordella pertussis*. They are referred to as AB$_5$-type proteins since they consist of one toxic ADP-ribosyltransferase subunit and five lectin domains that bind to gangliosides of gut or lung epithelia [102].

Other lectin domains are located at the top of pili or flagella. Also these lectin domains play a role in the attachment of the bacteria to epithelial cells. Such lectins are part of complex multiprotein architectures and are therefore difficult to express in soluble form and to crystallize. Only a limited number of structures of pili-related lectins are available: FimH and PapG from uropathogenic *E. coli* have been crystallized with mannose and αGal1–4Gal, respectively [103, 104], while GafD from the F17 pilus of enterotoxigenic *E. coli* has been crystallized with GlcNAc [56, 105]. Interestingly, all these domains share the same overall fold; however, they do not exhibit significant sequence identities and their respective carbohydrate-binding sites have different architectures.

Homomeric lectins have been observed in several microorganisms. Cyanovirin, a homomeric lectin from the cyanobacteria *Nostoc ellipsosporum,* has attracted considerable interest due to its high affinity for oligomannose structures and its related anti-HIV properties. The first three-dimensional structural model was obtained by NMR [106], while crystallography allowed for a detailed characterization of its complexes with large oligosaccharides [107]. The gram-negative bacterium *Pseudomonas aeruginosa,* which is responsible for severe airway infections of patients suffering from cystic fibrosis or immuno-suppression, produces two soluble lectins, one of which is specific for galactose and one for fucose. These lectins are associated with virulence factors that are proposed to play a role in the recognition, adhesion and toxicity towards epithelial cells and also involved in biofilm formation [108]. PA-IL, the galactose-specific lectin, has been crystallized as a tetramer in complex with calcium and galactose (Fig. 6) and displays a calcium-bridged binding mode for galactose that is very similar to the

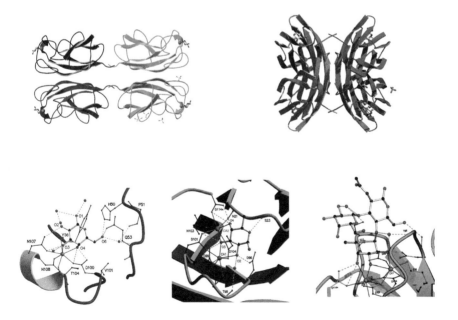

Figure 6. Graphical representation of the *Pseudomonas aeruginosa* calcium-dependent lectins. Top: tetramers of PA-IL (left) and PA-IIL (right). Bottom: binding site of PA-IL in complex with calcium and galactose (left), binding site of PA-IIL in complex with calcium and fucose (middle) and with calcium and Lewis a trisaccharide (right).

one observed in animal C-type lectins [109]. PA-IIL, the fucose-specific lectin, is specifically inhibited by human milk, which is unusually rich in fucosylated oligosaccharides and protects infants against microbial infections. Lewis a (Le^a), a trisaccharide determinant present in the human milk of most women, is the best ligand of PA-IIL, with an affinity constant of 2.2×10^{-7} M. The crystal structure of PA-IIL in complex with Le^a trisaccharide revealed the presence of two additional hydrogen bonds, hence explaining the higher affinity compared to the fucose monosaccharide [110].

4.2.1. Bacterial lectins – A case study

X-ray crystallography is a very powerful technique to determine the high-resolution structure of proteins and protein–ligand complexes. But more than that, the insights gained often reach far beyond the structure itself and can help to solve important biological questions. Take for example a study on bacterial toxins. Two researchers, Mike Lebens and Susann Teneberg from the University of Gothenburg, were intrigued by the different binding properties of two highly homologous toxins, cholera toxin and the heat-labile enterotoxin from enterotoxigenic *E. coli*. The receptor-binding B-subunits, CTB and LTB, respectively, share 83% sequence identity, yet cholera toxin binds with very high specificity only to its primary receptor, the GM1 ganglioside, whereas heat-labile enterotoxin is much more promiscuous and binds in addition to several other glycoconjugates [111–114]. In order to find out which of the amino acids are responsible for the broader binding specificity of heat-labile enterotoxin, the researchers constructed hybrids between the two carbohydrate-binding domains and tested their binding specificities [115]. Eventually, they created a hybrid that was indistinguishable in its binding properties from the native LTB, but had half of its residues from CTB. When substituting amino acids back one by one to the CTB sequence, the researchers were surprised: substitution of amino acid residue 4 (Ser→Asn) resulted in a hybrid with previously undescribed binding specificity to blood group antigens. Molecular modeling suggested binding to a site distinct from the primary GM1 binding site [116] – apparently, a new binding site had been created. But modeling of such complex systems is difficult and experimental verification needed. The crystal structure (at 1.9 Å resolution) indeed revealed the presence of a second binding site (Fig. 7), but with unexpected characteristics [82]. First of all, there were only very few direct protein–carbohydrate interactions; most of the hydrogen bonds were indirect, via water molecules. Second, even the interaction to the critical residue Asn4 turned out to be an indirect water-mediated interaction. And finally, except for Asn4, all amino acid residues in this new binding site were conserved also in native LTB. In fact, a single Ser4→Asn mutant of LTB showed equally strong binding to blood group antigens as did the structurally characterized hybrid [82]. This gave strong indications that the binding site must already have been present in the native toxin and the mutation only served to enhance the affinity such that it became detectable

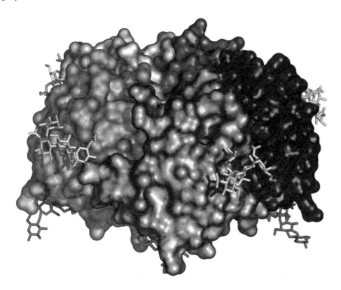

Figure 7. Graphical representation of the crystal structure of a hybrid between the lectin domains of cholera toxin and *E. coli* heat-labile enterotoxin. The carbohydrate-binding domains form a pentamer with two distinct binding sites per protomer, one for the primary receptor (represented by the GM1 pentasaccharide coloured in red) and one for blood group antigens (represented by the pentasaccharide coloured in yellow). The picture, which was kindly provided by Åsa Holmner Rocklöv, represents a collage of two structures (it has not been shown if the two different ligands can bind simultaneously).

by the binding assays. Indeed, a recent crystal structure of native LTB in complex with a blood group A antigen analog confirms this hypothesis [117]. The binding site is probably biologically relevant, as there are correlations between the severity of cholera and ETEC infections and the blood group of infected individuals, which can be explained on the basis of the crystal structure [117].

In this case, as in many others, the crystallographic analysis not only provided a detailed molecular picture of the binding characteristics, but also yielded new insights into the biology of the concerned bacterial infections and provided clues to previously puzzling epidemiological observations.

5. Hands-on Protocols and Practical Tips

5.1. Deglycosylation

As mentioned above (see Section 2.3), there are several alternative techniques to either partially or totally deglycosylate proteins. One can for example express the protein in a simpler or modified expression host or mutate the glycosylation sites.

The method to be discussed in more detail in this section is enzymatic deglyco-sylation. Several companies like SigmaAldrich and ProZyme have excellent the-oretical and practical information about deglycosylation on their webpages, which are well worth studying. Both *N*- and *O*-glycosylation[5] occur naturally in proteins, with *N*-glycosylation being the more common modification. *N*-glycans can generally be removed rather effectively by a single enzyme, PNGase F, whereas this is not the case for *O*-glycosylation, where several enzymes have to act in concert to exert the same effect. Deglycosylation can be tried both under native and denaturing conditions. While native conditions are in general prefer-able for crystallization purposes, deglycosylation under denaturing conditions is more effective and should for this reason always been done in parallel as a pos-itive control. To achieve higher efficiency even under native conditions, one may experiment with using higher temperatures (up to 37°C, instead of 0–4°C) and/or longer reaction times (1–5 days; preferably adding sodium azide to the mixture to prevent bacterial growth).

In some cases, native deglycosylation does not work well. In these cases, one may try to deglycosylate the protein under denaturing conditions and then refold the protein. Alternatively, one may use only slightly denaturing conditions, by applying various mild detergents (like β-octyl-glucoside, Chaps, Triton X-100, etc.). One can also try to incubate the reaction mixture in a sonicating water bath. Yet another option is to deglycosylate the protein only partially using exoglycosi-dases (neuraminidase, galactosidase, etc.) instead of endoglycosidases.

In practice, one often starts out with ca. 30 µl protein solution (concentrated to 0.5–5 mg/ml in water or a suitable buffer) and adds the glycosidase of choice (e.g. PNGase F, Endo H, Endo F1, F2, F3, neuraminidase or enzyme kits) in ratios varying from ca. 1:15 to 1:2,000 (w:w). The deglycosylation reaction is then monitored regularly by taking samples and analyzing them on SDS–PAGE or IEF gels. A final check is preferably done by mass spectrometry. With some of the glycosidases like neuraminidase, one should be very careful to fully remove the enzyme (e.g. by gelfiltration), since it crystallizes easily even in minute concentrations. Another way to ensure complete removal is to use glycosidases as fusion proteins (coupled, e.g. to glutathione-S-transferase; [118] and passing the mixture through a GST affinity column (Glutathione Sepharose) after the reaction is completed.

Of course, it is always well worth a try to crystallize the protein also in its fully glycosylated form. Some proteins even crystallize better in their glycosylated form, due to the involvement of the glycan chains in crystal contacts. Other pro-teins behave best when partially deglycosylated.

---

[5] *N*- and *O*-glycosylation refers to the covalent attachment of glycans to the amide nitrogen (N) of Asn and hydroxyl oxygen (O) of Ser or Thr, respectively. In some cases, also other amino acid residues can be glycosylated, but such modifications are rare.

5.2. Preparing the protein for crystallization

Before setting up crystallization experiments, one first needs to decide in which solution to keep the protein. Since the aim is to screen a large number of crystallization conditions and try out various different buffers, pHs and precipitants, it is usually a good idea to keep the protein solution as simple as possible. If the protein is stable in 5–10 mM of a typical biological buffer (Hepes, Tris, etc.) – fine! This is all that is needed. For initial screening, it is preferable to avoid phosphate buffer, though, as phosphate gives crystals with many inorganic cations.... If your protein is not stable in a solution consisting only of a diluted buffer, you might try to add some salt (NaCl) to the solution to keep the protein in solution. 100 mM of sodium chloride usually does no harm to the crystallization experiments. In case you are trying to crystallize a membrane protein, salt alone will not help to solubilize the protein and you will need to add detergents (Glycon and Anatrace have good selections and you might start out with a concentration of 2 CMC[6] of typical detergents used for crystallization such as β-octyl-glucoside or dodecyl-maltoside) (see Refs. [48, 119, 120] for further details on membrane protein crystallization). Finally, many proteins (notably cytosolic ones) are prone to oxidation. In these cases it can be valuable to add some di-thiothreitol (DTT) to the protein solution (ca. 5–50 mM).

Arguably the most crucial parameter for getting protein crystals is the protein purity. Check the protein on a gel before you start with crystallization experiments. As a rule of thumb, 1 μl of a 10 mg/ml sample should give only one band on a Coomassie-stained gel. Especially critical are impurities by protein isoforms or other protein variants as these are so similar that they are likely to be inserted in the same crystal lattice, but they may not be compatible with crystal growth.

Another critical factor is protein storage. In general, it is best to divide the protein into small aliquots of 50–100 μl (concentrated to ca. 10 mg/ml) and then shock-freeze the aliquots and store them at –80°C. Not all proteins, however, stand freezing (and even less freeze drying/lyophylization). If the protein precipitates upon freezing or looses activity, it may be better to store it at 4°C or on ice. Sometimes, it may even be necessary to use the protein straight from the purification column. In cases where the protein activity can be tested, this should be done in order to identify optimal protein storage conditions.

5.3. Protein crystallization

Once the protein is prepared, one needs to decide which crystallization conditions to test. For this, a large number of crystallization screens are available (from companies like Molecular Dimensions or Hampton Research), which

---

[6] CMC = critical micelle concentration.

have been developed based on successful crystallization reports. These screens provide for an easy start. In general, it is recommended to start with the most common screen, Screen nr. 1, and then continue with other screens based on the obtained results. It may, however, also make sense to take the characteristics of the protein you are trying to crystallize into consideration from the very beginning. Are you working with a membrane protein? Then, you should use a screen developed for membrane proteins. Are you working with a lectin? Then, you may check out the information compiled in the BMCD at http://wwwbmcd.nist.gov:8080/bmcd/bmcd.html (searching by keyword for "lectin"). Is there a common theme? In this case, you might like to devise your own crystallization screen. Maybe, one of the proteins in the database (either the crystallization database or the PDB) has especially high sequence identity to the protein you interested in? Then, the crystallization conditions for that particular protein may pave your way to success – best of course if you combine the two approaches and try out commercial sparse matrix screens as well as your own tailor-made ones!

If you are in the lucky situation that you have a dynamic or static light scattering device at hand, it can be a good idea to test how the protein behaves under different conditions (temperatures, buffers, etc.) before setting up the first crystallization experiments [121, 122]. Especially the temperature can be well worth testing. Conditions that indicate a monodisperse solution are much more likely to yield crystals than polydisperse solutions that contain various degrees of protein aggregates.

The next decision to take, concerns the crystallization method. Some techniques lend themselves better to screening then others. Very suitable are methods for which multi-well plates are available (either 96-well or 24-well plates). The plates do not have to be special-made for crystallization; just usual tissue culture plates will do. Personally, we prefer to start with the hanging-drop technique shown in Fig. 8.

The setups are really easy to prepare:
(1) Pipette 0.5–1 ml of the screening solution into the reservoir well (make sure that it is well-mixed in case of hand-made viscous solutions like PEG mixtures).
(2) Grease the rim of the wells with either silicon oil (e.g. NVH oil from Hampton Research) or vacuum grease.
(3) Place a clean cover slip (usually silanized) beside the plate (e.g. on the plate's lid, which provides a convenient and clean surface).
(4) Pipette first the protein solution (1–2 μl) onto the cover slip and then add the reservoir solution on top of it (usually the same volume) (DO NOT mix the solutions by pipetting up and down, because this can lead to protein denaturation).
(5) Take the cover slip and place it upside-down onto the well with the reservoir solution.

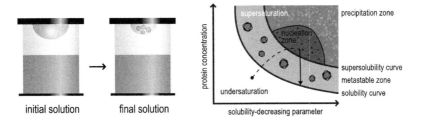

initial solution   final solution

Figure 8. Protein crystallization by the hanging-drop vapor-diffusion technique. The protein-containing drop is placed on a cover slip and hung upside down over a container with reservoir solution (left). Upon equilibration, the drop shrinks and the concentration in the drop rises until it matches the vapor pressure from the reservoir. If the conditions are just right, protein crystals will appear in the drop after some time (middle). Crystallization follows the phase diagram shown on the right, from low concentrations of protein and precipitating agent (undersaturation) over the solubility line into the supersaturated state. Crystal nuclei are formed at somewhat higher concentration than optimal for crystal growth. Upon crystallization, the protein concentration in the drop will again decrease until it reaches the solubility line. In an alternative to spontaneous crystallization (shown here), seeding techniques may be used, in which tiny crystal seeds are placed directly into the metastable zone. This picture was prepared by Alexandre Dmitriev.

(6) Possibly place plasticine into the corners of the lid to raise the lid slightly above the sealed cover slips when closing the lid over the plates.

(7) Store either at room temperature or at 4°C (or at any other temperature that may be suggested by dynamic light scattering experiments; note that a defined, constant temperature can be very valuable for reproducibility).

(8) Regularly check the setups and take detailed notes.

More information can be obtained in the excellent textbooks by Bergfors [2] and Ducruix and Giegé [123].

5.4. Good or bad precipitate?

To tell apart a crystal from aggregated protein precipitate is an easy task. However, to distinguish between promising and bad precipitate is much more difficult. This is where experience makes all the difference.

In order to get you on the right track for developing your own judgment skills, you might like to check out the excellent web site by Therese Bergfors at http://xray.bmc.uu.se/~terese/crystallization/library.html. Once you are able to distinguish promising from bad conditions, you are on the right way to optimize the crystallization experiments.

Good tips for optimization have been reviewed by D'Arcy [124] and Chayen [125] and, in a broader perspective, by Vincentelli *et al.* [9].

5.5. Protein or salt crystals?

Obtaining protein crystals is very exciting! But are those really *protein* crystals? Many experimenters have been very disappointed to find out that their precious crystals turned out to be salt. Here are some simple methods to test this:

(1) X-ray diffraction: If the crystal is big enough to test it in the X-ray beam, the diffraction pattern will show immediately if you crystallized protein or salt. Protein crystals exhibit many closely spaced diffraction spots, while for salt, with its small unit cell, the reflections are far apart (and much stronger!). If you do not see any spots, this can indicate a very weakly (non-)diffracting protein crystal. You should make sure, though, that you did not miss the spots by choosing a too small oscillation range for data collection (20° oscillation should do the trick).
(2) Brute force: Crush the crystal – if it is salt, it will be hard to break and you will hear a distinct clicking sound when snapping the needle on the cover slip. A protein crystal is soft and easily crumbles under a needle (but watch out: never use this method if you only have one single crystal!).
(3) Stains: Protein-staining dyes such as Methylene Blue (staining solutions can be purchased ready-to-use as "Izit" from Hampton Research) can be used in order to identify protein crystals in a non-destructive way. If the crystals turn blue, you can be sure that it is protein. If not, it might be salt, but you could also be dealing with very faintly staining protein crystals, especially if the crystals are small or thin.
(4) Check the reservoir: If you find crystals in the reservoir, the odds are high that the crystals in the drop are not protein either.
(5) Gel: Collect the crystals in an Eppendorf tube, wash them several times in mother liquor (by centrifuging the tube a couple of minutes at ca. 10,000 rpm and then removing the supernatant), then dissolve them and load them on a gel. If you see a distinct band characteristic for your protein – voilà, if not, you might either have crystallized salt, used too little material from the start or lost the protein in the procedure (usually, the method works, though).
(6) Negative control: Set up experiments with the same conditions (including salts and detergents or anything else you might have added to stabilize the protein, e.g. by using the filtrate solution from the concentration step), but without protein. If you get crystals, you know it is not protein.

5.6. Seeding techniques and other tricks

In some cases, it can be beneficial to induce crystal formation by introducing seeds into the crystallization solution. Three main techniques exist:

(1) *Macroseeding*:
Small crystals are introduced into a pre-equilibrated crystallization drop (e.g. with help of a loop or capillary). The crystal seeds should preferably first be washed several times in mother liquor (best in a solution with slightly lower concentration,

such that the surface layers start to dissolve and clean surfaces are available for crystal re-growth). The concentration of the drop, to which the seed crystal is to be added, should be in the metastable range (Fig. 8). Suitable conditions can be found either through parallel seeding experiments into drops with various concentrations or by first plotting out the phase diagram.

This technique works best if perfect small crystals can be obtained, but no crystals large enough to obtain high-resolution X-ray diffraction data.

(2) *Streak seeding*:
Streak seeding works best if large crystals are available, even if they are twinned or not useful for other reasons. For capturing the seeds, a hair or a cat's whisker is taken and gently stroked over the original crystal. Also acupuncture needles work very well for streaking. The hair or needle is then drawn through a series of pre-equilibrated crystallization setups. In this way, it will contain less and less seeds from one drop to the next, giving rise to streaks of crystals in the first drop(s) and single crystals in the last drop(s).

This method should definitely be tried for old crystallization setups, which yielded crystals only in some of the drops.

(3) *Microseeding*:
Microseeds are preferably generated by crushing a large crystal, e.g. with a needle, but they can also be obtained by crushing (many) small crystals with a glass rod or homogenizer. The drop containing the microseeds is then diluted with the mother liquor, to final dilutions of 1:10, 1:100, 1:1,000, and 1:10,000 (always using a fresh pipette tip when preparing the next dilution in the series) and vortexed in order to obtain a homogenous mixture of the seeds. Tiny amounts of these solutions are then added one by one to pre-equilibrated crystallization drops in order to test which dilution gives the best results. The seeding solutions can sometimes be stored for several days to months. Just be careful to vortex them again before using them the next time.

Good overviews over crystal seeding are also given by Bergfors [126] and Stura and Wilson [127]. Very nice illustrations of the seeding technique are shown in Ref. [128]. An interesting variation is so-called cross-seeding (seeding from one crystal form to another, which grows under a different condition) or to use other materials like porous silicon, minerals or crushed frozen hair for seeding [129, 130].

If seeding does not work, but small crystals can be obtained, one may also try to layer the experiments with various oils to slow down equilibration [131] or to transfer drops from a high to a lower precipitant concentration before crystals become visible macroscopically [132]. This method even works for very short crystallization times [133].

5.7. Co-crystallization

Co-crystallization experiments are performed in order to obtain the crystal structures of protein–ligand complexes. This method works exactly like a normal protein crystallization experiment. The only difference is that one has to add a ligand to the drop from the start. Even better if protein and ligand are mixed several hours before setting up the experiment (or over night), so that there is enough time for a complex to form (keep the protein on ice for this, so that it does not denature!). The ligand-to-protein ratio should be at least 1:1 (if equimolar binding is expected); however, usually better results are achieved when using a higher ligand-to-protein ratio (2:1 for strong binders up to 50:1 or more in cases of weak affinity).

5.8. Soaking

For large ligands, co-crystallization is the method of choice. Smaller ligands, however, do not need to be premixed with the protein, as the ligands can easily flow through the water channels of the protein crystals and then bind just as well as in solution. In such a case, one may apply the soaking technique, where one adds the ligand to drops that already contain protein crystals. The obvious advantage of this technique is that the consumption of ligand is much lower. The soaking time can be as short as 2 min, but might need to be adjusted and optimized together with the ligand concentration. Sometimes, the protein crystals crack upon soaking and in this case, one might rather use lower ligand concentrations and longer soaking times. It can also be advantageous to adapt the crystals slowly to the changed conditions by slowly adding higher concentrations of ligand or by transferring the crystals step by step to drops that contain gradually increasing ligand concentrations. An alternative can be to add a tiny grain of the solid compound to the far end of the drop and watch the crystal carefully. As soon as any disturbances become visible, the crystals need to be fished out of the drop with a loop and flash-frozen for data collection.

5.9. Cryoprotection

Fifteen years ago, most crystals were mounted in glass capillaries and data collection was performed at 4–20°C (above the freezing point). Nowadays, this is an exception. In more than 99% of the cases, protein crystals are now flash-frozen (e.g. in liquid nitrogen) and the X-ray data are collected at cryogenic temperatures [33]. In order to prevent ice-formation upon freezing, it is essential that the crystals are first cryoprotected by bathing them in solutions that contain additives such as glycerol, ethylene glycol, methylpentane diol (MPD), low molecular weight polyethylene glycol (e.g. PEG 400 or PEG 1000) or sugars. Also oils or cryo salts can be used [134]. A good overview over tested cryo conditions for the most common commercial screens are given in Ref. [135]. The procedure is similar to the soaking technique and these two methods can easily be combined in one step. Several transfer protocols may be used. Sometimes, it can be advisable to slowly increase the concentration of the cryoprotectant, in consecutive transfers. Alternatively — as preferred by us — the cryoprotectant may be added directly onto the drop, after which the crystal is immediately looped out and plunged into liquid nitrogen.

If the crystal does not diffract, one may test thawing it and then refreezing ("cryo-annealing") or other tricks like dehydration or cross-linking [136]. Often, this does not help (but no damage is done either), but in some cases amazing resurrections were reported and the such-treated crystals diffracted brilliantly after the treatment.

5.10. X-ray data collection

A good overview of parameters to be considered for optimal data collection and a discussion of data collection strategies are given by Evans [137] and Dauter [138], respectively.

5.11. Crystallographic questions?

One last practical tip, in case you run into unexpected trouble at any step of the crystallographic analysis: Do not hesitate to contact the CCP4 bulletin board (http://www.ccp4.ac.uk/ccp4bb.php) with questions. Even off-side topics (like the deglycosylation issue discussed above) are vented. This bulletin board is unique, as almost all protein crystallographers are subscribers to this electronic mailing list and willing to answer your pressing questions.

References

[1] H.M. Berman, J. Westbrook, Z. Feng, G. Gilliland, T.N. Bhat, H. Weissig, I.N. Shindyalov and P.E. Bourne, Nucl. Acids Res., 28 (2000) 235.

[2] T.M. Bergfors (Ed.), Protein crystallization: Techniques, strategies, and tips, International University Line, La Jolla, CA, USA, 1999.

[3] D. Blow, Outline of crystallography for biologists, Oxford University Press, Oxford, UK, 2002.

[4] J. Drenth, Principles of X-ray crystallography, Springer-Verlag, New York, USA, 1994.

[5] A. McPherson, Macromolecular crystallography, Wiley-Liss, Hoboken, NJ, USA, 2003.

[6] G. Rhodes, Crystallography made crystal clear, Academic Press, San Diego, CA, USA, 1993.

[7] Z.S. Derewenda, Methods, 34 (2004) 354.

[8] S. Yokoyama, Curr. Opin. Chem. Biol., 7 (2003) 39.

[9] R. Vincentelli, C. Bignon, A. Gruez, S. Canaan, G. Sulzenbacher, M. Tegoni, V. Campanacci and C. Cambillau, Acc. Chem. Res., 36 (2003) 165.

[10] W.A. Hendrickson, J.R. Horton and D.M. LeMaster, EMBO J., 9 (1990) 1665.

[11] S. Doublié, Methods Enzymol., 276 (1997) 523.

[12] S.A. Guerrero, H.-J. Hecht, B. Hofmann, H. Biebl and M. Singh, Appl. Microbiol. Biotechnol., 56 (2001) 718.

[13] M.-P. Strub, F. Hoh, J.-F. Sanchez, J.M. Strub, A. Böck, A. Aumelas and C. Dumas, Structure, 11 (2003) 1359.

[14] W.A. Hendrickson, Science, 254 (1991) 51.

[15] S.E. Ealick, Curr. Opin. Chem. Biol., 4 (2000) 495.

[16] Z. Dauter, M. Dauter, E. de La Fortelle, G. Bricogne and G.M. Sheldrick, J. Mol. Biol., 289 (1999) 83.

[17] U.A. Ramagopal, M. Dauter and Z. Dauter, Acta Crystallogr. D Biol. Crystallogr., 59 (2003) 1020.

[18] G.N. Sarma and P.A. Karplus, Acta Crystallogr. D Biol. Crystallogr., 62 (2006) 707.

[19] I. Usón and G.M. Sheldrick, Curr. Opin. Struct. Biol., 9 (1999) 643.

[20] I. Usón, B. Schmidt, R. von Bülow, S. Grimme, K. von Figura, M. Dauter, K.R. Rajashankar, Z. Dauter and G.M. Sheldrick, Acta Crystallogr. D Biol. Crystallogr., 59 (2003) 57.

[21] M.H. Nanao, G.M. Sheldrick and R.B.G. Ravelli, Acta Crystallogr. D Biol. Crystallogr., 61 (2005) 1227.

[22] R.B.G. Ravelli, H.-K. Schröder Leiros, B. Pan, M. Caffrey and S. McSweeney, Structure, 11 (2003) 217.

[23] P.H. Zwart, S. Banumathi, M. Dauter and Z. Dauter, Acta Crystallogr. D Biol. Crystallogr., 60 (2004) 1958.

[24] M.H. Nanao and R.B. Ravelli, Structure, 14 (2006) 791.

[25] G. Rosenbaum, K.C. Holmes and J. Witz, Nature, 230 (1971) 434.

[26] K. Moffat and Z. Ren, Curr. Opin. Struct. Biol., 7 (1997) 689.

[27] C. Riekel, M. Burghammer and G. Schertler, Curr. Opin. Struct. Biol., 15 (2005) 556.

[28] S. Arzt, A. Beteva, F. Cipriani, S. Delageniere, F. Felisaz, G. Förstner, E. Gordon, L. Launer, B. Lavault, G. Leonard, T. Mairs, A. McCarthy, J. McCarthy, S. McSweeney, J. Meyer, E. Mitchell, S. Monaco, D. Nurizzo, R. Ravelli, V. Rey, W. Shepard, D. Spruce, O. Svensson and P. Theveneau, Prog. Biophys. Mol. Biol., 89 (2005) 124.

[29] E. Girard, P. Legrand, O. Roudenko, L. Roussier, P. Gourhant, J. Gibelin, D. Dalle, M. Ounsy, A.W. Thompson, O. Svensson, M.-O. Cordier, S. Robin, R. Quiniou and J.-P. Steyer, Acta Crystallogr. D Biol. Crystallogr., 62 (2006) 12.

[30] Future looks bright for table-top synchrotron. Nature, 434 (2005) 8.

[31] H. Hope, Acta Crystallogr. B, 44 (1988) 22.

[32] E.F. Garman and T.R. Schneider, J. Appl. Cryst., 30 (1997) 211.

[33] E.F. Garman and S. Doublié, Methods Enzymol., 368 (2003) 188.

[34] Z. Dauter, Curr. Opin. Struct. Biol., 12 (2002) 674.

[35] A. Perrakis, R. Morris and V.S. Lamzin, Nat. Struct. Biol., 6 (1999) 458.

[36] S.X. Cohen, R.J. Morris, F.J. Fernandez, M. Ben Jelloul, M. Kakaris, V. Parthasarathy, V.S. Lamzin, G.J. Kleywegt and A. Perrakis, Acta Crystallogr. D Biol. Crystallogr., 60 (2004) 2222.

[37] P.D. Adams, N.S. Pannu, R.J. Read and A.T. Brunger, Acta Crystallogr. D Biol. Crystallogr., 55 (1999) 181.

[38] P.D. Adams, N.S. Pannu, R.J. Read and A.T. Brunger, Proc. Natl. Acad. Sci. U.S.A., 94 (1997) 5018.

[39] E. Dodson, G.J. Kleywegt and K. Wilson, Acta Crystallogr. D Biol. Crystallogr., 52 (1996) 228.

[40] G.J. Kleywegt, Acta Crystallogr. D Biol. Crystallogr., 56 (2000) 249.

[41] J. Jancarik and S.-H. Kim, J. Appl. Crystallogr., 24 (1991) 409.

[42] R. Page and R.C. Stevens, Methods, 34 (2004) 373.

[43] G.L. Gilliland, J. Cryst. Growth, 90 (1988) 51.

[44] G.L. Gilliland, M. Tung, D.M. Blakeslee and J.E. Ladner, Acta Crystallogr. D Biol. Crystallogr., 50 (1994) 408.

[45] H. Michel and D. Oesterhelt, Proc. Natl. Acad. Sci. U.S.A., 77 (1980) 1283.

[46] C. Hunte and H. Michel, Curr. Opin. Struct. Biol., 12 (2002) 503.

[47] M. Caffrey, J. Struct. Biol., 142 (2003) 108.

[48] M.C. Wiener, Methods, 34 (2004) 364.

[49] M.L. Pusey, Z.-J. Liu, W. Tempel, J. Praissman, D. Lin, B.-C. Wang, J.A. Gavira and J.D. Ng, Prog. Biophys. Mol. Biol., 88 (2005) 359.

[50] G.J. Kleywegt, K. Henrick, E.J. Dodson and D.M.F. van Aalten, Structure, 11 (2003) 1051.

[51] R.A. Engh and R. Huber, Acta Crystallogr. A, 47 (1991) 392.

[52] A.T. Brünger, Nature, 355 (1992) 472.

[53] G.J. Kleywegt, Acta Crystallogr. D Biol. Crystallogr., 52 (1996) 842.

[54] G.J. Kleywegt, M.R. Harris, J.-Y. Zou, T.C. Taylor, A. Wählby and T.A. Jones, Acta Crystallogr. D Biol. Crystallogr., 60 (2004) 2240.

[55] A. Minichino, J. Habash, J. Raftery and J.R. Helliwell, Acta Crystallogr. D Biol. Crystallogr., 59 (2003) 843.

[56] L. Buts, R. Loris, G.E. De, S. Oscarson, M. Lahmann, J. Messens, E. Brosens, L. Wyns, G.H. De and J. Bouckaert, Acta Crystallogr. D Biol. Crystallogr., 59 (2003) 1012.

[57] S.F. Gallego del, J. Gomez, S. Hoos, C.S. Nagano, B.S. Cavada, P. England and J.J. Calvete, Acta Crystallogr. F Struct. Biol. Cryst. Commun., 61 (2005) 326.

[58] T. Lütteke and C.W. von der Lieth, BMC Bioinformatics, 5 (2004) 69.

[59] T. Lütteke, M. Frank and C.W. von der Lieth, Nucleic Acids Res., 33 (2005) 242.

[60] W. Kühlbrandt and K.A. Williams, Curr. Opin. Chem. Biol., 3 (1999) 537.

[61] D.A.A. Myles, Curr. Opin. Struct. Biol., 16 (2006) 630.

[62] N. Niimura, S. Arai, K. Kurihara, T. Chatake, I. Tanaka and R. Bau, Cell. Mol. Life Sci., 63 (2006) 285.

[63] T. Ruiz, I. Erk and J. Lepault, Biol. Cell., 80 (1994) 203.

[64] T. Walz and N. Grigorieff, J. Struct. Biol., 121 (1998) 142.

[65] P.A. Thuman-Commike, FEBS Lett., 505 (2001) 199.

[66] L. Wang and F.J. Sigworth, Physiology (Bethesda), 21 (2006) 13.

[67] M. van Heel, B. Gowen, R. Matadeen, E.V. Orlova, R. Finn, T. Pape, D. Cohen, H. Stark, R. Schmidt, M. Schatz and A. Patwardhan, Q. Rev. Biophys., 33 (2000) 307.

[68] J. Frank, Annu. Rev. Biophys. Biomol. Struct., 31 (2002) 303.

[69] M.G. Rossmann, M.C. Morais, P.G. Leiman and W. Zhang, Structure, 13 (2005) 355.

[70] K. Mitra and J. Frank, Annu. Rev. Biophys. Biomol. Struct., 35 (2006) 299.

[71] T. Gonen, Y. Cheng, P. Sliz, Y. Hiroaki, Y. Fujiyoshi, S.C. Harrison and T. Walz, Nature, 438 (2005) 633.

[72] N.R. Chandra, N. Kumar, J. Jeyakani, D.D. Singh, S.B. Gowda and M. Prathima, Glycobiology, 16 (2006) 938.

[73] A.G. Murzin, S.E. Brenner, T. Hubbard and C. Chothia, J. Mol. Biol., 247 (1995) 536.

[74] J. Habash, J. Raftery, R. Nuttall, H.J. Price, C. Wilkinson, A.J. Kalb and J.R. Helliwell, Acta Crystallogr. D Biol. Crystallogr., 56 (2000) 541.

[75] J.L. Asensio, F.J. Canada, M. Bruix, A. Rodriguez-Romero and J. Jimenez-Barbero, Eur. J. Biochem., 230 (1995) 621.

[76] K.D. Hardman and C.F. Ainsworth, Biochem., 15 (1976) 1120.

[77] R. Loris, T. Hamelryck, J. Bouckaert and L. Wyns, Biochim. Biophys. Acta, 1383 (1998) 9.

[78] K. Drickamer and M.E. Taylor, Annu. Rev. Cell Biol., 9 (1993) 237.

[79] W.I. Weis and K. Drickamer, Annu. Rev. Biochem., 65 (1996) 441.

[80] K. Drickamer, Structure, 5 (1997) 465.

[81] M. Fernández-Alonso, F.J. Cañada, J. Jiménez-Barbero and G. Cuevas, J. Am. Chem. Soc., 127 (2005) 7379.

[82] Å. Holmner, M. Lebens, S. Teneberg, J. Ångström, M. Ökvist and U. Krengel, Structure, 12 (2004) 1655. Erratum in: Structure 15 (2007) 253.

[83] R. Ravishankar, M. Ravindran, K. Suguna, A. Surolia and M. Vijayan, Curr. Sci., 72 (1997) 855.

[84] J.V. Pratap, G.M. Bradbrook, G.B. Reddy, A. Surolia, J. Raftery, J.R. Helliwell and M. Vijayan, Acta Crystallogr. D Biol. Crystallogr., 57 (2001) 1584.

[85] P. Adhikari, K. Bachhawat-Sikder, C.J. Thomas, R. Ravishankar, A.A. Jeyaprakash, V. Sharma, M. Vijayan and A. Surolia, J. Biol. Chem., 276 (2001) 40734.

[86] Y. Guo, H. Feinberg, E. Conroy, D.A. Mitchell, R. Alvarez, O. Blixt, M.E. Taylor, W.I. Weis and K. Drickamer, Nat. Struct. Mol. Biol., 11 (2004), 591.

[87] H.G. Beisel, S. Kawabata, S. Iwanaga, R. Huber and W. Bode, EMBO J., 18 (1999) 2313.

[88] M.A. Bianchet, E.W. Odom, G.R. Vasta and L.M. Amzel, Nat. Struct. Biol., 9 (2002) 628.

[89] J.F. Sanchez, J. Lescar, V. Chazalet, A. Audfray, J. Gagnon, R. Alvarez, C. Breton, A. Imberty and E.P. Mitchell, J. Biol. Chem., 281 (2006) 20171.

[90] M. Wimmerova, E. Mitchell, J.F. Sanchez, C. Gautier and A. Imberty, J. Biol. Chem., 278 (2003) 27059.

[91] G. Cioci, E.P. Mitchell, V. Chazalet, H. Debray, S. Oscarson, M. Lahmann, C. Gautier, C. Breton, S. Perez and A. Imberty, J. Mol. Biol., 357 (2006) 1575.

[92] P. Paaventhan, J.S. Joseph, S.V. Seow, S. Vaday, H. Robinson, K.Y. Chua and P.R. Kolatkar, J. Mol. Biol., 332 (2003) 461.

[93] C. Birck, L. Damian, C. Marty-Detraves, A. Lougarre, C. Schulze-Briese, P. Koehl, D. Fournier, L. Paquereau and J.P. Samama, J. Mol. Biol., 344 (2004) 1409.

[94] M.E. Carrizo, S. Capaldi, M. Perduca, F.J. Irazoqui, G.A. Nores and H.L. Monaco, J. Biol. Chem., 280 (2005) 10614.

[95] M. Ban, H.J. Yoon, E. Demirkan, S. Utsumi, B. Mikami and F. Yagi, J. Mol. Biol., 351 (2005) 695.

[96] P.J. Walser, P.W. Haebel, M. Kunzler, D. Sargent, U. Kues, M. Aebi and N. Ban, Structure, 12 (2004) 689.

[97] J.M. Mancheno, H. Tateno, I.J. Goldstein, M. Martinez-Ripoll and J.A. Hermoso, J. Biol. Chem., 280 (2005) 17251.

[98] N. Kochibe and K.L. Matta, J. Biol. Chem., 264 (1989) 173.

[99] H. Ueda, H. Matsumoto, N. Takahashi and H. Ogawa, J. Biol. Chem., 277 (2002) 24916.

[100] C.S. Wright, J. Mol. Biol., 178 (1984) 91.

[101] C.S. Wright, J. Mol. Biol., 139 (1980) 53.

[102] E.A. Merritt and W.G. Hol, Curr. Opin. Struct. Biol., 5 (1995) 165.

[103] K.W. Dodson, J.S. Pinkner, T. Rose, G. Magnusson, S.J. Hultgren and G. Waksman, Cell, 105 (2001) 733.

[104] C.S. Hung, J. Bouckaert, D. Hung, J. Pinkner, C. Widberg, A. DeFusco, C.G. Auguste, R. Strouse, S. Langermann, G. Waksman and S.J. Hultgren, Mol. Microbiol., 44 (2002) 903.

[105] M.C. Merckel, J. Tanskanen, S. Edelman, B. Westerlund-Wikstrom, T.K. Korhonen and A. Goldman, J. Mol. Biol., 331 (2003) 897.

[106] C.A. Bewley, K.R. Gustafson, M.R. Boyd, D.G. Covell, A. Bax, G.M. Clore and A.M. Gronenborn, Nat. Struct. Biol., 5 (1998) 571.

[107] I. Botos, B.R. O'Keefe, S.R. Shenoy, L.K. Cartner, D.M. Ratner, P.H. Seeberger, M.R. Boyd and A. Wlodawer, J. Biol. Chem., 277 (2002) 34336.

[108] A. Imberty, M. Wimmerova, C. Sabin and E.P. Mitchell, Structures and roles of *Pseudomonas aeruginosa* lectins. In: C.A. Bewley (Ed.), Protein-carbohydrate interactions in infectious disease, The Royal Society of Chemistry, Cambridge, 2006, pp. 30–48.

[109] G. Cioci, E.P. Mitchell, C. Gautier, M. Wimmerova, D. Sudakevitz, S. Perez, N. Gilboa-Garber and A. Imberty, FEBS Lett., 555 (2003) 297.

[110] S. Perret, C. Sabin, C. Dumon, M. Pokorna, C. Gautier, O. Galanina, S. Ilia, N. Bovin, M. Nicaise, M. Desmadril, N. Gilboa-Garber, M. Wimmerova, E.P. Mitchell and A. Imberty, Biochem. J., 389 (2005) 325.

[111] J. Holmgren, M. Lindblad, P. Fredman, L. Svennerholm and H. Myrvold, Gastroenterology, 89 (1985) 27.

[112] J. Holmgren, P. Fredman, M. Lindblad, A.M. Svennerholm and L. Svennerholm, Infect. Immun., 38 (1982) 424.

[113] P.A. Orlandi, D.R. Critchley and P.H. Fishman, Biochem., 33 (1994) 12886.

[114] S. Teneberg, T.R. Hirst, J. Ångström and K.-A. Karlsson, Glycoconj. J., 11 (1994) 533.

[115] M. Bäckström, V. Shahabi, S. Johansson, S. Teneberg, A. Kjellberg, H. Miller-Podraza, J. Holmgren and M. Lebens, Mol. Microbiol., 24 (1997) 489.

[116] J. Ångström, M. Bäckström, A. Berntsson, N. Karlsson, J. Holmgren, K.-A. Karlsson, M. Lebens and S. Teneberg, J. Biol. Chem., 275 (2000) 3231.

[117] Å. Holmner Rocklöv, Molecular recognition of carbohydrates – Structural and functional characterisation of bacterial toxins and fungal lectins, PhD thesis, Chalmers University of Technology, Gothenburg, Sweden, 2005.

[118] F. Grueninger-Leitch, A. D'Arcy, B. D'Arcy and C. Chene, Protein Sci., 5 (1996) 2617.

[119] J. Abrahamson and S. Iwata, Crystallization of membrane proteins. In: T.M. Bergfors (Ed.), Protein crystallization: Techniques, strategies, and tips, International University Line, La Jolla, CA, USA, 1999, pp. 199–210.

[120] C. Hunte, G. von Jagow and H. Schagger (Eds.), Membrane protein purification and crystallization: A practical guide, Academic Press, San Diego, CA, USA, 2003.

[121] T.M. Bergfors, Dynamic light scattering. In: T.M. Bergfors (Ed.), Protein crystallization: Techniques, strategies, and tips, International University Line, La Jolla, CA, USA, 1999, pp. 29–38.

[122] A.R. Ferré-D'Amaré and S.K. Burley, Structure, 2 (1994) 357.

[123] A. Ducruix and R. Giegé (Eds.), Crystallization of nucleic acids and proteins. A Practical approach, Oxford University Press, Oxford, UK, 1992.

[124] A. D'Arcy, Acta Crystallogr. D Biol. Crystallogr., 50 (1994) 469.

[125] N.E. Chayen, Curr. Opin. Struct. Biol., 14 (2004) 577.

[126] T. Bergfors, J. Struct. Biol., 142 (2003) 66.

[127] E.A. Stura and I.A. Wilson, Seeding techniques. In: A. Ducruix and R. Giegé (Eds.), Crystallization of nucleic acids and proteins, Oxford University Press, Oxford, 1992, UK, pp. 99–125.

[128] D.E. McRee, Practical protein crystallography, Academic Press, San Diego, CA, USA, 1993.

[129] N.E. Chayen, E. Saridakis, R. El-Bahar and Y. Nemirovsky, J. Mol. Biol., 312 (2001) 591.

[130] A. D'Arcy, A. Mac Sweeney and A. Haber, Acta Crystallogr. D Biol. Crystallogr., 59 (2003) 1343.

[131] N.E. Chayen, Structure, 5 (1997) 1269.

[132] E. Saridakis and N.E. Chayen, Protein Sci., 9 (2000) 755.

[133] U. Krengel, R. Dey, S. Sasso, M. Ökvist, C. Ramakrishnan and P. Kast, Acta Crystallogr. F Struct. Biol. Cryst. Commun., 62 (2006) 441.

[134] K.A. Rubinson, J.E. Ladner, M. Tordova and G.L. Gilliland, Acta Crystallogr. D Biol. Crystallogr., 56 (2000) 996.

[135] M.B. McFerrin and E.H. Snell, Appl. Crystallogr., 35 (2002) 538.

[136] B. Heras and J.L. Martin, Acta Crystallogr. D Biol. Crystallogr., 61 (2005) 1173.

[137] P.R. Evans, Acta Crystallogr. D Biol. Crystallogr., 55 (1999) 1771.

[138] Z. Dauter, Acta Crystallogr. D Biol. Crystallogr., 55 (1999) 1703.

[139] C. Svensson, S. Teneberg, C.L. Nilsson, A. Kjellberg, F.P. Schwarz, N. Sharon and U. Krengel, J. Mol. Biol., 321 (2002) 69.

Lectins: Analytical Technologies
C.L. Nilsson (Editor)

Chapter 3

NMR Investigations of Lectin–Carbohydrate Interactions

Patrick Groves[a], Angeles Canales[a], M. Isabel Chávez[a, b], Małgorzata
Palczewska[c], Dolores Díaz[a], F. Javier Cañada[a] and Jesús Jiménez-Barbero[a]

[a]*Department of Protein Science, Centro Investigaciones Biológicas – CSIC, Ramiro de Maeztu 9, 28040 Madrid, Spain*
[b]*Instituto de Quimica, Universidad Nacional Autónoma de México, Ciudad Universitaria, 04510 Coyoacán (México DF), Mexico*
[c]*Department of Molecular and Cellular Biology, National Center for Biotechnology, CNB-CSIC, Campus Cantoblanco, 28049 Madrid, Spain*

1. Introduction

To walk into a room with a nuclear magnetic resonance (NMR) spectrometer is like walking into a candy store. There are thousands of different types and flavors of candy (experiments) to choose from, although the NMR experiments are hidden in the computer rather than displayed in big jars on the shelf. It is not always easy for the salesman (spectroscopist) to select the perfect mixed bag of candy. In this review, we provide a guide for the salesmen and parents on how to best formulate that mixed bag of candy, obtain the fastest results and get the best value for money. Each year, new flavors and types of candy become available and we will provide a glimpse into next year's treats.

Lectins are proteins that interact with carbohydrate ligands and often occur in oligomeric forms. Lectins are rarely suitable subjects for NMR structural studies due to their size. Carbohydrate ligands also provide challenges to the NMR spectroscopist. The poor dispersion of signals, compared to similar sized molecules, and the resulting overlap of resonances results in an ambiguity in data assignment. [13]C- and ^{15}N-labeling is regularly applied to protein and DNA samples to disperse overlapped resonances into two or more dimensions. However, carbohydrate synthesis usually precludes isotope labeling on the basis of cost. The inter-unit linkages in carbohydrates also provide a conformational flexibility, expressed as a diverse conformational ensemble, which broadens resonances. In other words, lectin–carbohydrate complexes present significant challenges to 3D structure determination for which NMR is a well-known tool. Fortunately, several NMR experiments are available that help us to overcome some of these problems and allow us to extract reliable data. Although highly detailed 3D lectin–carbohydrate structures still cannot be routinely obtained, a wide range of other useful information about the binding site and binding processes are available by NMR methods. This review is aimed at enlightening researchers without a specialized knowledge of the NMR spectroscopy field. We cover various experimental strategies that deliver distinctive data. We assume access to an instrument found in most biological departments: A 500 MHz instrument equipped with a multinuclear, gradient probe. As NMR spectrometers generate a large amount of data, the bottlenecks are generally the availability of samples and people to analyze the data. Therefore, if no instrument is available, there is likely to be a friendly NMR spectroscopist open to collaboration in a neighboring department or institute, or at a national center for NMR. We will indicate when more sophisticated equipment than a basic 500 MHz instrument is needed.

This review is organized into several sections. Following this introduction, we provide a short introduction on methods to obtain recombinant proteins suitable for NMR studies. Section 3 discusses one of the basic principles that govern what is observed by NMR, while Section 4 introduces the simplest NMR experiments that allow the measurement of chemical shifts of lectin and/or ligand during a titration to yield binding data. The assignment of the chemical shift data can provide an insight into structural changes, or the location of the binding site, although this is best done after obtaining a fuller 3D picture of the complex, as detailed in Section 5. Section 6 focuses on the ligand. Under the right conditions, it is possible to obtain data that specifically detail the bound conformation of the ligand. The final section mentions experimental methods to look out for in the coming years that promise to improve the existing strategies.

2. Protein Expression and Purification

Expression from a heterologous system is the most common way to obtain large amounts of protein material and *Escherichia coli* (*E. coli*) will probably provide the best protein expression system. However, many lectins, derived from plant or animal sources, can be incompatible with *E. coli*. Other options include yeast, baculovirus and cell-free systems. Protein yield is usually the prime focus for optimization, particularly for the production of isotope-labeled proteins.

2.1. Escherichia coli

E. coli is by far the most widely used host organism for the bacterial expression of proteins with expression yields of 0.5–0.8 g/L possible. However, this is only if the expressed protein is compatible with the *E. coli* expression and folding machineries. The choice of bacterial strain, medium formulation and the promoter and expression system are critical in maximizing protein yields.

Expression strains should be deficient in the most harmful natural proteases, maintain a stable level of the expression plasmid and confer the genetic elements relevant to the expression system. *E. coli* BL21 conforms to these specifications, is non-pathogenic and unlikely to survive in host tissues or cause disease [1]. There are strains derived from BL21 containing a variety of useful modifications, such as, recA negative strains for the stabilization of target plasmids containing repetitive (oligo)-nucleotide sequences (Novagen BLR strain), trxB/gor negative mutants for the enhancement of cytoplasmic disulfide bond formation (Novagen Origami and AD494 strains), lacY mutants enabling adjustable levels of protein expression (Novagen Tuner series) and mutants for the soluble expression of inclusion body prone and membrane proteins (Avidis C41(DE3) and C43(DE3) strains).

Many heterologously expressed proteins are expressed as fusion or chimeric proteins including a partner or "tag" linked to the target protein. Most tags provide one-step protein purification using specific affinity resins. However, the tag can also protect the target protein from intracellular proteolysis [2, 3], enhance solubility [4–6] or provide an expression reporter [7]. Good expression levels of an N-terminal tag can often be transferred to a poorly expressing target protein, probably as a result of mRNA stabilization [8]. Hammarstrom et al. [9] discuss the merits of different tags after testing 32 different proteins with seven different N-terminal tags. Common affinity tags are the polyhistidine tag (His-tag), which is compatible with immobilized metal affinity chromatography (IMAC), and the glutathione S-transferase (GST) tag for purification on glutathione-based resins. The His-tag is not always favored for NMR samples as it can be difficult to remove by proteolytic cleavage and its presence can chelate trace metals that cause spectrum and sample degradation.

The compatibility of codon usage between an expressed protein and host organism is an important factor in obtaining high levels of protein expression. All cells use a specific subset of the 61 available amino acid codons for the production of most mRNA molecules and the degree of application of each triplet varies between organisms. In case of major discrepancies, the point mutation of one codon to another can be considered. Alternatively, a specially prepared *E. coli* host enriched in rare tRNA is a good alternative, as for instance Rosetta (Novagene) which contains tRNA genes for most "problematic" rarely used codons encoding Arg, Ile, Gly, Leu and Pro.

2.2. Pichia pastoris

Methylothropic yeasts provide another protein expression system and *P. pastoris* is the most commonly used. Unlike *E. coli*, *P. pastoris* is a eukaryote so it can provide a variety of post-translational modifications necessary for proper expression/folding/stability or functionality of a protein. This feature is particularly relevant for target proteins containing multiple disulfide bonds or requiring glycosylation, phosphorylation or the absence of an amino-terminal methionine. *P. pastoris* expression systems also permit secretion of the authentic protein in a soluble form.

Glycosylation should be specifically considered when preparing a protein from yeast. Glycosylation can be advantageous for protein stability and/or solubility. To avoid it, we have to either choose intracellular expression, or the sequence of the secreted protein should be checked for the presence of possible glycosylation signals and modified by mutation where necessary.

Numerous strains of *P. pastoris* with a wide range of genotypes and commercially available vectors exist. Vectors without a secretion signal upstream of the multiple cloning sites can be used for either intracellular expression or extracellular expression by cloning the protein together with its own native secretion signal. Other vectors contain a signal sequence such as the alpha-mating factor pre-pro-leader sequence (-MF) prior to the cloning site, resulting in secretion of the target protein. *P. pastoris* is transformed by integration of the expression cassette into the chromosome at a specific locus. The resulting transformants are very stable with a rate of vector loss less than 1% per generation in the absence of selectable markers. The protein expression is under the control of an AOX (alcohol oxidase) promoter. In yeast cells, there are two AOX genes named AOX1 and AOX2, with the promoter of AOX1 being very highly up-regulated in the absence of glucose, while AOX2 shows a lower up-regulation. The AOX1 promoter can be used to drive the expression of recombinant protein to high levels, even with a single integrated copy of the expression cassette. Other benefits are that AOX1 can be switched off, as non-limiting amounts of carbon sources such as glycerol and glucose repress the promoter and minimize the possibility of selecting non-expressing mutants/contaminants during culture growth. At the same time, it allows production of proteins that may be toxic to *P. pastoris* cells. Transformants with a Mut$^+$ phenotype (AOX1 promoter) and Mut^s phenotype (AOX2 promoter) should be tested as it is hard to predict which phenotype will be better for target protein expression.

Flask cultures of *P. pastoris* can be optimized. In small-scale flask cultures, the *P. pastoris* cell density is lower and the extent of aeration more limited than in fermentor cultures. Baffled flasks can be used to increase the amount of available oxygen and growth rates. Media buffered to pH values between pH 3.0 and 6.0 can reduce the degree of proteolysis of the recombinant protein. Generally, expression of protein is performed at 30°C. An increase of temperature above 32°C normally stops protein expression, but lowering the temperature to 23–24°C can increase protein expression levels.

2.3. Cell-free system

In recent years, there have been several improvements in cell-free expression systems that provide expression yields of 5–10 mg/mL. The improvements include the optimization of lysate composition, introduction of semi-continuous reactions and energy regeneration systems. A number of reports have been published on cell-free systems from *E. coli* showing the possibility of producing large amounts of protein for NMR structural studies.

Several commercial cell-free systems are currently available and include the Rapid Translation System (RTS) platform (Roche) and Expressway (Invitrogen), the former offering a complete scalable system for small-scale PCR-mediated screening to large-scale (10 mL) production in a variety of cell lysates (*E. coli* and wheat germ bases). However, the expression of proteins that require SH-bond formation or membrane expression is not possible in either case. Furthermore, the commercial systems, while very quick and easy to conduct, are generally cost prohibitive. The productivity of the *E. coli* cell-free protein synthesis system (a coupled transcription–translation system) has been improved to 6 mg/mL of reaction mixture by several hours of dialysis for chloramphenicol acetyltransferase. The same method was successfully used for isotope labeling of a protein for NMR spectroscopy. The cell-free protein synthesis method is now routinely used for the uniform ^{15}N, uniform ^{13}C/^{15}N and amino acid-specific labeling of samples for NMR spectroscopy.

A eukaryotic cell-free system from wheat embryos has also been developed. The system was prepared from extensively washed embryos that were devoid of translation inhibitors. Translation continued for 4 h in a batch system, and >60 h in a dialysis system, yielding 1–4 mg of proteins.

2.4. Baculovirus system

Over the past 20 years, the baculovirus–insect cell expression system has become widely used for the production of recombinant proteins. A number of technological improvements have eliminated the original tedious procedures required to identify and isolate recombinant viruses. A wide variety of transfer vectors have been developed along with simplified recombinant virus isolation and quantification methods, advances in cell culture technology and the commercial availability of reagents. These enhancements have resulted in a virus-based expression system that is safe, easy to use and readily amenable to scale-up. A number of studies have shown enhanced protein production following cotransfection with baculoviruses expressing chaperone proteins, which are known to aid in the folding and modification of newly synthesized proteins. Significant increases in expression levels have also been reported by the addition of various DNA elements to the virus. This system is suitable for difficult to produce proteins but the cost of isotope labeling is prohibitive.

3. Fast or Slow Exchange

There is one principle of NMR spectroscopy, particularly for the observation of ligand-binding processes, which must be explained first. The question of fast or slow exchange strongly affects the obtained results and governs several of the experiments that will be described later.

Fig. 1 provides an analogy of slow and fast exchange. Fig. 1B and C illustrates what happens when kids take a break from their lessons. In slow exchange (Fig. 1B), we observe two, separate populations with the swing permanently occupied by a single child and the others hanging around with nothing exciting to do. In NMR terms, we also see two populations of resonances for the free and bound forms of the ligand although some of these resonances will be coincident. The relative size (integrals) of the two sets of populations/resonances represents the number of children/ligands that are on swings/receptors, or are not. Similarly, the relative integrals of the free and bound resonances relate to the proportions of free and bound molecules in the total population. Fig. 1B provides a fairly accurate, static picture of what is observed during the whole experimental time. In contrast, Fig. 1C represents a snapshot of the more dynamic, fast-exchange regime. Averaged over time, we see only one population that occupies the swing/lectin for part of the experimental time. In NMR terms, we see one set of peaks that are weight-averaged between free and bound chemical shifts. Both sets of data, whether derived from integrals or chemical shift positions, can be analyzed by standard equilibrium equations and software to yield binding data.

Another fundamental factor to consider is that the break times are of different lengths in different schools. In other words, we may observe slow exchange during a short break/experiment and fast exchange during a long break/experiment – even if the sample and exchange rates are unchanged. The data in NMR experiments typically evolve and are collected over the 20–2,000 ms range (and repeated for several min/h to improve signal intensity). This means that the off-rate, k_{off}, is around 10 s^{-1} or faster in order to occur in the fast-exchange regime.

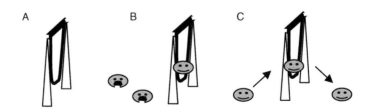

Figure 1. The lectin as a swing in the playground. (A) The swing is free when the kids are in school. (B) Slow exchange: The swing/lectin is permanently occupied by one child/ligand during break. (C) Fast exchange: All the kids get a turn.

However, this exchange rate will not be compatible with the shorter transferred NMR (TR-NMR) experiments described in Section 6. Typically, ligands with milli- to micro-molar affinities (10^3–10^6 M^{-1}) have suitable kinetics for TR-NMR experiments. Occasionally, ligands with nanomolar affinities (10^9 M^{-1}) have fast exchange kinetics.

Only a basic description of slow and fast exchange is given here (see Cavanagh et al. [10], for a full mathematical description). An intermediate exchange state can exist, which is an average or mixture of slow and fast exchange. It is difficult to analyze such data and it is better to perturb the experimental conditions to force either fast or slow exchange onto the system. This can be achieved by changing the temperature or pH, or adding a chemical that influences the strength of binding, for example, sodium chloride to weaken electrostatic interactions, or organic cosolvents, often tolerated by proteins at low concentrations, to change the solvent viscosity and/or interfere with hydrophobic interactions.

4. How to Track Chemical Shift Changes

What will we see after placing the sample into the magnet of the NMR instrument? Hopefully, we observe a spectrum comprising a series of resonances. Each proton (or other nucleus) in the sample should provide a signal although these can be identical, coincident or too broad or small to observe. The resonance positions (measured as a chemical shift) are sensitive to the environment of the nuclei. So amides and aromatics are generally seen to the left, carbohydrate ring protons in the middle and methyls on the right of the spectrum. The simplest, most basic NMR experiments available to the spectroscopist can yield interesting results. Some of the resonances in a spectrum will either shift or reappear in new positions when we titrate a carbohydrate ligand into a lectin. Peak broadening may also be observed.

The chemical shift (δ) of a particular resonance in the NMR spectrum depends on a number of factors, including the physical environment surrounding the nucleus that gives rise to the resonance. The process of ligand binding can subtly change the environment of the ligand nuclei, leading to changes in the chemical shifts, Fig. 2. (This is like the transfer of dirt between clothing and swing.) The resulting chemical shift perturbations need to be interpreted with caution regarding structure. The largest changes in chemical shift tend to occur for ligand resonances that come into close contact with aromatic rings or metal centers. Amide or hydroxyl resonances can also be perturbed strongly by the formation or destruction of hydrogen bonds, although carbohydrate hydroxyl protons can only be observed under optimized conditions [11, 12]. Structural interpretations of chemical shift data are best done in conjunction with 3D models of the complex.

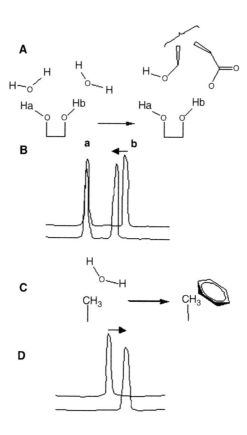

Figure 2. Predicted effects on ¹H resonance positions. (A) A carbohydrate fragment in aqueous solution binds to a protein. Ha becomes close to the –OH group of serine while Hb becomes close to the carboxylate group of Asp/Glu resulting in changes in chemical shift. (B) The environment of Ha is not too different in its bound form (no shift) while Hb is shifted upfield due to the stronger H-bonding and electron withdrawing effect of carboxylate over H₂O. (C) A methyl group changes environment from aqueous exposure to being placed alongside the edge of an aromatic ring (e.g., a Phe residue). (D) This would be expected to result in a downfield shift in the CH₃ peak.

4.1. ¹H Chemical shift perturbations

1D ¹H experiments dominate the spectrometer room as they are the simplest and most sensitive to perform. These types of experiment are perfect for titrations that require a large number of data points that will later be processed into binding curves and affinity constants. Using NMR to monitor titrations has several advantages. Often, the state of both lectin and carbohydrate can be followed. It is possible to monitor several resonances from the lectin and/or carbohydrate to yield several independent binding curves obtained from a single experiment. That last statement may sound strange. However, one shifted resonance can report ligand binding while another reports ligand-induced oligomerization. This status can be deduced from inconsistent binding curves obtained from the same titration. Ligand-binding data for glycopeptide antibiotics provide an example where different binding isotherms provide a complete picture of the different energy levels associated with bound monomer and dimer species [13, 14].

The basis of the titration experiment is that the free and complexed resonances in the protein and ligand have different chemical shifts. Fig. 3 illustrates an example of a 1D titration [15]. The partial spectra illustrate signals arising from the lectin, in this case a small protein mutant of AcAMP2. The assignments of the peaks in free and bound states are known from 2D NMR structural studies, such as detailed in Section 4. The resonance for Ser16HN shows classic "fast-exchange" behavior. The resonance shifts from the free to bound position as more ligand is added to the NMR tube. Fig. 3 also illustrates one of the drawbacks in such an NMR titration. The Lys26HN resonance shifts as much as Ser16HN in the presence of ligand. Unfortunately, Lys26HN overlaps with the Gly2HN, Gly12HN, Cys21HN and Cys15HN resonances at various stages of the titration. AcAMP2 is a small protein with relatively few resonances – spectra are progressively more crowded for larger proteins that contain more resonances. This places a limit on the usefulness of simple 1D ^{1}H NMR methods to follow such titrations. Very little change is noted for the ligand resonances in Fig. 3. At high concentrations, the ligand is mostly in a free state, and the signals are minimally perturbed.

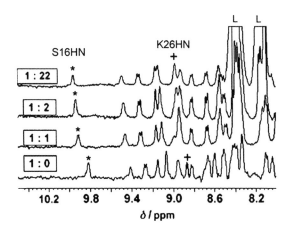

Figure 3. Titration of GlcNAc$_3$ into AcAMP2-Phe18Nal. The protein–ligand ratio is given to the left. The Ser16HN resonance (*) can be followed through the titration whereas the Lys26HN resonance (+), which shifts to a similar degree, cannot be followed due to overlap with other resonances. The assignment of K26HN was made based on 2D experiments on free and bound protein. The ligand resonances (L) cannot be clearly followed at low ligand concentrations and do not shift significantly at higher concentrations.

At lower concentrations, the ligand signals become lost in the field of protein signals. Besides which, at this lower concentration, the proportion of complex is high only if the affinity is high, which is not the case for this example.

4.2. Alternative experiments to obtain chemical shift perturbation data

The principal drawback to chemical perturbation studies is the overlapping resonances of lectin and ligand, particularly at low ligand–lectin ratios. One answer is to filter the signals to reduce the overlap. In the context of Fig. 1, this means specially wearing glasses that filter out most of the stains. Table 1 provides some possible approaches. Each example involves the labeling of either lectin or ligand. The choice of approach often depends on the ease and cost of preparing appropriately labeled compounds.

^{19}F is an NMR-sensitive nucleus, with a low occurrence in proteins and ligands. ^{19}F can be introduced into ligands as a reporter nuclei using fluorinated reagents, such as trifluoroacetyl chloride [16, 17] or by more complicated synthetic procedures [18]. In principle, the affinity of the ^{19}F-labeled reporter ligand can be measured directly and the affinity of other, non-fluorinated ligands can be measured by competition. It is important that the ^{19}F label is placed in a part of the ligand that will be sensitive to ligand binding and experiences chemical shift changes. This approach requires access to ^{19}F-labeled ligands and suitable NMR probes. However, libraries of fluorinated ligands can take advantage of the simpler ^{19}F spectra where neither the protein nor solvent contains ^{19}F [19]. The ^{19}F reporter can also be introduced into a recombinant protein [20] or a synthesized protein [15]. Indeed, ^{19}F-labels placed in the carbohydrate-binding residues of a hevein domain experienced more than one bound conformation, a heterogeneity that had not been observed using native protein and ^{1}H NMR techniques [15].

^{13}C-labeled ligands are usually expensive to synthesize. If the ligand can be obtained from bacteria or yeasts then ^{13}C-labeled media can be used to obtain labeled ligands. Carbohydrates can be obtained in yields of several grams per liter

Table 1. Ligand titrations can be followed by different NMR experiments.

System	Observe	Experiment
Unlabeled lectin, ligand	^{1}H	^{1}H 1D
^{19}F-labeled ligand, unlabeled lectin	^{19}F	^{19}F 1D
^{13}C-labeled ligand, unlabeled lectin	^{1}H	^{1}H,^{13}C-edited 1D
	^{1}H + ^{13}C	^{1}H,^{13}C HSQC 2D
^{2}H-labeled lectin, unlabeled ligand	^{1}H	^{1}H 1D
^{15}N-labeled lectin, unlabeled ligand	^{1}H + ^{15}N	^{1}H,^{15}N HSQC 2D

of bacterial culture [21–23]. The preparation of partly ^{13}C-labeled ligands through modifications or semisynthesis and ^{13}C-labeled reagents (e.g., ^{13}C-labeled acetic anhydride) allows competition protocols to be utilized. Filtered 1D [1]H(^{13}C) NMR experiments allow only the ^{1}H protons attached to ^{13}C to be observed in the spectrum (e.g., 1D ^{1}H,^{13}C-HSQC experiment). This experiment results in a 99% reduction in the background signals from the lectin and an efficient removal of the solvent signal as the natural abundance of ^{13}C is around 1.1%.

Another possibility is the perdeuteration of the protein, which is possible for recombinant proteins [20, 24]. This approach allows normal 1D ^{1}H NMR spectra of the ligand to be obtained. The spectrum of the lectin is effectively reduced by 95–98% in intensity as it contains less ^{1}H nuclei. The disadvantage of this approach is the cost of perdeuteration. Typically, Several rounds of protein expression on media consisting of increasing levels of deuterium are required in order to condition the bacteria to growing and expressing the protein of interest on high deuterium levels. Deuteration levels above 95% are difficult to achieve and isotope effects reduce protein yields.

So far, we have discussed 1D NMR methods. However, titrations can also be obtained by 2D NMR. The second dimension allows us to separate the NMR signals and can lead to the decongestion of overlapped signals for the price of increased experiment time to collect each titration data point. Correlation experiments that correlate nuclei that are separated by one or two chemical bonds, such as COSY, are relatively fast experiments – 30 min acquisition time if a good 1D ^{1}H spectrum can be attained in 15 s. The 2D spectrum often allows the resolution of resonances that are overlapped in the 1D spectrum. If one is interested in correlating the amide HN signal of the lectin residues with the Hα resonances, the experiment can run with a smaller spectral width in the second dimension, meaning higher resolution in the second dimension or the need for fewer experiments to be collected (shorter experiment time). The smaller spectral width means that many signals in the 2D will be aliased, overlapped and difficult to follow but the HN-Hα fingerprint region can be kept clear. The Lys26HN resonance of the AcAMP2 derivative shown in Fig. 3 could have been clearly followed by a series of 2D COSY experiments, as illustrated in Fig. 4. Significant shifts of both Gly2Hα, one Gly12Hα and Cys21Hα would have provided more binding isotherms from which to calculate the binding constants. The Hα region (3.8–5.8 ppm) is not useful in 1D titrations as it is crowded and overlapped with carbohydrate signals. However, the increased experimental time to obtain a 2D titration at one temperature would have precluded the temperature study carried out by Chavez et al. [15] in order to obtain information about the thermodynamics of binding.

Homonuclear COSY experiments are suitable for lectins obtained from natural sources up to 10 kDa. The linewidths of larger molecules exceed the coupling constants (typically in the 3–14 Hz range for $^3J_{HNH\alpha}$) that are to be measured. Recombinant proteins, where available, can be produced, relatively inexpensively,

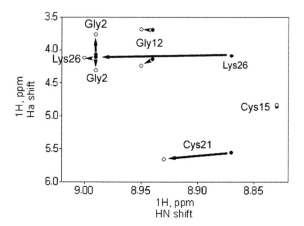

Figure 4. Simulation of peak patterns for HN-Hα correlations for Gly2, Gly12, Cys15, Cys21 and Lys26. Note, the scale of the two axis distort the real changes in chemical shift.

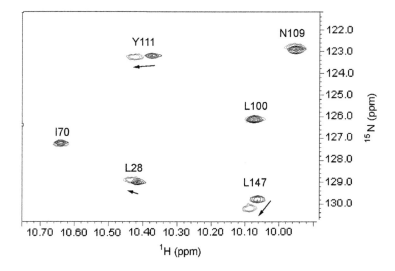

Figure 5. Chemical shift perturbation analysis using ^{15}N-labeled protein. Portion of overlaid ^{1}H,^{15}N HSQC spectra of free FGF-1 (black) and FGF-1 in complex with a heparin-like hexasaccharide analog (gray). Three residues (I70, L100 and N109) are insensitive to ligand binding. The ^{1}H shift experienced by Y111HN can be followed thanks to its separation from L28HN in the 15N dimension. L147 illustrates that ^{15}N nuclei can also be sensitive to ligand binding.

in ^{15}N-labeled forms. The ^{1}H,^{15}N HSQC experiment is relatively sensitive (and fast) and allows the filtering out of ligand signals and the majority of lectin signals. A signal is obtained for each residue in the protein (except prolines) plus some side-chain signals for Trp, Gln, Asn, His and Arg residues. Fig. 5 shows that while

Y111HN and L28HN may have similar ^{1}H shifts, their resonances can be cleanly separated thanks to the ^{15}N dimension. As $^1J_{HN}$ is about 90 Hz, this means that proteins above 10 kDa can be easily observed.

Sensitivity is a factor in perturbation experiments. An affinity of 10 μM (10^{-5} M) can only be accurately determined if the receptor concentration is maintained at 100 μM or lower with the ligand titrated over a 0–250 μM range. While a 1D ^{1}H NMR spectrum of 1 mM ligand at good signal–noise ratio is achievable in a few scans, a 10-fold dilution requires a 100-fold increase in scans to achieve the same S:N. This places strong limits on what can be accurately measured by these NMR methods, especially if 2D methods are to be deployed. A second consideration, particularly with naturally obtained materials, is that a 0.5 mL sample of 30 kDa lectin at 100 μM concentration requires 1.5 mg of protein. However, as NMR is a non-destructive method, some of this material can be recovered by repurification.

Despite the drawbacks of chemical shift perturbations, they remain a valid and often used technique for studying lectin–carbohydrate interactions [11, 15, 25]. The availability of ^{15}N-labeled proteins allows the chemical shift perturbations observed in ^{1}H,^{15}N correlated experiments (HSQC) to be mapped against protein structure if an assignment is available [26]. This approach overcomes many of the problems that occur in simple 1D ^{1}H NMR spectra and titrations can be obtained at protein concentrations of 0.1 mM. HSQC spectra provide a good dispersion of signals, essentially a signal for each backbone amide in the protein plus a few side-chain HN signals, that allow individual peaks to be unambiguously tracked during titrations. This allows a better definition of the ligand-binding site in the protein and/or better idea of conformational changes that occur within the protein [27].

One should be aware that lectins do not always bind carbohydrate ligands in simple 1:1 complexes [28–30]. Therefore, titrations rarely report the binding of one ligand to one protein molecule. Binding can be complicated by multiple ligand sites in a protein that have different affinities or induce different degrees of chemical shift perturbation, or by changes in protein aggregation state. In these cases, the ^{1}H,^{15}N HSQC protocol is particularly useful as it provides a global change in the protein.

5. How to Obtain 3D Structures of Lectin–Carbohydrate Complexes

NMR spectroscopy has developed since the 1980s into a tool capable of determining the 3D structures of lectins [31]. However, the number of lectin structures determined by NMR spectroscopy compared to crystallography is low. One reason is that although lectins may comfortably fall into the <30 kDa upper limit of well-established NMR structure determination protocols, this limit is often broken by protein oligomerization.

5.1. Lectin structure by homonuclear NMR methods

Hevein is a naturally occurring, monomeric protein that binds chitin with a micromolar affinity. Its small size (43 amino acid residues, ~4.7 kDa mass) falls comfortably within the range capable of homonuclear NMR methods. This methodology is based on the identification of the spin systems and assignment of individual amino acid resonances and then determining unambiguous contacts between sequential amino acids, as introduced in the early 1980s, Fig. 6 [32–34]. The NMR data can be acquired with standard 2D NMR experiments such as COSY, TOCSY, NOESY and ROESY have been shown for hevein, AcAMP2, pseudohevein, wheat germ agglutinin domain B (WGA-B) and an Eucommia antifungal peptide in ligand-free forms [35–39] and ligand-bound forms [29, 37, 38, 40]. Hevein is small enough to be produced by automatic peptide synthesis. This has allowed the production of mutant proteins, including truncated heveins [25] or heveins containing unnatural amino acids in the ligand-binding site of the protein [15]. As it is difficult to incorporate unnatural amino acids by the usual protein expression methods, this allows the effects of the unnatural amino acids to be measured in the context of a protein–ligand model.

Homonuclear methods are generally limited to small proteins of less than 100 amino acids (and the incorporation of unnatural amino acids by peptide synthesis to less than 50 residues). Hevein domains are within these limits and they can be studied in the absence of other domains that are unessential for ligand binding or even cause dimerization. This strategy was used for WGA-B (~5 kDa), one of the four domains in WGA, which otherwise is found as part of a ~35 kDa dimer [38, 41]. The dispersion of resonances in carbohydrate ligands can result in ambiguous data that makes it difficult to determine the contacts between lectin and ligand. Fortunately, lectins tend to contain aromatic residues at the ligand-binding site that help with this problem if the ligand binds tightly enough that it can be studied at

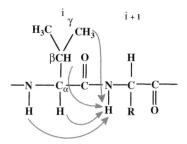

Figure 6. The resonances of individual amino acid residues can be determined using correlation spectroscopy (COSY, TOCSY). However, the carbonyl residue effectively blocks the correlation through the amide bonds linking two amino acids. These links can be determined from characteristic NOEs that are indicated by arrows.

concentrations close to 1:1. The temperature or solution conditions (salt, organic cosolvent) can be adapted to achieve this. This predominance of aromatic residues in carbohydrate binding sites has been theoretically investigated [42].

Several other aspects should be noted. The heterogeneity of protein resonances has several potential sources, which can also affect other experiments such as the chemical shift perturbations. Natural proteins can be produced as closely related splice products or post-translationally modified that are difficult to separate. For example, a C-type lectin from human dendritic cells is produced as two major and six minor splice products [43]. Many post-translational effects, such as nitrosylation or phosphorylation, are transitory modifications that can be lost during protein purification. Other effects, as well as the effects of purification media, can result in heterogeneity [44]. Heterogeneity at the N- or C-terminus can be due to incorrect removal of signal peptides [45] or fusion tag. Oxidation of cysteine or methionine residues can occur [46]. Asn and Gln residues can undergo deamidation processes [47]. Protein obtained by heterologous expression can be partly misfolded – particularly if the protein contains Cys–Cys crossbridges. Protein obtained by the heterologous expression from yeasts can be subjected to unwanted glycosylation. However, multiple sets of protein signals can be observed even if you have a pure, heterogeneous protein sample if cis-trans isomerization of proline residues [48, 49], or if a loop region [50] adopts multiple conformations, in the slow-exchange regime.

While NMR is the tool that provides the raw data, new computer tools are under development in order to assist in resonance assignment, model building and verification.

5.2. Medium-sized lectins, 10–30 kDa

The 1990s saw the advent of heteronuclear protein structure determination. Again, the methodology is well defined in the literature [31]. This methodology requires the ^{15}N- and/or ^{13}C-labeling of the protein receptor. Ligands are rarely produced in ^{13}C-labeled forms. Magnetic field strengths in the 600–900 MHz range are typically used for protein complexes above 15 kDa.

Chemical shift mapping is a useful method for revealing ligand-binding sites in proteins [51, 52]. Generally, the backbone assignment of the protein is required as a minimum starting point. The titration of a ligand can be followed by ^{1}H,^{15}N HSQC experiments. Not only this can yield data but also effect of the ligand on chemical shift can be plotted against residue number, or plotted onto a three-dimensional structure of the protein. The identification of the ligand-binding site is complementary to the transferred NMR methods described below. If the binding site is well determined, and the bound conformation of the ligand known, then molecular modeling methods can provide a good complex structure by docking methods, which utilize lectin structures, most often determined by crystallography methods, that are deposited in the Protein Database.

One experiment not normally described as part of this heteronuclear approach is the half-filtered NOESY experiment [53]. This experiment typically utilizes a sample in which one component of a complex is ^{13}C-labeled. The experiment reports NOEs between ^{1}H attached to unlabeled component and ^{1}H attached to the ^{13}C-labeled component. These NOEs specifically refer to the interface between a receptor and a ligand (or at a dimer interface if unlabeled and labeled proteins are mixed). In the case of lectin–carbohydrate complexes, the lectin will provide the ^{13}C-labeled component.

5.3. Simplify the problem

Very often, within a family of proteins, you find a member that does not oligomerize, for example, calbindin D9k within the S100 family [54]. This protein might provide a better candidate for NMR studies, at least providing a homologous model from which mechanisms can be inferred. However, care should be taken that the protein is monomeric at NMR concentrations (millimolar) as well as biological concentrations (often micromolar).

A second strategy is to mutate a lectin to reduce its ability to oligomerize. This assumes that ligand binding and oligomerization occur in separate sites and it might not be suitable for complexes in which there is a strong cooperativity between ligand binding and dimerization. Mutations can be made randomly, based on predicted 3D structures, or by detection of a naturally occurring isoform [55].

Another strategy is to modify the ligand. This strategy is applicable where an oligosaccharide provides several protein-binding sites. Fibroblast growth factor (FGF) binds heparin by effectively sandwiching the polysaccharide between induced protein dimer interfaces. A monomer complex can be formed by the design of a heparin oligosaccharide that contains sulfate groups on only one face of the ligand [28]. The chemical synthesis of this short hexasacharide is lengthy but simplifies the problem related to the size of the complex.

6. Transferred NMR Methods

If we return to Fig. 1, TR-NMR methods can be explained as an observation of the paint that is transferred from the swing to the kids shorts in Fig. 1C. TR-NMR methods only work in the fast-exchange regime. It should be noted that a sample could give results for one type of TR-NMR experiment (fast exchange) but no results for a second TR-NMR type (slow exchange) due to their different experimental (mixing) times.

TR-NMR methods allow information about the bound ligand to be obtained in the presence of a low concentration of a large receptor. Lectins are suitable receptors in such studies while the milli- to micro-molar affinity of carbohydrate ligands tend to bind lectins with suitable binding kinetics providing fast exchange. Here, we will describe four TR-NMR methods.

The success of TR-NMR methods relies on the binding kinetics of the ligand and relative concentrations of the lectin and ligand. Because of this, we cannot make too many conclusions about the binding of two different ligands to a receptor based on their TR-NMR spectra. There are no exact formulae to guide sample preparation and the optimal ligand–lectin ratio is often different for the different types of experiment [56]. In this respect, TR-NMR samples and experimental conditions involve several rounds of optimization.

The basis of TR-NMR experiments is that lectins and their small carbohydrate ligands have distinct physical and spectroscopic properties. DOSY exploits the different size and diffusion properties of receptors and ligands. TR-NOESY exploits the fact that bound ligands develop NOEs faster than free ligands. STD exploits the wider spectrum of the receptor compared to the ligand. For TR-RDCs, we expect the ligand to orient differently in the magnetic field in its free or bound states.

The TR-NMR sample typically contains a relative low concentration of lectin and an excess of ligand. Experimental parameters are chosen where the physical properties of the lectin produce significant data that, due to its low concentration, are nonetheless unobservable. The more concentrated ligand alone also does not produce interesting data under the selected conditions. However, the reasons for this differ for each transferred NMR experiment, as explained below. In the complex, the bound ligand effectively borrows the physical properties of the lectin that can be observed when they are transferred to the sharp resonances of the free ligand form (essentially, the paint from the swing, Fig. 1C). The experiment relies on the exchange of all the ligand molecules between free and bound states during the experiment. For this reason, the binding kinetics, particularly off-rate relative to the experiment time, are important.

6.1. Diffusion ordered spectroscopy (DOSY)

The aggregation state of a sample can be difficult to judge with molecules sticking together to make a mess that is difficult to disentangle by NMR spectroscopy. This is often the case when experimental systems are concentrated to allow NMR spectroscopic analysis. The monomeric conditions that biologists work with can be different at the higher concentrations used by NMR. Fortunately, we can use DOSY to define what we are looking at – single molecules, dimers, complexes, etc. [57].

DOSY can be successfully used to measure the relative size of lectins and their ligands [30, 57, 58]. Several reviews to recent developments and other practical applications of DOSY have been recently published [59, 60]. The experimental parameters of different DOSY experiments vary considerably, depending on the subject. Longer diffusion constants (values of Δ and δ) are recommended for the observation of protein aggregation [57]. This protocol is sensitive for protein concentrations above 1 mg/mL and in the 0–100 kDa range. The original protocol uses samples in D_2O buffers. Baseline distortions in standard DOSY experiments lead to variable results in H_2O buffers. New pulse sequences to overcome this problem are under development.

DOSY allows the measurement of the diffusion coefficient of a molecule and therefore its aggregation state [61]. With matched ligand library samples in the presence and absence of a receptor, one can detect changes in the diffusion properties of a ligand that binds to the receptor. Such diffusion editing allows the determination of the full 1D ^1H NMR spectrum of a bound ligand present in a ligand library or compound mixture. Several disadvantages to this methodology prevent it from becoming a major method for ligand screening. Firstly, the method relies on matched samples, which increases the errors in data analysis. Secondly, the experiment is sensitive at low ligand–receptor ratios – this means high concentrations of receptor are required as sensitivity and errors prevent a lowering of ligand concentrations. Thirdly, the low ligand–receptor ratio leads to strong line-broadening effects on the bound ligands so it is better to use smaller receptors. Fourthly, the experiment may work well if a single compound binds specifically to the lectin but the experiment is less sensitive if several compounds in the library compete for the same binding site. On the positive side, ligands that bind in slow exchange can be detected with this method, and differentiated from fast-exchange ligands because of an opposite phase.

6.2. NOESY: Transfer NOESY (TR-NOESY) and 1D selective NOESY

From the NMR viewpoint, TR-NOESY is a standard NOESY experiment [56]. The difference lies in the sample conditions and choice of experimental parameters. Samples typically contain low, micromolar concentrations of lectin and high, millimolar concentrations of ligand. The NOESY parameters are selected such that few, if any, crosspeaks are observed for matched samples containing either protein or ligand alone. In the complexed sample, the ligand spends part of it's time bound to the lectin, where strong NOEs develop and these can be observed after transfer to the sharp, free ligand resonances. Small NOEs can develop for the uncomplexed ligand and it is important to obtain reference data for these samples. Where the NOEs for the starting conformation cannot be obtained by NOESY, ROESY can be used instead. The sizes (signs) of ROEs are almost independent of molecular size.

TR-NOESY parameters for 1 mM ligand include a mixing time of 25–500 ms, depending on the size of the protein. Longer values can be subject to spin-diffusion – a relaying of information over distances greater than 5 Å, which can lead to strain in calculations of the bound ligand. The optimal range of mixing times might be determined by a series of 1D NOE experiments. Indeed, if the ligand contains a small number of well-resolved resonances, 1D NOESY can provide data more quickly, or at higher intensity. Furthermore, the collection of NOE data over a range of mixing times allows NOE buildup curves to be constructed and analyzed to give a more accurate determination of inter-proton distances. A long relaxation delay of 2–5 s is usually employed. The TR-NOESY experiment may be complemented by the STD technique to determine the binding epitope [62, 63].

6.3. Transfer residual dipolar couplings (TR-RDC)

RDCs can be obtained by placing the sample in a medium that causes a physical or electromagnetic alignment with the magnetic field. Bicelles, stretched acrylamide gels and lanthanides have all been used to achieve alignment [64–66]. Dipolar couplings are averaged to zero when a molecule rotates freely in solution. However, they can be partly recovered by alignment in a magnetic field. If we consider a ^{1}H,^{13}C bond as a vector, then the RDC effectively reports the angle (in reality, the solution is one of four angles) that the vector makes with the magnetic field. Other coupled nuclei can be examined although one-bond ^{1}H,^{13}C and ^{1}H,^{15}N RDCs provide the greatest sensitivity.

The set of RDCs for a molecule can change after interaction with another molecule. This has been followed up in the study of protein–protein contacts [66, 67]. However, Koenig [68] proposed the study of ligands in the presence and absence of a receptor. The RDC values for the studied ligand were quite small, as were the observed RDC changes that are proportional to the amount of ligand that is bound to the protein. This means that the two sets of data were similar and the differences not much larger than the errors. Nonetheless, the differences still allowed the bound structure of the ligand to be determined.

There are a number of problems with the measurement of RDCs. The first is that the RDCs depend on a weak interaction between protein and the medium – which can lead to a distortion in molecular structure or stronger, irreversible interaction with the medium [65]. Bicelle systems are essentially 5–10% detergent systems – concentrations that typically denature proteins. Despite this, the strategy is sound and Prestegard's group recently obtained better sensitivity by using a lectin that had been modified with a membrane anchor [69]. As the protein is more strongly aligned, the bound ligand is also more strongly aligned than if it binds to a soluble protein. This means that the RDC datasets for the free and bound ligand are more different, with an improvement in resolution and angle determination.

7. Emerging NMR Methodologies

The application of TROSY-based methods allows the structural biomolecular complexes above 30 kDa to be studied by NMR [70]. However, a complete, TROSY-based structure determination of a high molecular weight protein would demand several preparations incorporating different isotope labels resulting in a considerable cost in time and money. The majority of reports on TROSY present improved experiments with few complete studies. Nolis and Parella [71] have illustrated that TROSY-based methods can overcome the spectral crowding associated with carbohydrate samples.

A "middle" way involves the backbone assignment of a protein using the strong coupling that exists between adjacent nuclei to walk along the protein backbone. Nevertheless, spectra are still crowded and this can cause problems. Three-dimensional structures can be obtained by obtaining RDC data for the backbone resonances [72]. The result is a scaffold of secondary structure elements with flexible side chains that hopefully find their correct conformation during the modeling process. The measurement of chemical shift anisotropies [73] and incorporation of CRINEPT for biomolecules above 200 kDa [74] may provide further approaches to collecting structural data on lectin complexes.

8. Summary

NMR spectroscopy provides a route to the understanding of lectin–carbohydrate interactions. NMR is not as powerful as X-ray crystallography in solving the structures of large, multimeric proteins. However, small monomeric proteins are suitable and the described transferred NMR methods offer an imprint of the ligand-binding site. The ideal situation is when a 3D structure of a lectin is available and a combination of transferred NMR and modeling provide a structure of the complex and such projects often provide a saving in materials and time compared to solving the crystal structure of the complex. We hope this broad overview of NMR spectroscopy is of value to the lectin biochemist.

References

[1] H. Chart, H.R. Smith, R.M. La Ragione and M.J. Woodward, J. Appl. Microbiol., 89 (2000) 1048.

[2] A. Jacquet, V. Daminet, M. Haumont, L. Garcia, S. Chaudoir, A. Bollen and R. Biemans, Protein Exp. Purif., 17 (1999) 392.

[3] A. Martinez, P.M. Knappskog, S. Olafsdottir, A.P. Doskeland, H.G. Eiken, R.M., Svebak, M. Bozzini, J. Apold and T. Flatmark, Biochem. J., 306 (1995) 589.

[4] G.D. Davis, C. Elisee, D.M. Newham and R.G. Harrison, Biotechnol. Bioeng., 65 (1999) 382.

[5] R.B. Kapust and D.S. Waugh, Protein Sci., 8 (1999) 1668.

[6] H.P. Sorensen, H.U. Sperling-Petersen and K.K. Mortensen, Protein Exp. Purif., 32 (2003) 252.

[7] G.S. Waldo, B.M. Standish, J. Berendzen and T.C. Terwilliger, Nat. Biotechnol., 17 (1999) 691.

[8] I. Arechaga, B. Miroux, M.J. Runswick and J.E. Walker, FEBS Lett., 547 (2003) 97.

[9] M. Hammarstrom, N. Hellgren, S. van Den Berg, H. Berglund and T. Hard, Protein Sci., 11 (2002) 313.

[10] J. Cavanagh, W.J. Fairbrother, A.G. Palmer and N.J. Skelton, Protein NMR spectroscopy: Principles and practice, Academic Press, San Diego, 1996.

[11] H.C. Siebert, S. Andre, J.L. Asensio, F.J. Canada, X. Dong, J.F. Espinosa, M. Frank, M. Gilleron, H. Kaltner, T. Kozar, N.V. Bovin, C.W. von Der Lieth, J.F. Vliegenthart, J. Jimenez-Barbero and H.J. Gabius, Chembiochem., 1 (2000) 181–195.

[12] H.C. Siebert, S. Andre, J.F. Vliegenthart, H.J. Gabius and M.J. Minch, J. Biomol. NMR, 25 (2003) 197.

[13] Y.R. Cho, A.J. Maguire, A.C. Try, M.S. Westwell, P. Groves and D.H. Williams, Chem. Biol., 3 (1996) 207.

[14] H. Shiozawa, B.C.S. Chia, N.L. Davies, R. Zerella and D.H. Williams, J. Am. Chem. Soc., 124 (2002) 3914.

[15] M.I. Chavez, C. Andreu, P. Vidal, N. Aboitiz, F. Freire, P. Groves, J.L. Asensio, G. Asensio, M. Muraki, F.J. Canada and J. Jimenez-Barbero, Chemistry, 11 (2005) 7060.

[16] P. Midoux, J.P. Grivet, F. Delmotte and M. Monsigny, Biochem. Biophys. Res. Commun., 119 (1984) 603.

[17] M.V. Krishna Sastry, M.J. Swamy and A. Surolia, J. Biol. Chem., 263 (1988) 14826.

[18] M. Michalik, M. Hein and M. Frank, Carbohydr. Res., 327 (2000) 185.

[19] T. Tengel, T. Fex, H. Emtenas, F. Almqvist, I. Sethson and J. Kihlberg, Org. Biomol. Chem., 2 (2004) 725.

[20] P.G. Gettins, Int. J. Biol. Macromol., 16 (1994) 227.

[21] E. Samain, V. Chazalet, R.A. Geremia, J. Biotechnol., 72 (1999) 33.

[22] T. Antoine, A. Heyraud, C. Bosso and E. Samain, Angew. Chem. Int. Ed. Engl., 44 (2005) 1350.

[23] C. Dumon, C. Bosso, J.P. Utille, A. Heyraud and E. Samain, Chembiochem., 7 (2006) 359.

[24] C.G. Shibata, J.D. Gregory, B.S. Gerhardt and E.H. Serpersu, Arch. Biochem. Biophys., 319 (1995) 204.

[25] N. Aboitiz, M. Vila-Perello, P. Groves, J.L. Asensio, D. Andreu, F.J. Canada and J. Jimenez-Barbero, Chembiochem, 5 (2004) 1245.

[26] C. Peng, S.W. Unger, F.V. Filipp, M. Sattler and S. Szalma, J. Biomol. NMR, 29 (2004) 491.

[27] M. Scharpf, G.P. Connelly, G.M. Lee, A.B. Boraston, R.A. Warren and L.P. McIntosh, Biochemistry, 41 (2002) 4255.

[28] J. Angulo, R. Ojeda, J.L. de Paz, R. Lucas, P.M. Nieto, R.M., Lozano, M., Redondo-Horcajo, G. Gimenez-Gallego and M. Martin-Lomas, Chembiochem., 5 (2004) 55.

[29] J.L. Asensio, F.J. Canada, H.C. Siebert, J. Laynez, A. Poveda, P.M. Nieto, U.M. Soedjanaamadja, H.J. Gabius and J. Jimenez-Barbero, Chem. Biol., 7 (2000) 529.

[30] P. Groves, M.O. Rasmussen, M.D. Molero, E. Samain, F.J. Canada, H. Driguez and J. Jimenez-Barbero, Glycobiology, 14 (2004) 451.

[31] G.M. Clore and A.M. Gronenborn, Prog. Biophys. Mol. Biol., 62 (1994) 153.

[32] K. Nagayama and K. Wuthrich, Eur. J. Biochem., 114 (1981) 365.

[33] G. Wagner, A. Kumar and K. Wuthrich, Eur. J. Biochem., 114 (1981) 375.

[34] M. Billeter, W. Braun and K. Wuthrich, J. Mol. Biol., 155 (1982) 321.

[35] N.H. Andersen, B. Cao, A. Rodriguez-Romero and B. Arreguin, Biochemistry, 32 (1993) 1407.

[36] J.C. Martins, D. Maes, R. Loris, H.A. Pepermans, L. Wyns, R. Willem and P. Verheyden, J. Mol. Biol., 258 (1996) 322.

[37] J.L. Asensio, H.C. Siebert, C.W. von Der Lieth, J. Laynez, M. Bruix, U.M. Soedjanaamadja, J.J. Beintema, F.J. Canada, H.J. Gabius and J. Jimenez-Barbero, Proteins, 40 (2000) 218.

[38] J.F. Espinosa, J.L. Asensio, J.L. Garcia, J. Laynez, M. Bruix, C. Wright, H.C. Siebert, H.J. Gabius, F.J. Canada and J. Jimenez-Barbero, Eur. J. Biochem., 267 (2000) 3965.

[39] R.H. Huang, Y. Xiang, G.Z. Tu, Y. Zhang and D.C. Wang, Biochemistry, 43 (2004) 6005.

[40] P. Verheyden, J. Pletinckx, D. Maes, H.A. Pepermans, L. Wyns, R. Willem and J.C. Martins, FEBS Lett., 370 (1995) 245.

[41] C.S. Wright, J. Mol. Biol., 215 (1990) 635.

[42] M.C. Fernandez-Alonso, F.J. Canada, J. Jimenez-Barbero and G. Cuevas, J. Am. Chem. Soc., 127 (2005) 7379.

[43] J.A. Willment, S. Gordon and G.D. Brown, J. Biol. Chem., 276 (2001) 43818.

[44] M. Svensson, A. Hakansson, A.K. Mossberg, S. Linse and C. Svanborg, Proc. Natl. Acad. Sci. U S A., 97 (2000) 4221.

[45] M. Palczewska, P. Groves and J. Kuznicki, Protein Exp. Purif., 17 (1999) 465.

[46] C. Jacob, G.I. Giles, N.M. Giles and H. Sies, Angew. Chem. Int. Ed. Engl., 42 (2003) 4742.

[47] K.J. Reissner and D.W. Aswad, Cell Mol. Life Sci., 60 (2003) 1281.

[48] J. Kordel, S. Forsen, T. Drakenberg and W.J. Chazin, Biochemistry, 29 (1990) 4400.

[49] Y. Chi, T.K. Kumar, I.M. Chiu and C. Yu, J. Biol. Chem., 275 (2000) 39444.

[50] M.F. Garcia-Mayoral, L. Garcia-Ortega, M.P. Lillo, J. Santoro, A. Martinez del Pozo, J.G. Gavilanes, M. Rico and M. Bruix, Protein Sci., 13 (2004) 1000.

[51] E.R. Zuiderweg, Biochemistry, 41 (2002) 1.

[52] J. Clarkson and I.D. Campbell, Biochem. Soc. Trans., 31 (2003) 1006.

[53] G. Otting and K. Wuthrich, Q. Rev. Biophys., 23 (1990) 39.

[54] D.B. Zimmer, E.H. Cornwall, A. Landar and W. Song, Brain Res. Bull., 37 (1995) 417.

[55] T. Miura, M. Takahashi, H. Horie, H. Kurushima, D. Tsuchimoto, K. Sakumi and Y. Nakabeppu, Cell Death Differ., 11 (2004) 1076.

[56] J. Jiménez-Barbero and T. Peters, (Eds.), NMR spectroscopy of glycoconjugates, Wiley-VCH, Weinheim, 2002.

[57] P. Groves, M. Palczewska, M.D. Molero, G. Batta, F.J. Cañada and J. Jiménez-Barbero, Anal. Biochem., 331 (2004) 395.

[58] M. Politi, P. Groves, M.I. Chavez, F.J. Canada and J. Jimenez-Barbero, Carbohydr. Res., 341 (2006) 84.

[59] J.C. Cobas, P. Groves, M. Martín-Pastor and A. De Capua, Curr. Anal. Chem., 1 (2005) 289.

[60] L.O. Sillerud and R.S. Larson, Methods Mol. Biol., 316 (2006) 227.

[61] F. Chevalier, J. Lopez-Prados, P. Groves, S. Perez, M. Martín-Lomas and P.M. Nieto, Glycobiology, 16 (2006) 969.

[62] B. Meyer and T. Peters, Angew Chem. Int. Ed., 42 (2003) 864.

[63] A. Bernardi, D. Arosio, D. Potenza, I. Sanchez-Medina, S. Mari, F.J. Canada and J. Jimenez-Barbero, Chem. Eur. J., 10 (2004) 4395.

[64] J.H. Prestegard, H.M. al-Hashimi and J.R. Tolman, Q. Rev. Biophys., 33 (2000) 371.

[65] J.H. Prestegard, C.M. Bougault and A.I. Kishore, Chem. Rev., 104 (2004) 3519.

[66] A.M. Bonvin, R. Boelens and R. Kaptein, Curr. Opin. Chem. Biol., 9 (2005) 501.

[67] D.C. Williams Jr., M. Cai, J.Y. Suh, A. Peterkofsky and G.M. Clore, J. Biol. Chem., 280 (2005) 20775.

[68] B.W. Koenig, ChemBioChem., 3 (2002) 975.

[69] T. Zhuang, H. Leffler and J.H. Prestegard, Protein Sci., 15 (2006) 1780.

[70] K. Pervushin, R. Riek, G. Wider and K. Wuthrich, Proc. Natl. Acad. Sci. USA, 94 (1997) 12366.

[71] P. Nolis and T. Parella, J. Magn. Reson., 176 (2005) 15.

[72] J.H. Prestegard, K.L. Mayer, H. Valafar and G.C. Benison, Methods Enzymol., 394 (2005) 175.

[73] F. Cisnetti, K. Loth, P. Pelupessy and G. Bodenhausen, Chemphyschem., 5 (2004) 807.

[74] R. Riek, K. Pervushin and K. Wuthrich, Trends Biochem. Sci., 25 (2000) 462.

Chapter 4

Applications of Isothermal Titration Calorimetry to Lectin–Carbohydrate Interactions

Tarun K. Dam and C. Fred Brewer

Department of Molecular Pharmacology, Albert Einstein College of Medicine, Bronx, New York 10461, USA

1. Introduction

1.1. Background

Lectins are a group of carbohydrate-binding proteins that are found in animals, plants, and lower organisms [1]. The biological functions of animal lectins include glycoprotein trafficking and clearance, immune defense, malignancy, and apoptosis [2]. Although less is known about the biological functions of plant lectins, they were first to be identified, isolated, and characterized in terms of their carbohydrate-binding properties. As a consequence, plant lectins have been used in a number of biological applications including the generation of mutant cell lines for genetic, functional, and biosynthetic studies of carbohydrates in cells; the isolation of carbohydrates and glycoprotein receptors from cells by affinity chromatography; use in histochemistry studies; and in biophysical studies of lectin–carbohydrate interactions.

The biological activities of lectins appear to be primarily due to their carbohydrate-binding properties [2]. Thus, determination of the mechanisms of binding and specificities of lectins for carbohydrates and glycoconjugates provides insight into their biological properties. Historically, the relative affinities and specificities of lectins for carbohydrates were first addressed using hemagglutination inhibition, equilibrium dialysis, and quantitative precipitation inhibition techniques [3]. While these methods are still employed, additional techniques are often used including gradient affinity chromatography, nuclear magnetic resonance, and surface plasmon resonance. However, all of the above methods either suffer from being indirect measurements of binding constants or have special conditions required for the measurements. For example, hemagglutination inhibition and quantitative precipitation inhibition techniques are indirect binding methods, while equilibrium dialysis typically requires radioactivity or a chromophore in the molecules. Affinity chromatography suffers from potential matrix interactions with ligands in solution, and covalent attachment of a receptor or ligand to the matrix. Nuclear magnetic resonance measurements require the kinetics of binding between two molecules to be in the so-called intermediate to slow exchange condition. Surface plasmon resonance mandates covalent attachment of one of the binding molecules on a solid-state chip. Equilibrium constants are derived from kinetic on–off rate measurements of a ligand in a flowing system. Thus, all of the techniques have limitations in determining quantitative binding constants for carbohydrate–lectin interactions in solution.

1.2. Advantages and disadvantages

The major advantage of isothermal titration calorimetry (ITC) is that it provides direct determination of the association constant (K_a) for binding of unmodified molecules in solution by measuring their heat of binding. ITC also provides the stoichiometry of binding and all thermodynamic binding parameters (described below). Thus, ITC is the preferred method for determining K_a as well as other important binding parameters for lectin–carbohydrate interactions. As a consequence, ITC has gained increasing use in studies of lectin–carbohydrate interactions since its commercial introduction in the early 1990s. A disadvantage of ITC is the amounts of carbohydrate and lectin required, typically 1 ml of 100 μM to 1 mM carbohydrate and 2 ml of 10 to 100 μM lectin. The range of affinity constants that can be measured is in the mM to nM region.

There are many reviews on the thermodynamics of biomolecular interactions [4–13], but relatively few reviews on the thermodynamics of lectin–carbohydrate interactions [14–16]. The present review provides selective examples of the application of ITC to lectin–carbohydrate interactions including multivalent interactions.

1.3. Thermodynamic binding parameters

ITC directly determines the thermodynamics of binding of two molecules in solution at a constant temperature. A series of data points of the amount of heat released (exothermic) or absorbed (endothermic) per mole of injectant (ligand) is plotted as a function of the molar ratio (L_T/M_T) of ligand (L_T) and macromolecule (M_T) after each injection to generate the binding isotherm. Thermodynamic binding parameters are determined by nonlinear least-squares analysis of the binding isotherm. ΔH, the change in enthalpy (kcal/mol) on binding; K_a, the association constant (M^{-1}); and n, the number of binding sites per monomer of receptor are the adjustable parameters in the fits. From the equation

$$\Delta G = -RT \ln K_a \tag{1}$$

ΔG, the free energy of binding (kcal/mol), can be calculated. And from the equation

$$\Delta G = \Delta H - T\Delta S \tag{2}$$

$T\Delta S$, the entropy of binding (kcal/mol), can be determined. Determination of the temperature dependence of the enthalpy and entropy changes allows evaluation of the changes in heat capacity (ΔC_p):

$$\Delta H(T_1) = \Delta H(T_0) + \Delta C_p(T_1 - T_0) \tag{3}$$

$$\Delta S(T_1) = \Delta S(T_0) + \Delta C_p(T_1 - T_0)/T_0 \tag{4}$$

Hence, ITC measurements allow simultaneous determination of the thermodynamic binding parameters K_a, ΔG, ΔH, $T\Delta S$, and n in a single experiment [15, 17], as well as ΔC_p from temperature dependent measurements.

2. Characterization of the Binding Specificity and Binding Sites of Lectins

2.1. Concanavalin A

ITC provides a means of investigating the specificity, molecular contacts, and size of the binding sites of lectins with carbohydrates. ITC studies of concanavalin A (ConA), a Glu/Man specific plant lectin, illustrate some of these applications.

2.1.1. Evidence of an extended binding site
Binding of ConA, a Man/Glc specific plant lectin from the Jack bean *Canavalia ensiformis*, to a series of mono- and oligosaccharides including the "core" trimannoside, 3,6-di-*O*-(α-D-mannopyranosyl)-α-D-mannopyranoside, a branched chain trisaccharide moiety found in all *N*-linked carbohydrates, was investigated using ITC [18]. K_a and $-\Delta H$ values for methyl α-D-mannopyranoside (MeαMan) showed that the specificity of the lectin was greatest for this monosaccharide. These results confirmed hemagglutination inhibition and equilibrium dialysis experiments in the literature. However, the K_a (4.9×10^5 M^{-1}) and ΔH (-14.4 kcal/mol) values for the

core trimannoside (**1** in Fig. 1) binding to ConA are considerably greater than those of MeαMan ($K_a = 8.2 \times 10^3 \, M^{-1}$; $\Delta H = -8.2 \, \text{kcal/mol}$) and constituent disaccharides, Man(1,3)Man and Man(1,6)Man, which represent the two arms of the 3,6-trimannoside. The 60-fold increase in K_a value and $-6.2 \, \text{kcal/mol}$ increase

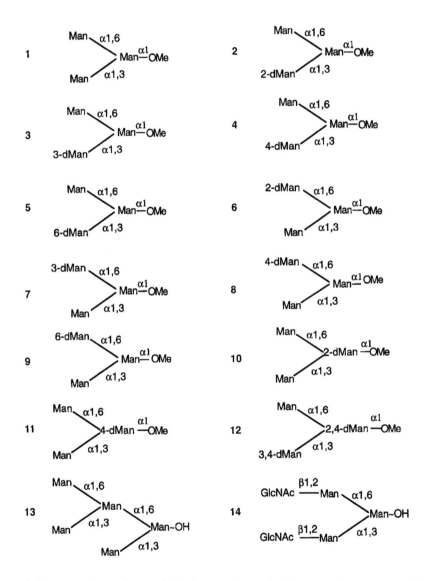

Figure 1. Structures of core trimannoside **1**, deoxy analogues **2–12**, Man 5 oligomannose carbohydrate **13**, and biantennary complex carbohydrate **14**. Man, GlcNAc, 2-dMan, 3-dMan, 4-dMan, and 6-dMan represent mannose, *N*-acetylglucosamine, 2-deoxymannose, 3-deoxymannose, 4-deoxymannose, and 6-deoxymannose residues, respectively.

in ΔH for the core trimannoside relative to MeαMan provides evidence that ConA possesses an extended binding site that recognizes the two nonreducing Man residues of the core trimannoside [18]. Swaminathan et al. [19] determined the ΔH for ConA–trimannoside interaction at pH 5.2 that was similar to that reported by Mandal et al. [18]. Importantly, the ITC derived n values, the number of binding sites per monomer of protein, were close to one for all of the above carbohydrates indicating that they bind as monovalent ligands [18].

2.1.2. Mapping contacts of the binding site with the core trimannoside

A variety of synthetic analogs of the core trisaccharide binding to ConA were investigated using ITC. Data for analogs possessing an α-glucosyl or α-galactosyl residue substituted at either the $\alpha(1,6)$ or $\alpha(1,3)$ position of the trimannoside (**15–18**, respectively, in Fig. 2) indicated that the $\alpha(1,6)$ residue of the parent trimannoside occupied the so-called "monosaccharide site" and the $\alpha(1,3)$ residue, a weaker secondary site [20]. ITC was also used to investigate binding of deoxy analogs of the $\alpha(1,3)$Man residue of the trimannoside (Fig. 1) [20]. The results demonstrated that only the 3-deoxy analog of the trimannoside on the $\alpha(1,3)$Man

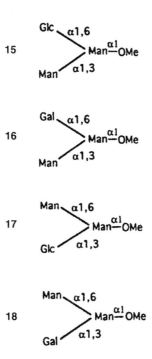

Figure 2. Glucose (Glc) and galactose (Gal) substituted core trimannoside (Man is mannose).

arm (**3**) (Fig. 1) bound with ~10-fold lower affinity and 3.4 kcal/mol lower enthalpy than the parent trimannoside (**1**). This suggested that the 3-hydroxyl of the α(1,3)Man arm makes specific hydrogen-bonds with the protein in the extended binding site. ΔH of the 3-deoxy analog (**3**) (-11 kcal/mol) was, however, higher than that of MeαMan (-8.4 kcal/mol). This indicated the presence of another contact site between the trimannoside and Con A, most likely the central Man residue. Thus, it was concluded that ConA had an extended binding site that included a high affinity site that recognized the 3-, 4-, and 6-hydroxyl groups of the α(1,6)Man residue of the trimannoside, a lower affinity that bound the 3-hydroxyl of the α(1,3)Man residue, and a third site that involved the "core" Man residue [20].

X-ray crystal data for ConA complexed with the core trimannoside [21] were later obtained that supported the ITC findings. These results demonstrate that ITC can provide structural information about carbohydrate–lectin interactions in solution, as well as thermodynamic data. Additional ITC studies using a complete set of deoxy analogs (Fig. 1) as well as di- and trideoxy analogs of the trimannoside [22] showed agreement with the X-ray crystallographic data [21].

2.1.3. Comparison of ITC and X-ray crystal data

Fig. 3 shows a view of the H-bonding interactions between the hydroxyl groups of the trimannoside and the binding site of ConA, derived from the X-ray crystal data [21]. The 3-, 4-, and 6-hydroxyl groups of the α(1,6)Man residue of the trimannoside bind in the same manner as MeαMan in the crystal complex does with ConA [23]. These results agree with the ITC data for the α(1,6)3-deoxy (**7**), α(1,6)4-deoxy (**8**), and α(1,6)6-deoxy (**9**) trimannoside analogs in Fig. 1 [22]. All of the analogs showed reduced K_a and ΔH values compared to the parent trimannoside. The X-ray crystallographic data also showed binding of the 3-OH of the α(1,3)Man residue to the N-H and side chain O of Thr 15, and the 4-OH of the α(1,3)Man residue to the side chain -OH of Thr 15 (Fig. 3). Additionally, the 2-OH and 4-OH of the central core Man residue are hydrogen-bonded to the protein. These observations are in agreement with the thermodynamic data. K_a and ΔH values for the α(1,3)3-deoxy (**3**), α(1,3)4-deoxy (**4**), "core"2-deoxy (**10**) and "core"4-deoxy (**11**) analogs were all less than the core trimannoside [22]. Thus, ITC provided structural data for the binding of carbohydrates to ConA.

2.1.4. Nonlinearity of $\Delta\Delta H$ and $\Delta\Delta G$ values of deoxy analogs of the core trimannoside

The $\Delta\Delta H$ values for the monodeoxy analogs of the core trimannoside are nonlinear [22]. For example, the combined $\Delta\Delta H$ value for the 3-OH and 4-OH of the α(1,3)Man residue of trimannoside for **3** and **4** (Fig. 1), respectively, and the 2-OH and 4-OH of the central Man residue for **10** and **11** (Fig. 1), respectively, is ~-8.8 kcal/mol. This can be compared to the difference in ΔH of -6.2 kcal/mol

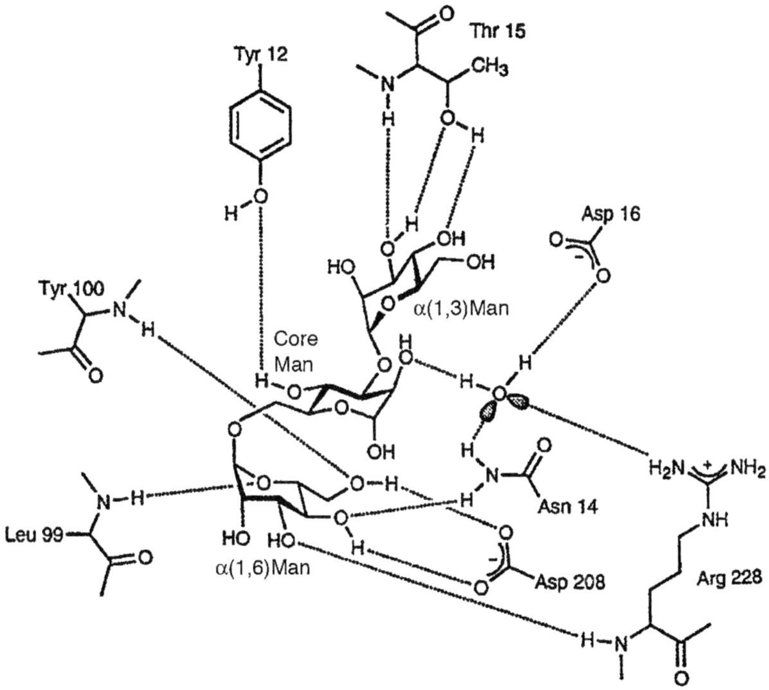

Figure 3. View of the X-ray crystal structure of core trimannoside (**1** with no anomeric methoxy group) bound to ConA [21].

between **1** and MeαMan that reflects binding of the α(1,3)Man and the central Man residues of **1**. In addition, the sum of the $\Delta\Delta H$ values for the 3-, 4-, and 6-OH of the α(1,6)Man residue (**7**, **8**, and **9**), the 3- and 4-OH of the α(1,3)Man residue (**3** and **4**), and the 2- and 4-OH of the central Man residue (**10** and **11**) is −17.5 kcal/mol which was greater than the ΔH for **1** of −14.4 kcal/mol. The sum of the $\Delta\Delta H$ values for the hydroxyl groups of **1** obtained from the monodeoxy analogs does not correspond to the observed ΔH of **1**. In all of these cases, the sum of the $\Delta\Delta H$ values for specific hydroxyl groups on specific Man residues of **1** obtained from corresponding monodeoxy analogs is greater than the measured ΔH for that residue(s). This nonlinear relationship in $\Delta\Delta H$ is also present in the di- and trideoxy analogs. The same nonlinearity is also present in the $\Delta\Delta G$ values of the monodeoxy analogs.

$\Delta\Delta H$ and $\Delta\Delta G$ values for the monodeoxy analogs of **1** also did not scale with the number of H-bonds at each position as determined from X-ray crystallography. $\Delta\Delta H$ values for the monodeoxy analogs are not proportional to the type or number of H-bonds of specific hydroxyl groups of **1**. This is of interest since it has been suggested that the free energy associated with elimination of a H-bond between an uncharged donor/acceptor pair is 0.5–1.5 kcal/mol, and 3.5–4.5 kcal/mol between a neutral-charged pair. The ITC data indicate no such relationship in the free energy difference ($\Delta\Delta G$) of monodeoxy analogs that possess a loss of one or more H-bonds such as **7** versus **8** and **9**.

The presence of nonlinear relationships in the $\Delta\Delta H$ and $\Delta\Delta G$ values for the deoxy analogs of the core trimannoside indicate other contributions to these terms such as protein and solvent effects. The magnitude of the $\Delta\Delta H$ and $\Delta\Delta G$ values represent not only the loss of the H-bond(s) involved, but also differences in the solvent and protein contributions to binding of **1** and the deoxy analogs. Thus, ITC measurements of deoxy analogs of a carbohydrate binding to a lectin do not provide direct measurements of the free energy and enthalpy of binding of the H-bonds involved.

2.1.5. Binding of a biantennary complex type pentasaccharide

A biantennary complex pentasaccharide that possesses a terminal $\beta(1,2)$-GlcNAc residue on each arm of the core trimannoside (**14** in Fig. 1) possesses an affinity ~4-fold greater than that of the core trimannoside as determined by ITC [18]. However, ΔH was -10.6 kcal/mol for the complex oligosaccharide as compared to -14.4 kcal/mol for trimannoside **1**. These results indicate that the increase in affinity of the complex oligosaccharide is due to entropic effects ($T\Delta S$) and not due to increases in $-\Delta H$. The disaccharide, GlcNAc$\beta(1,2)$Man, which constitutes the two branched chains of the oligosaccharide also shows a relatively low $-\Delta H$ but a favorable $T\Delta S$ contribution to the free energy of binding. ITC data for $\alpha(1,2)$ dimannoside (Man$\alpha(1,2)$Man) and $\alpha(1,2)$ trimannoside (Man$\alpha(1,2)$Man$\alpha(1,2)$Man) binding to ConA both show enhanced entropic contributions to binding [18]. These results are consistent with a sliding and recapture mechanism between Man residues of the oligosaccharides at the "monosaccharide" binding site of the lectin for their enhanced affinities, similar to that proposed from nuclear magnetic relaxation dispersion studies [24]. Furthermore, a subsequent X-ray crystallographic study supports these conclusions [21].

2.2. Dioclea grandiflora lectin (DGL)

The seed lectin from *Dioclea grandiflora* lectin (DGL), a Man/Glc binding protein from Northeastern Brazil, is a member of a group of lectins including ConA from the subtribe Diocleinae. DGL possesses a high degree of sequence homology with ConA differing in 52 out of 237 residues [25]. Six of the seven residues that have been implicated as ligands for the Ca^{2+} and transition metal ion sites are conserved, and the amino acid residues surrounding the carbohydrate-binding site of ConA are also conserved in DGL [25]. DGL also binds with high affinity to the "core" trimannoside. However, its affinity for complex carbohydrate **14** (Fig. 1) is different from that of ConA [26]. In addition, ConA and DGL possess different biological activities such as histamine release from rat peritoneal mast cells [27]. ITC studies reveal the molecular basis of trimannoside and complex carbohydrate binding to DGL and important similarities and differences in carbohydrate recognition by these two highly homologous lectins.

2.2.1. Binding of the core trimannoside

DGL binds to the trimannoside **1** with a ΔH of -16.2 kcal/mol and a K_a of 1.2×10^6 M^{-1} [28]. The ΔH is -8.0 kcal/mol greater and K_a 270-fold greater for **1** than the corresponding values for MeαMan. These results indicate that DGL possesses an extended binding site for the trimannoside, similar to that of ConA. These results were confirmed by the X-ray crystal structure of DGL complexed with the trimannoside [29].

2.2.2. Comparison of ITC data for DGL and ConA

The α(1,3)4-deoxy analog (**4**) (Fig. 1) shows a loss in $-\Delta H$ ($\Delta \Delta H$) of 2.4 kcal/mol and a ~4-fold reduction in K_a for ConA, but a loss in $\Delta \Delta H$ of 1.6 kcal/mol and a 2-fold reduction in K_a for DGL binding [28], indicating slight difference in the mode of binding at this position. ITC results for deoxy analogs **7**, **8**, and **9** of the α(1,6) arm of **1** (Fig. 1) indicate binding of the 3-, 4-, and 6-OH groups to DGL, as observed in ConA [28]. However, data for deoxy analog **6** indicate binding of the 2-OH to DGL although the crystal structure of the DGL complex with **1** shows no lectin–carbohydrate interactions at this site [29]. The reduction in binding by the α(1,6)Man 2-deoxy analog appears to reflect indirect binding to the protein of the 2-hydroxyl of the parent trimannoside via a water molecule. The overall pattern of $\Delta \Delta H$ data for DGL is similar to that for ConA, however, the magnitude of the $\Delta \Delta H$ data for certain analogs of the α(1,6) arm is different [28]. The 3-, 4-, and 6-deoxy α(1,6)Man analogs possess $\Delta \Delta H$ values that are nearly twice as great for DGL (~6.1 kcal/mol) as for ConA (~3.2 kcal/mol). The two lectins are nearly identical in terms of the residues involved in binding and their hydrogen-bonds to **1**. Hence, differences in the $\Delta \Delta H$ values of the two lectins for the 3-, 4-, and 6-deoxy α(1,6)Man analogs are not due to differences in the direct lectin–carbohydrate hydrogen-bonding interactions. The ~2.9 kcal/mol difference in average $\Delta \Delta H$ values for the 3-, 4-, and 6-deoxy α(1,6) analogs binding to DGL (6.1 kcal/mol) and to ConA (3.1 kcal/mol) is nearly the same as the ~2.7 kcal/mol difference in $\Delta \Delta H$ (with regards to trimannoside) values for the 2-deoxy α(1,6) analog (**6**) binding to DGL (3.4 kcal/mol) versus to ConA (0.7 kcal/mol). This indicates a common mechanism underlying the differences in the thermodynamics of binding of all four α(1,6) deoxy analogs to the two lectins. In this regard, it is interesting that the average $\Delta \Delta H$ values for the 3-, 4-, and 6-deoxy α(1,6) analogs have nearly the same magnitude for each protein. The relatively constant $\Delta \Delta H$ values of the 3-, 4-, and 6-deoxy α(1,6) analogs for each lectin occur despite the different number and type of hydrogen-bonds at each position in the parent trimannoside. This further indicates a common thermodynamic mechanism of binding of the 3-, 4-, and 6-deoxy $\alpha(1,6)$ analogs to each lectin.

The K_a and ΔH values obtained with $\alpha(1,3)3,4$-deoxy, "core"2,4-deoxy analog **12** (Fig. 1) for ConA and DGL are similar to those of MeαMan [28]. These results are consistent with the tetradeoxy analog being equivalent to MeαMan since the former is devoid of participating hydroxyl groups on $\alpha(1,3)$Man and "core" Man residues.

Nonlinearity and a lack of scaling of $\Delta\Delta H$ and $\Delta\Delta G$ values with the number and nature of hydrogen-bonds were observed with the deoxy trimannosides binding to DGL similar to that for ConA [28].

2.2.3. Comparisons of ITC and X-Ray crystal data for DGL and ConA

There are four amino acid differences within an area surrounding the binding sites in the two lectins for the core trimannoside: residue 21 (Asn in DGL, Ser in ConA), residue 168 (Asn in DGL, Ser in ConA), residue 205 (Glu in DGL, His in ConA), and residue 226 (Gly in DGL, Thr in ConA). These residues are indirectly involved in ligand binding. They interact with a network of hydrogen-bonded water molecules which, in turn, interact with **1**.

The greatest deviation in water molecule organization appears near the shift in residues 222–227. In ConA, the side chain of Thr 226 is oriented to make hydrogen-bond interactions with the ordered water molecule network, and the smaller side chain of Ser 168 accommodates a water molecule between itself and Thr 226. In DGL, residue 226 is a Gly and ordered water molecules fill the space of the missing side chain. This network of hydrogen-bonded water molecules interacts with the hydroxyl oxygens at positions 2 and 3 of the $\alpha(1,6)$ arm of **1** and thermodynamic data indicate that the strength and specificity of DGL and ConA binding differ at these positions in **1** [28]. Thus, differences observed in the $\Delta\Delta H$ values for DGL and ConA binding to the 2-, 3-, 4-, and 6-deoxy $\alpha(1,6)$Man analogs may be due to altered structural water molecules in this region of the binding sites of the lectins.

Alternatively, DGL and ConA can undergo different conformational transitions on binding **1** and the deoxy analogs that may contribute to the observed differences in $\Delta\Delta H$ values of the analogs. In any case, differences in the $\Delta\Delta H$ values of the 2-, 3-, 4-, and 6-deoxy $\alpha(1,6)$Man analogs (**6–9** in Fig. 1) binding to DGL and ConA are not due to direct protein–ligand interactions.

2.2.4. Comparison of the binding of a biantennary complex type pentasaccharide to DGL and ConA

Hemagglutination inhibition experiments and affinity column chromatography show that DGL binds biantennary complex carbohydrate **14** (Fig. 1) much more poorly than ConA [26]. Using ITC, the K_a value of **14** for DGL is 4.7×10^4 M^{-1}

as compared to a K_a of 1.2×10^6 M^{-1} for ConA. The ΔH values of the two lectins for **14** are very different, with $\Delta H = -10.6$ kcal/mol for ConA and $\Delta H = -4.6$ kcal/mol for DGL. These results indicate that although both lectins possess high affinities for **1**, they possess different affinities for **14**.

The X-ray crystal structure of ConA complexed with **14** shows that the $\beta(1,2)$-GlcNAc residue on the $\alpha(1,6)$ arm of the pentasaccharide fits into an extended groove of ConA, and makes hydrogen-bond contacts on both sides of the sugar rings [30]. The interactions of the $\beta(1,2)$-GlcNAc residue on the $\alpha(1,6)$ arm are with Thr 226 and Ser 168 of ConA. Superposition of the X-ray crystal structure of DGL bound to the trimannoside onto that of the corresponding ConA complex reveals that binding contacts between DGL and the pentasaccharide are prevented due to key amino acid differences at residue 226 (Thr in ConA, Gly in DGL) and residue 168 (Ser in ConA, Asn in DGL), and the shift in the backbone of residues 222–227.

Superposition of the trimannoside structure in DGL with the structure of complex carbohydrate **14** bound to ConA [30] shows that the core trimannoside moiety of **14** bound to ConA deviates by less than 0.5 Å from the position of **1** bound to DGL, indicating similar binding of this moiety in both complexes. However, the $\beta(1,2)$-GlcNAc residue on the $\alpha(1,6)$ arm of **14** modeled into DGL reveals contacts different from those observed in the ConA complex. In DGL, the side chain is missing from residue 226 and thus no hydrogen-bond to the 3-hydroxyl of the $\beta(1,2)$-GlcNAc residue can exist. The backbone carbonyl oxygen of Gly 224 is too far away to make a hydrogen-bond with the 4-hydroxyl of the $\beta(1,2)$-GlcNAc residue, and the side chain of Asn 168 is too large to allow a hydrogen-bond to exist in the 7-hydroxyl of the $\beta(1,2)$-GlcNAc residue [29]. These differences in the interactions of the $\beta(1,2)$-GlcNAc residue on the $\alpha(1,6)$ arm of **14** in DGL appear to explain the 30-fold lower affinity as well as the lower ΔH value of DGL relative to ConA. The interactions of $\beta(1,2)$-GlcNAc residue on the $\alpha(1,3)$ arm of **14** in DGL suggest little interference in binding of this region of the complex.

The contact differences for the $\beta(1,2)$-GlcNAc residue on the $\alpha(1,6)$ arm of **14** in DGL and ConA provide an explanation for the failure of DGL to bind GlcNAc$\beta(1,2)$Man, while ConA binds well to the disaccharide [26].

2.3. Other Diocleinae lectins

ConA and DGL belong to the subtribe Diocleinae. Seven other lectins from the same subtribe were investigated for their binding interactions with **1** as well as the deoxy analogs in Fig. 1 by ITC. The seven lectins were isolated from *C. brasiliensis, C. bonariensis, Cratylia floribunda, D. rostrata, D. virgata, D. violacea, and D. guianensis*. Despite their phylogenetic proximity and apparently conserved sequences, these Diocleinae lectins possess different biological activities such as histamine release from rat peritoneal mast cells [27], lymphocyte proliferation and interferon-γ production [31], peritoneal macrophage stimulation and inflammatory reaction [32], as well as induction of paw edema and peritoneal cell immigration in rats [33]. Hemagglutination and thermodynamic studies show important similarities and differences in the carbohydrate-binding properties of the lectins. The different carbohydrate specificities of the Diocleinae lectins correlate with histamine release from rat peritoneal mast cells, as discussed below.

2.3.1. Binding sites of Diocleinae lectins

ITC data for the seven Diocleinae lectins binding to MeαMan show differences in their affinities as well as ΔH and $T\Delta S$ values [34]. *C. brasiliensis* possesses the highest K_a value (1.3×10^4 M^{-1}), while *D. rostrata* the lowest (1.7×10^3 M^{-1}). *C. brasiliensis*, *D. guianensis*, *D. violacea*, and *D. virgata* have ΔH values between -5.8 and -4.9 kcal/mol, while *C. bonariensis*, *C. floribunda*, *D. rostrata*, ConA, and DGL possess $-\Delta H$ values between -6.9 and -8.9 kcal/mol. The K_a values of the lectins for binding MeαMan did not correlate with their respective $-\Delta H$ values, indicating compensating entropy terms.

All seven Diocleinae lectins possess enhanced K_a and $-\Delta H$ values for the core trimannoside relative to MeαMan [34]. The $-\Delta H$ values for all seven lectins binding to trimannoside are -5 to -7 kcal/mol greater than that for MeαMan, similar to the differences observed for DGL and ConA. These data suggest similar extended binding sites for all nine Diocleinae lectins.

ITC experiments were carried out with all seven Diocleinae lectins using deoxy analogs of the trimannoside (Fig. 1). K_a and ΔH values of **7**, **8**, and **9** are lower than that of **1** for all seven lectins which support the involvement of the 3-, 4-, and 6-hydroxyl groups of the α(1,6)Man of **1**, but not the 2-hydroxyl, in binding to all seven Diocleinae lectins [35]. These findings suggest that the α(1,6)Man residue of **1** occupies the "monosaccharide binding site" in all of the lectins [3].

Deoxy analogs **2**, **3**, **4**, and **5** were used to determined the involvement of the 2-, 3-, 4-, and 6-hydroxyl groups of α(1,3)Man of **1**, respectively, in binding to the Diocleinae lectins [35]. Only analog **3** exhibited a loss in K_a and ΔH relative to **1**, suggesting that the 3-hydroxyl of the α(1,3)Man of **1** binds to the seven Diocleinae lectins.

The X-ray crystal structures of ConA and DGL complexed with the core trimannoside [21, 29] show that the 3- and 4-hydroxyl of the α(1,3)Man of **1** are involved in H-bonds with conserved residues of the two lectins. Binding of ConA [22] to **4** shows a 5-fold reduction in K_a value and a loss in $\Delta\Delta H$ value of 2.1 kcal/mol relative to **1**. DGL binding to **4** shows a 2-fold reduction in K_a and a loss in $\Delta\Delta H$ of 1.6 kcal/mol relative to **1** [28]. Although the X-ray crystal structures show binding of the 4-hydroxyl of the α(1,3)Man of **1** with both lectins, the thermodynamic data for **4** suggests only very weak interactions of the 4-hydroxyl with the two proteins. The ITC data for **4** binding to the seven Diocleinae lectins also show only weak interactions of the 4-hydroxyl group of the α(1,3)Man of **1** with the Diocleinae lectins [35].

The X-ray crystal structures of the ConA [21] and DGL trimannoside complexes [29] also show the involvement of the 2- and 4-hydroxyl groups of the core Man of trimannoside. Unlike the 4-hydroxyl group, 2-hydroxyl group shows only water mediated binding to the lectins. ITC data show $\Delta\Delta H$ values of 2.3 and 3.4 kcal/mol, respectively, for ConA and DGL binding to **11** [22, 28]. However, the $\Delta\Delta H$ values for ConA and DGL binding to **10** were 1.0 kcal/mol and 1.4 kcal/mol, respectively [22, 28].

A similar pattern of larger $\Delta\Delta H$ values for **11** as compared to **10** exists with the seven Diocleinae lectins [28, 35]. Only the *C. grandiflora* and *D. violacea* lectins possess $\Delta\Delta H$ values for **10** large enough to be evidence for H-bonding of the respective hydroxyl group of **1** to the lectins. The presence or absence of H-bonding of the 2-hydroxyl group of the core Man of **1** to the remaining Diocleinae lectins will have to await X-ray crystallographic analysis of the trimannoside complexes. The absence of $\Delta\Delta H$ values greater than 1.0 kcal/mol for certain deoxy analogs such as **10** with five of the Diocleinae lectins may be taken as evidence that such H-bonds are either energetically weak or absent in the corresponding solution complexes of **1** with the lectin.

2.3.2. Differences in the binding of biantennary complex oligosaccharide

The affinity of DGL for biantennary complex oligosaccharide **14** (Fig. 1) is weaker than that of ConA [34]. A structural explanation for this difference was provided by Rozwarski et al. [29]. All of the Diocleinae lectins showed correlated binding affinities toward **14** and its constituent disaccharide, GlcNAcβ1-2Man. Hemagglutination inhibition results indicated that **14** had higher inhibition potencies with *C. brasiliensis, D. guianensis*, and *D. virgata* as compared to the other Diocleinae lectins. This parallels the binding activities of the lectins toward GlcNAcβ1-2Man [34]. ITC data show greater K_a values of *C. brasiliensis, D. guianensis*, and *D. virgata* for **14** relative to the other four lectins. Among the nine Diocleinae lectins, ConA shows the highest K_a value for **14**, while DGL shows the lowest K_a [34]. *C. brasiliensis, D. guianensis*, and *D. virgata* possessed greater $-\Delta H$ values for **14** of the seven lectins, while $-\Delta H$ values were much lower for other four Diocleinae lectins.

An enthalpy–entropy compensation plot of the data for **14** yielded different slopes for the above two groups of Diocleinae lectins [34]. The lectins from *C. brasiliensis, D. guianensis, D. virgata*, and ConA fell on a line with a slope of 1.44 (correlation coefficient 0.85), while a line with a slope of 0.85 (correlation coefficient 0.98) was observed for the lectins from *C. bonariensis, C. floribunda, D. rostrata, D. violacea*, and DGL. A similar plot of the lectins binding to **1** shows a single line with a slope of 1.21 (correlation coefficient 0.97). These results indicate different mechanisms of binding of the four higher affinity lectins for **14**, as compared to the five lower affinity lectins. Although all nine Diocleinae lectins show conserved high affinities binding for **1**, four of the lectins show higher affinities for **14**, while the other five lectins show lower affinities. Thus, binding differences among this group of Diocleinae lectins exist toward biantennary complex carbohydrate **14**.

Gomes and coworkers [27] investigated histamine release from rat peritoneal mast cells induced by lectins from the Diocleinae subtribe. ConA and the lectins from *C. brasiliensis, D. guianensis*, and *D. virgata* induced high levels of histamine release, whereas the lectins from *D. grandiflora, C. bonariensis, C. floribunda, D. rostrata*, and *D. violacea* induced lower levels. A significant correlation was shown to exist between the level of histamine release and the affinity constants of the lectins for **14** [34]. The strong histamine inducing lectins ConA, *C. brasiliensis, D. guianensis*, and *D. virgata* exhibited relatively high affinities for **14**, while the low histamine inducing lectins from *D. grandiflora, C. bonariensis, C. floribunda, D. rostrata*, and *D. violacea* possessed lower affinities for **14**. It appears, therefore that induction of histamine release from rat peritoneal mast cells requires relatively strong binding of a Diocleinae lectin to a biantennary complex carbohydrate and/or structurally homologous epitope present on the cell surface.

2.3.3. Differences in the thermodynamic binding parameters of deoxy trimannoside analogs

ITC derived thermodynamic data in combination with available X-ray crystallographic data show that all nine Diocleinae lectins interact with essentially the same set of hydroxyl groups of the trimannoside. However, ITC data show a range of $\Delta\Delta H$ values for certain deoxy analogs that have corresponding hydroxyl groups involved in binding to the nine Diocleinae lectins [35]. These include $\Delta\Delta H$ values ranging from 6.5 to 7.7 kcal/mol for **7**, **8**, and **9** binding to *D. rostrata* to the much lower values of ~3 kcal/mol for ConA [22]. Thus, there is variation in these $\Delta\Delta H$ values of the nine Diocleinae lectins. This is the case for not only the same deoxy analog with different lectins, but also for different deoxy analogs that have corresponding hydroxyl groups that bind to the same protein. For example, analogs **3**, **7**, and **11** possess different $\Delta\Delta H$ values in binding to DGL and ConA, respectively, even though their respective hydroxyl groups of **1** show hydrogen-bond to both lectins. Another example is a comparison of data for ConA and *C. brasiliensis*. The X-ray crystal structure of the *C. brasiliensis* lectin shows only two amino acid changes relative to ConA [36]. Gly-58 and Gly-70 in *C. brasiliensis* are replaced by Asp and Ala in ConA, respectively. Neither of the residues is near the carbohydrate-binding sites in both proteins, and only small changes in the quaternary structures of the two lectins were noted. However, these two amino acid changes result in differences in the $\Delta\Delta H$ values of both lectins binding to analogs **7**, **8**, and **9** (~5 kcal/mol for *C. brasiliensis* versus ~3.0 kcal/mol for ConA). In addition, although the K_a values of the two lectins for **1** are similar (3.7×10^5 M^{-1} for *C. brasiliensis* and 4.9×10^5 M^{-1} for ConA), the *C. brasiliensis* lectin possesses a ΔH of -12.4 kcal/mol for **1**, while ConA possesses a ΔH of -14.4 kcal/mol for

1 [28]. It is also of note that ConA and the *C. brasiliensis* lectin are reported to have different lectin-induced nitric oxide production in murine peritoneal cells in vitro [37]. Differences in the thermodynamics of binding in a homologous group of lectins where the binding residues are conserved indicate important roles of nonconserved residues away from the carbohydrate-binding site in these proteins. Changes in the hydration of the lectins or subtle differences in their conformations due to minor alterations of amino acid residues away from the carbohydrate-binding sites may be responsible for differences in their binding thermodynamics [35]. Indeed, Siebert et al. [38] have shown the effects of single-site mutations on conformational features of lectins.

Changes in the binding thermodynamics of lectins due to changes in amino acids away from the carbohydrate-binding site have been reported. Galectin-1 from Chinese hamster ovary cells was reported to undergo significant changes in ΔH and $T\Delta S$ but not ΔG in binding to LacNAc when single or multiple mutations were introduced in the N-terminal region of the protein [39]. ΔH for the parent galectin-1 binding to LacNAc is -6.6 kcal/mol, while a Cys-2 to Ser-2 mutant possesses a ΔH for LacNAc of -2.8 kcal/mol. Both lectins possess similar K_a values for the disaccharide. Interestingly, a monomeric four-substituted mutant with amino acid substitutions at the 2-, 4-, 5-, and 6-positions also shows essentially no change in K_a, but possesses a ΔH of -0.6 kcal/mol, hence making binding an entropy-driven process. Interestingly, all of these mutations are ~ 20 Å from the carbohydrate-binding site of the galectin.

3. Binding of Multivalent Carbohydrates to Lectins

3.1. Binding of multivalent carbohydrates to ConA and DGL

3.1.1. Thermodynamic binding parameters

The mechanism of binding of synthetic multivalent carbohydrates to ConA and DGL was investigated by ITC. Synthetic glycosides bearing multiple terminal trimannoside residues in Fig. 4 show increased affinities for ConA and DGL relative to trimannoside **1** (Fig. 1) [40]. Bi-, tri-, and tetravalent analogs **19**, **20**, and **21**, respectively, in Fig. 4 show 6-, 11-, and 35-fold higher K_a values for ConA, respectively, and 5-, 8-, and 53-fold higher K_a values, respectively, for DGL [40] (Table 1 in Ref. 40). These K_a values agree with relative inhibition values obtained by hemagglutination inhibition measurements [41].

ITC data indicate that trimannoside **1** binds to ConA and DGL with *n* values close to 1.0 [28]. These results agree with X-ray crystal data that demonstrate single carbohydrate-binding sites on each monomer of ConA and DGL [29]. Thus, trimannoside **1** is a monovalent ligand for ConA and DGL. The theoretical values of *n* for binding of bi-, tri- and tetravalent carbohydrates to ConA and DGL are 0.5 (1.0/2), 0.33 (1.0/3) and 0.25 (1.0/4), respectively, based on the structural valence

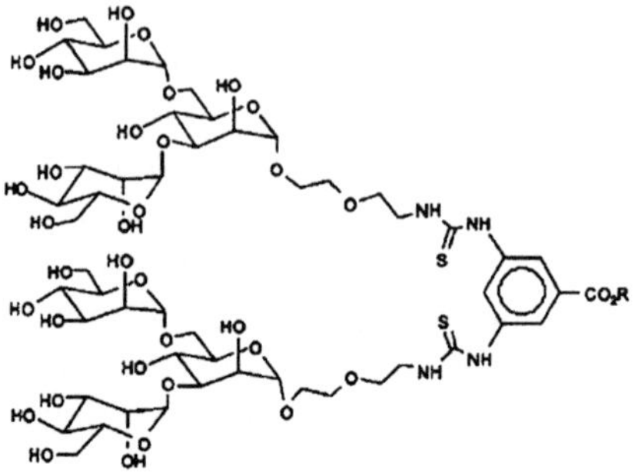

Figure 4. Structures of bi (**19**), tri (**20**) and tetraantennary (**21**) analogs of the core trimannoside (**1**).

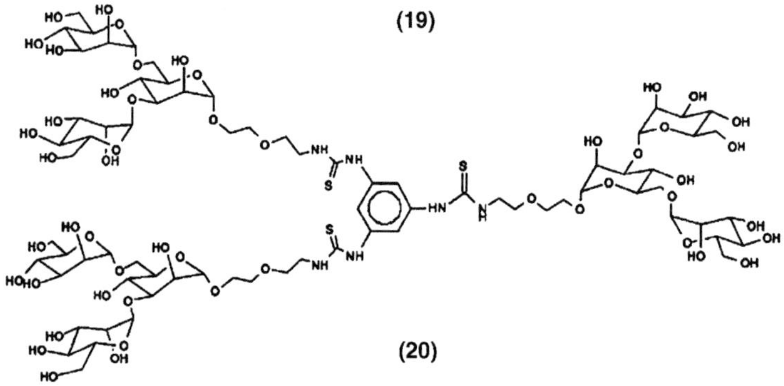

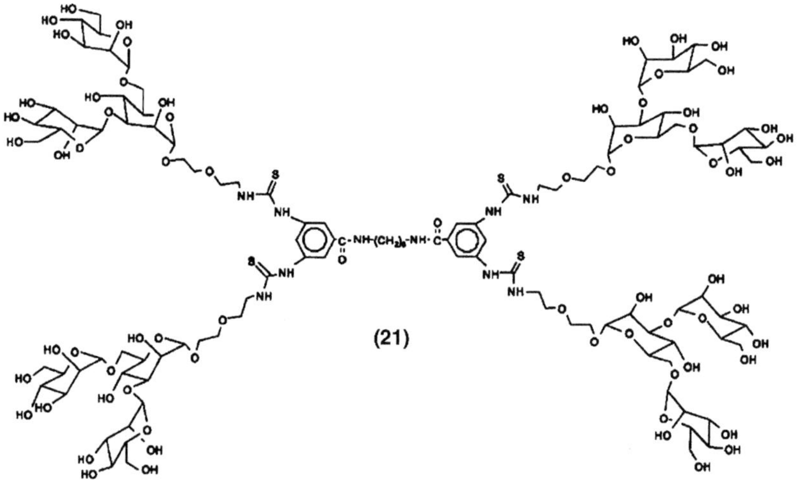

of the respective analogs. The values obtained from ITC experiment are consistent with predicted structural valence values except for trivalent analog **20** [40]. The *n* values obtained with analog **20** are 0.51 with ConA and 0.40 with DGL. These values show that the structurally trivalent analog is functionally bivalent for binding to ConA and a mixture of bi- and trivalent binding to DGL. These results indicate that the functional valence of a multivalent carbohydrate can differ from its structural valence.

ITC data of multivalent analogs with relatively high affinities show that the observed ΔH is approximately, the sum of the ΔH values of the individual epitopes of an analog. The ΔH values of bivalent Man analogs binding to ConA were almost 2-fold greater than that of MeαMan [40]. Similar observations were made with DGL [40]. The ΔH value of -26.2 kcal/mol for bivalent analog **19** binding to ConA is almost than twice the ΔH value of -14.7 kcal/mol for **1**. The same is found for **19** binding to DGL. The ΔH value for tetravalent analog **21** binding to ConA is -53.0 kcal/mol, which is approximately four times the ΔH value of **1** (-14.7 kcal/mol) [40]. The ΔH value for **21** binding to DGL is -58.7 kcal/mol, which is also nearly four times the ΔH value of **1** (-16.2 kcal/mol). The ΔH values of analog **20** are more complicated since the analog is bivalent for ConA ($n = 0.51$), and a mixture of bi- and trivalent for DGL ($n = 0.40$). These results suggest that the ΔH values of high affinity multivalent carbohydrates are approximately the sum of the ΔH values of the individual binding epitopes of the analogs. Similar observations have been made for the binding of a trivalent system of receptor and ligand derived from vancomycin and D-Ala-D-Ala [42].

Although ΔH scales proportionally to the number of binding epitopes in higher affinity multivalent carbohydrates, $T\Delta S$ does not. Instead, $T\Delta S$ is more negative than if it proportionally scaled to the number of binding epitopes in the carbohydrates. For example, tetraantennary analog **21** possesses four trimannosyl binding epitopes, and its ΔH value binding to ConA is -53 kcal/mol, which is approximately four times the ΔH of -14.7 kcal/mol for **1** [40]. However, the corresponding $T\Delta S$ value for **21** is -43.3 kcal/mol, not -28.4 kcal/mol if scaled with the $T\Delta S$ value of -7.1 kcal/mol for **1** [40]. The resulting ΔG for **21** would also be much greater if $T\Delta S$ scaled with valence since the difference between ΔH and $T\Delta S$ would be greater. However, the observed ΔG value(s) for **21** are much smaller. The same is true for the other multivalent carbohydrates investigated [40].

The observation that ΔH scales for multivalent carbohydrate analogs binding to ConA and DGL but $T\Delta S$ does not is characteristic of the binding of multivalent ligands to separate receptor molecules. In the current case, **21** binds to four separate ConA or DGL molecules. The spacer length between two carbohydrate epitopes of any of the multivalent analog in Fig. 4 is not long enough to span two binding sites of ConA or DGL. Therefore, the carbohydrate epitopes of the multivalent analogs (Fig. 4) interact with different lectin molecules. This is distinguished from binding of a multivalent carbohydrate to a single lectin molecule possessing multiple binding sites. In the latter instance, the increase in affinity is much greater. For example, binding of a triantennary complex carbohydrate to the hepatic asialoglycoprotein receptor that possess three subsites results in a $\sim 10^9$ M^{-1} inhibition constant relative to the $\sim 10^3$ M^{-1} inhibition constant of the corresponding monovalent oligosaccharide [43]. More dramatic is the increase in affinity to $\sim 10^{17}$ M^{-1} of a trivalent derivative of vancomycin binding to a trivalent derivative of D-Ala-D-Ala in which the affinity of the corresponding monovalent analogs is $\sim 10^6$ M^{-1} [42]. In the latter study, thermodynamic measurements showed that both ΔH and $T\Delta S$ scaled proportionally to the number of binding epitopes in the ligand. As a result, the resulting ΔG value is much greater than that for ConA or DGL binding to the multivalent carbohydrate analogs in Fig. 4. Thus, ITC measurements can distinguish between binding of a multivalent carbohydrate to separate lectin molecules or to an extended binding site on a single lectin molecule.

3.1.2. Thermodynamic basis for affinity enhancements of multivalent analogs
The enhanced affinities (avidities) of the multivalent carbohydrates in Fig. 4 for ConA and DGL are associated with their epitopes binding to separate lectin molecules. For example, the observed K_a for tetravalent analog **21** is the average of the four microscopic K_a values of its four epitopes binding to a separate lectin molecule [40]. Since ΔH is constant at each epitope and each epitope is approximately the same as that of **1**, then increases in the overall macroscopic K_a values (ΔG) of the four epitopes require more favorable $T\Delta S$ contributions of the individual four epitopes of the tetravalent analog compared with **1** [40]. This was demonstrated in reverse ITC experiments for the di- and trivalent Man analogs in Fig. 4 binding to ConA [44] (below).

3.1.3. Multivalent carbohydrate bind with negative cooperativity and a gradient of microscopic affinity constants
ITC derived n value and other thermodynamic binding parameters clearly indicate that each epitope of the multivalent analogs possesses microscopic binding parameters including microscopic K_a, ΔH, and $T\Delta S$ values. Tetravalent analog **21** (Fig. 4) has a total of four microscopic K_a values (K_{a1}, K_{a2}, K_{a3}, and K_{a4}) associated with its four epitopes. The macroscopic K_a of this analog determined by ITC is an average of the four microscopic K_a. Scatchard and Hill plot analyses of ITC raw data reveal that the microscopic K_a values are not equal and that $K_{a1} > K_{a2} > K_{a3} > K_{a4}$ [41]. This indicates that the multivalent analog binds to the lectin with decreasing affinity and negative cooperativity as observed in Scatchard and Hill plots [41].

The physical basis for the decreasing K_a values of epitopes of multivalent ligands is due to reduction in their functional valence as they bind an increasing number of lectin molecules. For example, Fig. 5 shows the various microequilibrium constants for **21** as its four epitopes sequentially bind one, two, three, and four

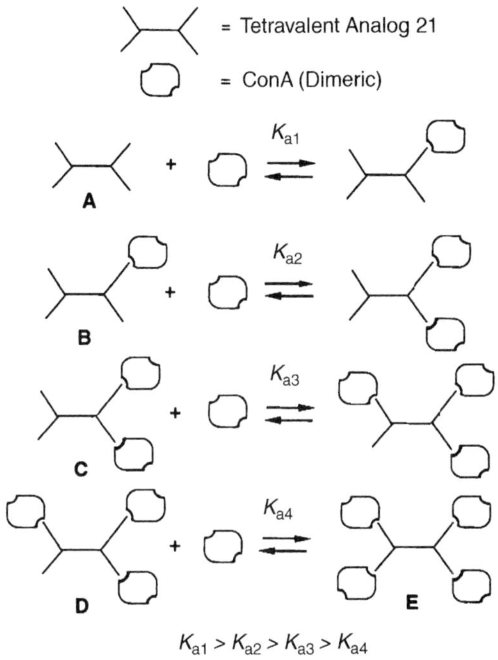

Figure 5. Four microequilibrium constants of tetravalent analog **21** binding to dimeric ConA represented by K_{a1}, K_{a2}, K_{a3}, and K_{a4}.

molecules of ConA. The functional valence of unbound **21** (species A) is four, the functional valence of **21** with one bound lectin molecule (species B) is three, the functional valence of **21** with two bound lectin molecules (species C) is two, and the functional valence of **21** with three bound lectin molecules (species D) is one. Sequential occupancy of the four epitopes of the analog results in a gradual decrease in the overall valence and binding affinity of **21** that is consistent with Hill plots of the raw ITC data with ConA and DGL [41].

Another factor that may play a role in the negative cooperativity shown by multivalent analogs binding to ConA and DGL is the formation of noncovalent cross-linked complexes between lectins and multivalent carbohydrates. The equilibria in Fig. 5 are simplified in that each lectin molecule, represented as a monomer in the scheme, is actually a dimer under the conditions of the experiment. Hence, each molecule of ConA or DGL is capable of binding and cross-linking the multivalent carbohydrates in the present study. However, ITC binding data of asialofetuin (ASF), a multivalent glycoprotein, shows similar negative cooperative effects in its interaction with monomeric galectins that are not expected to possess cross-linking activities [45].

3.1.4. Range of microscopic affinity constants for multivalent carbohydrates binding to ConA and DGL

Based on Fig. 5, Eq. (5) was derived to describe the relationship between the observed macroscopic free energy of binding and the microscopic free energies of binding of the various epitopes of a multivalent carbohydrate binding to a lectin [41].

$$\Delta G(\text{obs}) = \frac{\Delta G_1 + \cdots + \Delta G_n}{n} \tag{5}$$

Eq. (5) states that the observed macroscopic ΔG value ($\Delta G(\text{obs})$), determined by ITC, of a multivalent carbohydrate is the average of the microscopic ΔG values of the individual epitopes, n the number of epitopes of the multivalent ligand [41]. This equation correctly estimates the difference in microscopic ΔG values of the two epitopes of bivalent analog **19** binding to ConA [44]. In this case, Eq. (6) for binding of **19** is

$$\Delta G(\text{obs}) = \frac{\Delta G_1 + \Delta G_2}{2} \tag{6}$$

Eq. (6) shows that $\Delta G(\text{obs})$ obtained from an ITC experiment allows calculation of ΔG_1, the first epitope of the divalent carbohydrate, assuming that ΔG_2 for the second epitope is the same as that of a monovalent ligand analog. This latter assumption was shown to be true from a reverse ITC experiment that allows direct determination of ΔG_1 and ΔG_2 [44]. The difference between ΔG_1 and ΔG_2 calculated from Eq. (6) using $\Delta G(\text{obs})$ from an ITC experiment agreed well with that determined from the reverse ITC [44]. Eq. (5) can also be used to estimate the spread in microscopic ΔG values for the tetraantennary analog **21** that binds to ConA and DGL. Eq. (7) describes the relationship between the macroscopic $\Delta G(\text{obs})$ and four microscopic ΔG values for binding of **18** to DGL.

$$\Delta G(\text{obs}) = \frac{\Delta G_1 + \Delta G_2 + \Delta G_3 + \Delta G_4}{4} \tag{7}$$

ΔG_1 in Eq. (7) is associated with the binding of the first carbohydrate epitope of tetraantennary analog **21**, ΔG_2 with the second, ΔG_3 with the third, and ΔG_4 with the fourth. The ITC determined macroscopic $\Delta G(\text{obs})$ for binding of **21** to DGL is $-10.6\,\text{kcal/mol}$ [40], while ΔG_4 in Eq. (7) can be taken as the $\Delta G(\text{obs})$ for binding monovalent trimannoside **1** to DGL which is $-8.3\,\text{kcal/mol}$ [40]. Since $\Delta G(\text{obs})$ is the average of the four microscopic ΔG values, then

$$\Delta G_1 - \Delta G(\text{obs}) = \Delta G(\text{obs}) - \Delta G_4 \tag{8}$$

assuming that ΔGobs $- \Delta G_2 \sim \Delta G_3 - \Delta G$obs (i.e., there is a symmetrical distribution of microscopic ΔG values on either side of ΔG(obs)). The numerical value of ΔG_1 calculated from Eq. (8) is -12.9 kcal/mol, which is 4.6 kcal/mol greater than ΔG_4. This difference between ΔG_1 and ΔG_4 translates to a difference in microscopic K_a values, K_{a1} and K_{a4}, of approximately 2,800 fold. In absolute terms, K_{a1} is approximately 0.3 nM, while K_{a4} is approximately 0.8 mM. Thus, the microscopic K_{a1} of the first unbound epitope of tetraantennary analog **21** binding to DGL is 2,800-fold greater than K_{a4} for binding of the fourth. For **21** binding to ConA, this difference between K_{a1} and K_{a4} is nearly 1,200-fold [40]. This indicates a decreasing gradient of microscopic binding constants of the four epitopes of **21** binding to ConA and DGL. These differences have been postulated to be due to kinetic effects on the off rates of the various fractionally bound complexes of the multivalent carbohydrates [40]. The microscopic off rate (k_{-1}) for K_{a1} in Fig. 5 ($K_{a1} = k_1/k_{-1}$) would be expected to be slower than the microscopic off rate for K_{a2}, etc., due to binding and recapture of the first bound lectin molecule by the remaining unbound trimannoside residues of the tetravalent analog before complete dissociation of the complex.

3.1.5. Determination of microscopic affinity constants for multivalent carbohydrates by reverse ITC

In a reverse ITC experiment lectin is titrated into a solution of carbohydrate. Such experiments were performed with ConA titrated into solutions of **19** and **20** [44]. Data were fitted with one- or two-site models depending on the functional valence of the carbohydrate determined from "normal" ITC data [40]. Importantly, selection of other values for the number of epitopes failed to provide fits of the data.

ConA was titrated into a solution of **1** as a control and fitted with a one-site model [44]. The *n* value for **1** in the reverse experiment is 0.99 that agrees with the *n* value of the normal titration experiment. K_a for **1** in the reverse titration is 6.3×10^5 M^{-1} as compared to 3.9×10^5 M^{-1} for the normal titration [44]. ΔH for **1** in the reverse experiment is -13.1 kcal/mol as compared to -14.7 kcal/mol for the normal titration. Thus, the results of the reverse ITC of ConA binding to **1** agree with the previously reported normal ITC data.

The reverse ITC profile of ConA binding to **19** was fitted using a two-site model [44] since data from the normal ITC experiment demonstrated two binding sites for this analog [40]. The data provided thermodynamic data for the two individual epitopes of **19** [44]. The *n* value for the first epitope (n_1) is 0.97 and for the second (n_2) is 0.94. This indicates that both epitopes of **19** are fully bound to ConA.

Analog **20** is functionally bivalent for binding to ConA [44]. Importantly, the reverse ITC data of ConA binding to **20** could only be fit with a two-site model and not a three-site model [44]. The values of n for the two sites are 1.05 (n_1) and 1.09 (n_2), which are consistent with binding of two of the three epitopes of **20**.

The reverse ITC data provided two microscopic K_a values for the two epitopes of **19** binding to ConA [44]. K_{a1} is 1.6×10^7 M^{-1} and K_{a2} is 8.8×10^5 M^{-1}. Hence, the microscopic affinity constant of the first epitope is 18-fold greater than that of the second. Interestingly, the latter value is close to the affinity constant for **1**. The observed macroscopic ΔG value for **19** binding to ConA in the normal ITC experiment is -8.7 kcal/mol. The microscopic ΔG values for ConA binding to the two epitopes of **19** are $\Delta G_1 = -9.8$ kcal/mol and $\Delta G_2 = -8.1$ kcal/mol, respectively [44]. The average value of ΔG_1 and ΔG_2 is -9.0 kcal/mol that is similar to the macroscopic ΔG of -8.7 kcal/mol for **19** binding to ConA [44]. Thus, the observed macroscopic K_a agrees with the average of the two microscopic K_a values for **19**. The microscopic K_a values of the two sites of **20** are $K_{a1} = 4.6 \times 10^{-7}$ M^{-1} and $K_{a2} = 8.6 \times 10^{-5}$ M^{-1}. Thus, there is a 53-fold higher affinity of the first binding site of **20** relative to its second binding site for ConA. The affinity constant of the second site on **20** is similar to that of **19**, while the affinity constant of the first site on **20** is 2.5-fold greater than that of **19**. This may be due to the greater structural valence of **20** relative to **19** although their functional valences are the same.

The observed macroscopic ΔG value for **20** binding to ConA in the normal ITC experiment is -9.0 kcal/mol [40]. The microscopic ΔG values for ConA binding to the two epitopes of **20** are $\Delta G_1 = -10.4$ kcal/mol and $\Delta G_2 = -8.1$ kcal/mol [44]. The average of ΔG_1 and ΔG_2 is -9.3 kcal/mol that is similar to the observed macroscopic ΔG value of -9.0 kcal/mol for **20** [40]. Thus, the observed macroscopic K_a for **20** agrees with the average of the two microscopic K_a values reported by reverse ITC.

The microscopic enthalpies of binding of the two epitopes (ΔH_1 and ΔH_2) of **19** are essentially the same, and similar to that for **1** [44]. Similar results were found for **20**. These findings agree with the conclusion reached in using "normal" ITC measurements [40] that the two functional binding epitopes of **19** and **20** possess essentially equal microscopic ΔH values and are additive in generating the respective macroscopic ΔH values of the two analogs.

The microscopic entropy of binding values of the two epitopes ($T\Delta S_1$ and $T\Delta S_2$) of **19** and **20** were calculated from corresponding microscopic ΔG_1 and ΔG_2 values and microscopic ΔH_1 and ΔH_2 values of **19** and **20**, respectively [44]. $T\Delta S_1$ for the first epitope of **19** is 1.5 kcal/mol more favorable relative to $T\Delta S_2$ of the second. Similarly, there is 1.2 kcal/mol more favorable entropy of binding value for the first epitope of **20** compared to the second. These results provide direct determination of the favorable entropy effects in the enhanced affinities of bivalent carbohydrates. Furthermore, the results from reverse ITC experiments are consistent with the progressively decreasing K_a values and the increasing negative cooperativity of multivalent carbohydrates binding to ConA and DGL as demonstrated by Scatchard and Hill plots [41].

3.2. Binding of asialofetuin (ASF) to galectins

3.2.1. K_a values for ASF
Binding of galectins-1, -2, -3, -4, -5, and -7, and truncated, monomer versions of galectins-3 and -5 to ASF, a naturally occurring 48 kDa glycoprotein that possesses nine LacNAc epitopes, has been studied by ITC [45]. The observed K_a values for ASF binding to the galectins and the two truncated forms are 50- to 80-fold greater than that of the monovalent disaccharide LacNAc. Hill plot analysis shows that ASF binds to all of the galectins with negative cooperativity that is consistent with a gradient of microscopic K_a values for the various epitopes of ASF. Noncovalent cross-linking does not appear to contribute to the negative cooperativity since similar results were obtained with monomeric forms of the galectins (truncated galectins-3 and -5).

3.2.2. Range of microscopic K_a values for ASF binding to galectins
The nine LacNAc chains of ASF possess nine microequilibrium constants represented by K_{a1}, K_{a2}, ..., and K_{a9} for binding to the galectins. Hill plot analysis of the binding data of ASF with the galectins [45] showed evidence of increasing negative cooperativity, similar to that observed in Hill plots of the ITC data for the binding of multivalent carbohydrates **19–21** to ConA and DGL [41]. Thus, a similar analysis was performed to determine the range of microscopic binding constants of ASF with the galectins.

Since the ITC derived ΔG(obs) value for ASF binding to the galectins is the average of the nine microscopic ΔG values as shown in the following equation:

$$\Delta G(\text{obs}) = \frac{\Delta G_1 + \Delta G_2 + \Delta G_3 + \Delta G_4 \cdots + \Delta G_9}{9} \tag{9}$$

and assuming that there is a symmetrical distribution of decreasing microscopic ΔG values on either side of the average ΔG value (i.e., $\Delta G_8 - \Delta G_7 \sim \Delta G_3 - \Delta G_2$, etc.), then the value of ΔG(obs) is approximately equal to ΔG_5 in Eq. (9). It follows that if ΔG_9, which represents binding of the last unbound epitope of ASF to a galectin, is nearly equal to ΔG for LacNAc binding to a galectin, then the difference between ΔG(obs) and ΔG_9 is half the difference between ΔG_1 and ΔG_9. Thus, the 2.8 kcal/mol increase in ΔG(obs) for ASF binding to galectin-3 versus ΔG for LacNAc binding [45] indicates that the difference between ΔG_1 and ΔG_9 for ASF binding to galectin-3 is $\sim$5.6 kcal/mol. Since 2.8 kcal/mol is a 78-fold increase in affinity of ASF for galectin-3 over LacNAc, the difference in affinity of the first unbound LacNAc epitope of ASF for galectin-3 is 78 $\times$ 78 or approximately 6,000-fold increase in affinity over LacNAc with a gradient of decreasing affinities to that of LacNAc for the last unbound epitope of ASF.

ITC data show that all of the galectins including truncated galectin-5 possess K_a(obs) values that are 50–78 fold greater than that of LacNAc for the respective galectin [45]. This indicates that the first unbound epitope of ASF binds to all of the galectins with 3,000- to 6,000-fold higher affinity than LacNAc, and that there is a gradient of decreasing affinities of the remaining epitopes of ASF. In terms of absolute affinity constants, a 6,000-fold increase in affinity of galectin-3 for the first unbound epitope of ASF is eqivalent to a 10 nM affinity constant, using the K_a value for LacNAc as an estimate for the last microscopic K_a value (K_{a9}). This estimate for K_{a1} is for all of the galectins binding to the first unbound epitope of ASF. Thus, a large range of decreasing microaffinity constants exists for the nine epitopes of ASF binding to the galectins.

3.2.3. Implications of gradient binding of galectins to multivalent glycoproteins
The implications of a gradient of decreasing affinity constants of ASF for the galectins have important implications. For example, relatively low concentrations of the galectins can be expected to bind to only a few high affinity sites on multi-valent glycoprotein receptors. In this regard, as few as three galectin-1 molecules bound to ASF are observed to lead to homogeneous cross-linking of the molecules into large insoluble aggregates [46]. In addition, binding and cross-linking of mul-tivalent glycoprotein receptors on the surface of a cell by a divalent lectin such as galectin-1 with individual affinity sites on each glycoprotein of $\sim$10 nM would result in an overall avidity for galectin-1 of $\sim$$10^{16}$ M^{-1}. Hence, cross-linking by a dimeric lectin would be essentially irreversible under these conditions. Furthermore, cross-linking of glycoprotein receptors by lectins by such a mecha-nism could lead to supramolecular assemblies of homogeneous cross-linked receptors [47] or heterogeneous cross-linked receptors [48]. Such assemblies can trigger cell surface signal transduction mechanisms, similar to those observed for galectin-1 binding to T cell receptors that leads to apoptosis (cf. [49]).

The concept of gradient of microscopic binding constants of ligands to mul-tivalent receptors also has implications for other types of receptor systems. The concept of "spare receptors" is well known in the pharmacology literature [50]. Maximum dose–activity responses are observed at relatively low fractional occupancy of the receptors. Such "spare receptor" systems may exhibit enhanced affinity for a specific ligand through a clustering mechanism, similar to that observed for ASF binding to the galectins [45]. The enhancement in affinity of a ligand could be as much as 1,000- to 10,000-fold or greater by clus-tering of the receptors. Occupancy of a portion of the clustered or "spare recep-tors" by antagonist would diminish the total number of unbound receptors, but may not reduce the "avidity" of the remaining receptors [50]. The "efficacy" of the agonist effect may be related to cross-linking or activating a fraction of the receptors that is necessary for full activity. As an example, fractional occupancy of a few epitopes of a multivalent glycoprotein receptor by galectin molecules can lead to cross-linking interactions and subsequent signal transduction effects such as apoptosis, as observed in the binding of galectin-1 to human T cells glycoprotein receptors [51].

Acknowledgments

This work was supported by Grant CA-16054 from the National Cancer Institute, Department of Health, Education and Welfare, and Core Grant P30 CA-13330 from the same agency (C. F. B.).

References

[1] H. Lis and N. Sharon, Chem. Rev., 98 (1998) 637–674.
[2] A. Varki, R. Cummings, J. Esko, H. Freeze, G. Hart and J. Marth, Essentials of Glycobiology, Cold Springs Harbor, Laboratory Press, New York, 1999, pp. 653.
[3] I.J. Goldstein and R.D. Poretz. In: I.E. Liener, N. Sharon and I.J. Goldstein (Eds.), The Lectins, Academic Press, New York, 1986, pp. 35–244.
[4] E. Freire, O.L. Mayorga and M. Straume, Anal. Chem., 62 (1990) 950A–959A.
[5] E.J. Toone, Curr. Opin. Struct. Biol., 4 (1994) 719–728.
[6] H.F. Fisher and N. Sing, Methods Enzymol., 259 (1995) 194–221.
[7] B.M. Baker and K.P. Murphy, Methods Enzymol., 295 (1998) 294–315.
[8] L. Indyk and H.F. Fisher, Methods Enzymol., 295 (1998) 350–364.
[9] A. Cooper, Curr. Opin. Chem. Biol., 3 (1999) 557–563.
[10] J. Jimenez-Barbero, J.L. Asensio, F.J. Canada and A. Poveda, Curr. Opin. Struct. Biol., 9 (1999) 549–555.
[11] J.E. Ladbury and B.Z. Chowdhry, Chem. Biol., 3 (1996) 791–801.
[12] A. Cooper and C.M. Johnson. In: C. Jones, B. Mulloy and A.H. Thomas (Eds.), Methods in molecular biology: microscopy, optical spectroscopy, and macroscopic techniques, Humana Press, Totowa, NJ, 1994, pp. 109–124.
[13] J.E. Ladbury and R. Peters, Biotechnology, 12 (1994) 1083–1085.
[14] T. Christensen and E.J. Toone, Methods Enzymol., 362 (2003) 486–504.
[15] T.K. Dam and C.F. Brewer, Chem. Rev., 102 (2002) 387–429.
[16] T.K. Dam and C.F. Brewer, Methods Enzymol., 379 (2004) 107–128.
[17] T. Wiseman, S. Williston, J.F. Brandt and L.-N. Lin, Anal. Biochem., 179 (1989) 131–137.
[18] D.K. Mandal, N. Kishore and C.F. Brewer, Biochemistry 33 (1994) 1149–1156.
[19] C.P. Swaminathan, N. Surolia and A. Surolia, J. Am. Chem. Soc., 120 (1998) 5153–5159.
[20] D.K. Mandal, L. Bhattacharyya, S.H. Koenig, R.D. Brown, III, S. Oscarson and C.F. Brewer, Biochemistry 33 (1994) 1157–1162.
[21] J.H. Naismith and R.A. Field, J. Biol. Chem., 271 (1996) 972–976.

[22] D. Gupta, T.K. Dam, S. Oscarson and C.F. Brewer, J. Biol. Chem., 272 (1997) 6388–6392.

[23] Z. Derewenda, J. Yariv, J.R. Helliwell, A.J. Kalb, E.J. Dodson, M.Z. Papiz, T. Wan and J. Campbell, EMBO J., 8 (1989) 2189–2193.

[24] C.F. Brewer and R.D. Brown, III, Biochemistry 18 (1979) 2555–2562.

[25] M. Richardson, F.D.A.P. Campos, R.A. Moreira, I.L. Ainouz, R. Begbie, W.B. Watt and A. Pusztai, Eur. J. Biochem., 144 (1984) 101–111.

[26] D. Gupta, S. Oscarson, T.S. Raju, P. Stanley, E.J. Toone and C.F. Brewer, Eur. J. Biochem., 242 (1996) 320–326.

[27] J.C. Gomes, R.R. Rossi, B.S. Cavada, R.A. Moreira and J.T.A. Oliveira, Agents Actions, 41 (1994) 132–135.

[28] T.K. Dam, S. Oscarson and C.F. Brewer, J. Biol. Chem., 273 (1998) 32812–32817.

[29] D.A. Rozwarski, B.M. Swami, C.F. Brewer and J.C. Sacchettini, J. Biol. Chem., 273 (1998) 32818–32825.

[30] D.N. Moothoo and J.H. Naismith, Glycobiology, 8 (1998) 173–181.

[31] M. Barral-Netto, S.B. Santos, A. Barral, L.I.M. Moreira, C.F. Santos, R.A. Moreira, Oliveira, J.T.A. and B.S. Cavada, Immunol. Invest., 21 (1992) 297–303.

[32] D. Rodrigues, B.S. Cavada, J.T.A. Oliveira, R.D.A. Moreira and M. Russo, Braz. J. Med. Biol. Res., 25 (1992) 823–826.

[33] C.A.M. Bento, B.S. Cavada, J.T.A. Oliveira, R.A. Moreira and C. Barja-Fidalgo, Agents Actions, 38 (1993) 48–54.

[34] T.K. Dam, B.S. Cavada, T.B. Grangeiro, C.F. Santos, F.A.M. de Sousa, S. Oscarson and C.F. Brewer, J. Biol. Chem., 273 (1998) 12082–12088.

[35] T.K. Dam, B.S. Cavada, T.B. Grangeiro, C.F. Santos, V.M. Ceccatto, F.A.M. de Sousa, S. Oscarson and C.F. Brewer, J. Biol. Chem., 275 (2000) 16119–16126.

[36] J. Sanz-Aparicio, J. Hermoso, T.B. Grangeiro, J.J. Calvete and B.S. Cavata, FEBS Lett., 405 (1997) 114–118.

[37] J.L. Andrade, S. Arruda, T. Barbosa, L. Paim, M.V. Ramos, B.S. Cavada and M. Barral-Netto, Cell. Immunol., 194 (1999) 98–102.

[38] H.C. Siebert, R. Adar, R. Arango, M. Burchert, H. Kaltner, G. Kayser, E. Tajkhorshid, C.W. von der Lieth, R. Kaptein, N. Sharon, J.F. Vliegenthart and H.-J. Gabius, Eur. J. Biochem., 249 (1997) 27–38.

[39] D. Gupta, M. Cho, R.D. Cummings and C.F. Brewer, Biochemistry, 35 (1996) 15236–15243.

[40] T.K. Dam, R. Roy, S.K. Das, S. Oscarson and C.F. Brewer, J. Biol. Chem., 275 (2000) 14223–14230.

[41] T.K. Dam, R. Roy, D. Pagé and C.F. Brewer, Biochemistry, 41 (2002) 1351–1358.

[42] J. Rao, J. Lahiri, L. Isaacs, R.M. Weis and G.M. Whitesides, Science, 280 (1998) 708–711.

[43] Y.C. Lee, R.R. Townsend, M.R. Hardy, J. Lonngren, J. Arnarp, M. Haraldsson and H. Lonn, J. Biol. Chem., 258 (1983) 199–202.

[44] T.K. Dam, R. Roy, D. Pagé and C.F. Brewer, Biochemistry, 41 (2002) 1359–1363.

[45] T.K. Dam, H.-J. Gabius, S. Andre, H. Kaltner, M. Lensch and C.F. Brewer, Biochemistry, 44 (2005) 13564–12571.

[46] D. Gupta and C.F. Brewer, Biochemistry, 33 (1994) 5526–5530.

[47] C.F. Brewer, Trends Glycosci. Glycotechnol., 9 (1997) 155–165.

[48] N. Ahmad, H.-J. Gabius, S. André, H. Kaltner, S. Sabesan, R. Roy, B. Liu, F. Macaluso and C.F. Brewer, J. Biol. Chem., 279 (2004) 10841–10847.

[49] C.F. Brewer, M.C. Miceli and L.G. Baum, Curr. Opin. Struct. Biol., 12 (2002) 616–623.

[50] A. Goldstein, L. Aronow and S.M. Kalman, 2nd edition, Wiley, New York, 1974, pp. 101–104.

[51] K.E. Pace, C. Lee, P.L. Stewart and L.G. Baum, J. Immunol., 163 (1999) 3801–3811.

Lectins: Analytical Technologies
C.L. Nilsson (Editor)

Chapter 5

Bioanalytical Studies Based on Lectin–Carbohydrate Interactions Measured by Ellipsometry and Surface Plasmon Resonance Techniques

Jana Masárová[a], Fredrik Winquist[b] and Bengt Danielsson[a]

[a]*Pure and Applied Biochemistry, Lund University, SE-221 00 Lund, Sweden*
[b]*Applied Physics, Linköping University, SE-581 83 Linköping, Sweden*

1. Introduction

By definition, a biosensor is an analytical device, which exploits a biological detection or recognition system for a target molecule or macromolecule, in conjunction with a physiological transducer, which converts the biological recognition event into a usable output signal. The sensor is comprised of three essential components: the detector, which recognizes the physical stimulus, the transducer, which converts the stimulus to a useful, invariably electronic output; and the output system itself, which involves amplification, display, etc. in an appropriate format [1]. The term biosensor is generally applied to those devices that use a combination of biological receptor compounds (antibody, enzyme, lectin, nucleic acid, carbohydrate, etc.) and physical or physico-chemical transducer directing, in most cases, "real time" observation of a specific biological event (e.g. antibody–antigen, lectin–carbohydrate interactions) [2] as depicted in Fig. 1.

Biosensors may be divided into six basic groups, depending on the transduction process: electrochemical (potentiometry, amperometry, voltammetry), electrical (surface conductivity, capacitance), thermal (calorimetry, enzyme thermistor), magnetic (paramagnetism), piezoelectric (thickness-shear mode, surface acoustic waves, acoustic plate mode, Love wave) and optical (fluorescence, luminescence, reflection, scattering, surface plasmon resonance (SPR)) [3–5].

Certain optical biosensors are very attractive because they allow direct label-free and real-time detection of biological events. They are frequently used for medical purposes and they are very promising for assessment of many diseases

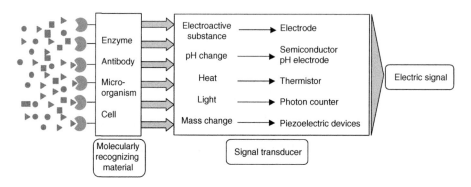

Figure 1. A schematic representation of possible biosensor constructions.

enhancing quality of human life. In the following sections, we deal with two optical techniques (SPR and ellipsometry) and their utilization in lectinology.

1.1. Surface plasmon resonance technique

SPR is a powerful optical biosensing technique for non-label bioaffinity interaction analysis (antigen–antibody, lectin–saccharide and receptor–hormone). It is a phenomenon that occurs during optical illumination of a metal surface and it can be harnessed for real-time biomolecular interaction analysis [6]. Light passing from a denser medium (higher refractive index) into a less dense medium (such as through a prism) is refracted toward the plane of the interface. Above a critical angle of incidence total internal refraction occurs and no light passes into the less dense medium. Although the beam at this state does not lose any net energy across the interface, it leaks an electrical field intensity called evanescent wave into the less dense medium with the same wavelength as that of the incident light. The energy of the wave decreases exponentially with the distance from the interface, decaying over a distance of about one wavelength from the surface (Fig. 2). If the evanescent wave is allowed to interact with a conducting medium, such as a metal, its p-polarized component (in the plane of incidence) can penetrate the metal film (50 nm gold in the Biacore chip) and excite electromagnetic waves that propagate in the metal at the interface with the sample solution called surface plasmons (analogous to the photons). For a non-magnetic metal the surface plasmons will also be p-polarized creating an enhanced evanescent wave field as they propagate on the surface of the metal layer penetrating a short distance (ca. 500 nm for the Biacore) into the less dense medium.

Plasmons are generated only when the energy and momentum of the light vector in the metal plane correspond to those of the surface plasmons. This SPR

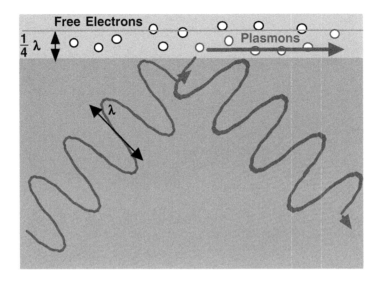

Figure 2. Surface plasmon resonance.

condition occurs at a certain angle of the incident light that depends on different factors, one being the refractive index of the sample solution. At this angle the intensity of the reflected light is at minimum, which is technically easy to detect. In the Biacore instruments this is done with a robust system without moving parts. The polarized light comes from a 760-nm light emitting diode that is focused into a wedge-shaped light beam illuminating the sensor surface under total internal reflection (Fig. 3). The reflected light is monitored with a diode array over a range of angles corresponding to refractive index changes in the range of 1.33–1.40 (pure water is 1.3333). SPR results in a reduction of the intensity of the light reflected from the sensor surface at a specific angle that can be calculated from the diode array response with high precision using interpolation algorithms. The unit for the SPR signal is resonance unit (RU) and 1000 RU corresponds to a shift in resonance angle of 0.1°. A surface protein concentration of 1 ng/mm^2 gives a response of ca. 1000 RU. A more detailed account of SPR-based bioaffinity inter-action analysis can be found in Ref. [7].

Six companies currently manufacture biosensor hardware for SPR technology: Biacore AB (Uppsala, Sweden), Affinity Sensors (Franklin, MA, USA), Windsor Scientific Limited (Berks, UK), BioTul AG (Munich, Germany), Nippon Laser and Electronics Lab (Hokaido, Japan), and Texas Instrument (Dallas, TX, USA). Some of these allow monitoring interaction processes in real-time and in continu-ous mode using highly automated device enabling convenient routine analysis. Biacore AB released the first commercial instrument in 1990 and from that time

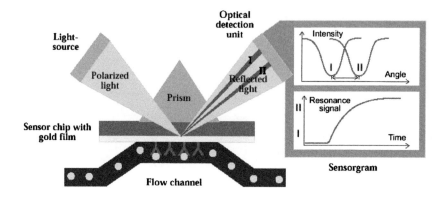

Figure 3. Principle of Biacore measurement. Fixed wavelength light is directed at the sensor surface and binding events are detected as changes in the particular angle where SPR creates extinction of light.

there has been a progressing increase in number of articles dealing with SPR, approximately 90% of which cite the Biacore instrument [8]. Below is a list of instrumental concepts mentioned in literature:

- Biacore AB – Biacore T100, Biacore 3000, Biacore A100, Flexchip, Biacore C, Biacore X
- Biosensing instrument Inc. (947 E. Redfield Rd., Tempe, AZ 85283) – BI-SPR 1000, BI-SPR 1100
- Windsor Scientific Limited – Autolab SPR
- Hofmann Sensorsysteme (Lorchnmuhle 1, D-96346 Wallenfels) – Plasmonic®
- GenOptics (France) – SPRi-Plex™ array system, SPRi-Lab™ array system
- Reichert, Inc. – SR7000 Surface plasmon resonance refractometer
- IBIS Technologies BV (PO Box 1242, NL-7550 BE Hengelo) – IBIS I, IBIS II, IBIS iSPR
- Nomadics, Inc. – Spreeta™ Evaluation Module
- DKK-TOA Corporation (Cambridge, UK) – SPR-20
- GWC Technologies – SPRimager®II
- Analytical μ-Systems (Sinzing, Germany) – Biosuplar

Lectin–glycoconjugate interaction determined by SPR was published in the literature practically from the beginning of the technique commercialization [9]. While earlier works were focused on the determination of affinity constants and kinetic parameters of the lectin–glycoconjugate interactions [10], the current articles deal with the implementation of the SPR technique in biology/biochemistry studies [11–13]. In all cases, the SPR experiments have to match the following conditions:

- Buffers and reagents are chemically pure
- Decide which binding partner will be immobilized on the sensor surface
- Choose the appropriate surface
- Choose the most suitable immobilization method
- Run the lectin binding experiment as well as control experiment
- Analysis of the data

The quality of SPR data is directly proportional to the quality of the reagents. The buffers used should be freshly prepared and degassed prior to the application. In general, the biosensor data should be viewed as a biophysical instrument and, as with any high-resolution technique, the better the reagents the better the results [14].

The immobilization of the biomolecule is the next very important step in biosensor analysis. The choice of the immobilized binding partner is dependent on the particular experiment, but it should meet the following criteria:

- Specificity
- Regeneration stability
- Purity
- Size

For immobilization of the selected binding partner, there are many commercially available surfaces that allow various chemistries to be applied (Table 1). The immobilization method should not affect the activity of the immobilized molecule and the prepared active surface should be stable enough. The covalent immobilization can be done by using different functional groups in the macromolecules. The amine groups of the lectin molecule are often used for immobilization [15–17]. The carboxymethylated surface activated with a mixture of *N*-hydroxysuccinimide (NHS) and *N*-ethyl-*N'*-(dimethylaminopropyl)carbodiimide (EDC) is utilized for the amine coupling. If the lectin (or glycoconjugate) molecule does not contain a suitable number of the accessible amine groups, other functionalities can be used for its attachment to the surface, e.g. thiol–disulphide exchange, biotin–avidin interaction [13, 18] or aldehyde coupling.

An alternative to the covalent binding of the lectin to the sensor surface is capturing. The advantage of capturing is the reversibility of such immobilization and the preservation of the lectin activity, because of the fresh lectin surface prepared for each analysis. The disadvantage is the consumption of the lectin sample. The lectin can be captured to the surface by the specific monosaccharide. In this case, the reduced number of the lectin binding sites should be considered in the further analysis. We have developed a capturing method for concanavalin A (Con A)

Table 1. An example of the variety of the commercially available surfaces for immobilization of biomolecules.

2D surfaces
  2D surface with primary and secondary amine functionalities
  Bare gold. Not derivatized
  Biotin, covalently immobilized on a 2D saccharide monolayer
  2D carboxymethyldextran surface
  Dendritic polyglycerol, carboxymethylated
  Tetraethylene glycol monolayer, carboxymethylated
  Hydrazomodified 2D carboxyl surface
  Hydrophobic planar alkyl layer
  Ni^{2+} ions complexed on a 2D chelating surface
  Dense 2D saccharide layer. Not modified
  Protein A, covalently immobilized on a 2D carboxymethyldextran surface
  Streptavidin, covalently immobilized on a 2D carboxymethyldextran surface
  Tetraethylene glycol monolayer
  Disulfide modified 2D carboxymethyldextran surface
3D surfaces
  Linear polycarboxylate hydrogel, NHS activated
  Dextran hydrogel
  Gelatin hydrogel
  Hydrazide modified linear polycarboxylate hydrogel
  Heparin. Thin hydrogel
  Ni^{2+} ions complexed in a linear polycarboxylate hydrogel
  Ni^{2+} ions complexed in a linear carboxymethyldextran hydrogel
  Protein A, immobilized in a linear polycarboxylate hydrogel
  Protein A, immobilized in a carboxymethyldextran hydrogel
  Pectin hydrogel
  Polyethyleneglycol
  Poly-L-lysine
  Streptavidin, immobilized in a linear polycarboxylate hydrogel
  Streptavidin, immobilized in a carboxymethyldextran hydrogel
  Disulfide modified linear polycarboxylate hydrogel
  Disulfide modified carboxymethyldextran hydrogel

The data are adopted from http://www.xantec.com.

immobilization. It utilizes the fact that every protein molecule contains a hydrophobic region which has a capacity to bind the other molecule via Van der Waals' forces. The highly hydrophobic antimicrobial peptide polymyxin B was covalently immobilized on the activated carboxymethylated surface via amine coupling [19]. This oriented the hydrophobic region of the peptide from the surface and allowed strong hydrophobic binding of the Con A. The required amount of the Con A for one immobilization was lower than in the covalent attachment of Con A (Fig. 4). Four times less amount of Con A was adsorbed on the PmB surface (corresponding to 465 RU) compared to covalent immobilization of Con A (1895 RU). Fig. 5 shows the image of non-modified C1 chip (left part), the image of C1 chip with covalently bound PmB ("PmB surface," right part) and the image of Con A adsorbed on PmB surface (middle part). As can be seen, the lower amount of adsorbed Con A is sufficient for the creation of coherent lectin layer on the surface.

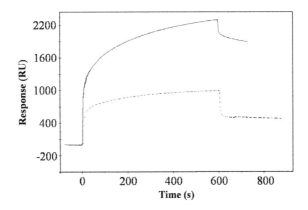

Figure 4. Surface plasmon resonance sensorgrams of covalent immobilization of concanavalin A (—) and concanavalin A adsorption on polymyxin B covered surface (- - -). Both immobilization methods were carried out on commercial Biacore C1 chips.

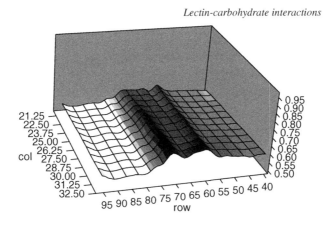

Figure 5. Surface plasmon resonance imaging of C1 chip. The *z*-axis indicates the amount of sample bound on the surface. The left part (rows 95–80) corresponds to non-modified surface; the right part (rows 55–40) shows polymyxin B layer covalently bound on the activated C1 chip; the middle part (rows 75–60) represents concanavalin A layer adsorbed on polymyxin B surface.

On the other hand, the affinity of the yeast mannan was higher than is usual for lectin–carbohydrate interaction. A more detailed investigation of the kinetic parameters of Con A–mannan interaction shows in particular a difference in the kinetic association constant k_a (Table 2). The association was much faster in the case of Con A adsorbed on PmB. This could be due to the hydrophobic nature of the PmB–Con A interaction resulting in a better access of the carbohydrate-binding domain of the lectin to mannan. Crystallographic studies have shown that the carbohydrate methyl-α-D-mannopyranoside is bound to Con A via a network of hydrogen bonds as well as hydrophobic bonds with amino acids Tyr-12, Asn-14, Leu-99, Tyr-100, Asp-208 and Arg-208 [20]. The higher kinetic dissociation constant k_d of adsorbed Con A–mannan interaction can be caused by the engagement of hydrophobic residues (Tyr and/or Leu) of Con A for its adsorption on PmB surface. No interaction between only PmB and mannan was observed. Because of very good storage and operational stability, such capturing can be used either in the biosensor applications or in the purification of the glycoconjugates by affinity chromatography [19].

The further SPR study of the lectins–cyclic peptides interactions confirmed the predominant hydrophobic nature of it with a small contribution of the overall charge of the lectins. In this study, nine lectins, including Con A, with different binding specificities as well as physico-chemical properties (Table 3) were investigated. They were captured on the surface by four non-ribosomal cationic peptides with and without hydrophobic tail and with various charges (Table 4).

All nine lectins showed affinity to the peptides. The strength of lectin–peptide interaction depended mainly on the lectin hydrophobicity. The strongest binding was recorded for PmB which has the highest overall charge as well as a hydrophobic tail. The strength of PmB–lectin interaction was determined within the range from $K_A = 2.7 \times 10^8$ (ECA) to 9.0×10^{11} (AAA). The lectins were adsorbed on all peptides adsorbed by the two-step interaction. The lectins approached the peptides in the first stage and interaction occured after conformational change of lectin structure (second stage). All adsorbed lectins were further able to specifically bind glycoconjugates. This indicates that a different site of the lectins was occupied by peptides. The glycoconjugates had non-typically high association rate constant and kept the same

Table 2. Binding characteristics of Con A immobilized covalently on C1 chip and adsorbed on PmB covered C1 chip determined by its interaction with yeast mannan.

Con A immobilization method	Binding capacity of immobilized Con A[a] (RU)	K_D (M)	k_a (M^{-1} s^{-1})	k_d (s^{-1})
Covalent	205 ± 23	1.1×10^{-7}	4.3×10^3	4.5×10^{-4}
Adsorption	189 ± 15	1.8×10^{-9}	3.0×10^6	5.3×10^{-3}

[a] Calculated from sensorgrams as R_{max} value for the same yeast mannan concentration (1 mg/ml).

Table 3. The carbohydrate specificity and physico-chemical properties of the studied lectins.

Lectin	Abbrev.	Carbohydrate specificity	M_r	p*I*
Canavalia ensiformis	Con A	α-Man, α-Glc	104,000	4.5–5.5
Maackia amurensis	MAA	α-SA	140,000	4.7
Aleuria aurantia	AAL	α-Fuc	72,000	9
Erythrina cristagalli	ECA	α/β-GalNAc, α/β-Gal	54,000	nd
Ulex europaeus	UEA I	α-Fuc	63,000	
Dolichos biflorus	DBA	α-GalNAc	120,000	5.5
Triticum vulgare	WGA	α-GlcNAc	36,000	9
Lycopersicon esculentum	LEA	α-GlcNAc	100,000	nd
Lens culinaris	LCA	α-Man	49,000	8.5

nd = not determined

Table 4. The polymyxins used for the capturing of the lectins.

Peptide	Abbrev.	Sequence and structure	Charge
Polymyxin B	PmB	BTBB(BFdLBBT)	+5
Polymyxin E (colistin)	PmE	BTBB(BLdLBBT)	+5
Polymyxin B nonapeptide	PmBN	TBB(BFdLBBT)	+4
Polymyxin E methane sulfonate (colymycin M)	PmEMS	BTBB(BLdLBBT)	0

pattern as published [19]. These findings can be helpful for developing immobilization/capturing strategies of the lectins and because lectins are common adhesion agents in bacterial infection, in understanding the mechanism of the activity of antimicrobial peptides and finally in the area of antibacterial agent development.

Another advantage of such capturing can be its further utilization in the kinetic analysis of the lectin–carbohydrate interaction because of the low lectin surface capacity achieved by this method. In general, the low surface binding capacity should be used for the kinetic analysis and the high surface binding capacity for the concentration measurements (*BIAtechnology Handbook*, pp. 5–10).

As mentioned above, most SPR articles dealing with lectin–glycoconjugate interaction are focused on qualitative and kinetic analysis. In the following text we describe examples of kinetic analysis of lectin–carbohydrate interactions.

1.1.1. Kinetic analysis of lectin–carbohydrate interactions
Earlier work was oriented to the determination of the kinetics of plant lectin–glycoprotein or plant lectin–oligosaccharide binding. In this work, the glycoconjugate was immobilized on the surface [8, 9, 21–23]. In very few papers,

lectin was preferred as the immobilized molecule, although this is more appropriate in order to analyze the sensorgrams [24, 25]. Generally, the lectins have two or more binding sites. If they are present in the solution flowed over the immobilized glycoconjugate, they have the potential to cross-link two or more glycomolecules on the surface. This would result in an apparent higher affinity and the kinetics cannot be described with a simple interaction model [14]. On the other hand, if the lectin is immobilized on the sensor surface in low amounts, the mass transport effect is negligible and simple monomolecular interaction model can be applied in the kinetic analysis.

1.1.1.1. Theory
The binary binding system is described by the following equation:

$$L + C \; \underset{k_d}{\overset{k_a}{\rightleftharpoons}} \; LC$$

where L is the symbol for the lectin, C the symbol for the carbohydrate molecule, k_a the rate constant of the association, and k_d the rate constant of the dissociation of the lectin–carbohydrate complex (LC).

The association rate is

$$\frac{d[LC]}{dt} = k_a [L][C] \tag{1}$$

The dissociation rate is

$$-\frac{d[LC]}{dt} = k_d [LC] \tag{2}$$

At the equilibrium, association and dissociation rates are equal:

$$k_a [L][C] = k_d [LC] \tag{3}$$

The equilibrium association K_A and dissociation K_D constants can be then expressed as

$$K_A = \frac{k_a}{k_d} = \frac{[LC]}{[L][C]} \tag{4}$$

$$K_D = \frac{k_d}{k_a} = \frac{[L][C]}{[LC]} \tag{5}$$

The SPR technique allows for measuring the binding events in real time. Thus, when the analyte is flowed over the immobilized ligand, the formation of ligand–analyte complex is described by the following equation:

$$\frac{d[LC]}{dt} = k_a[L][C] - k_d[LC] \tag{6}$$

Note that Eq. (6) is valid if the mass transport of the analyte is much faster than the interaction-controlled association. This means that the analyte at the surface is maintained at the same concentration as in the bulk phase. It is achieved when the concentration of the bound ligand is kept low and the flow rate is relatively high during the kinetic measurements.

The concentration of the free ligand (lectin [L]) is expressed as the difference between the total amount of the lectin bound on the surface $[L_0]$ and the amount of the lectin–carbohydrate complex [LC]:

$$[L] = [L_0] - [LC]$$

Then Eq. (6) becomes

$$\frac{d[LC]}{dt} = k_a[C]([L_0] - [LC]) - k_d[LC] \tag{7}$$

If the total amount of lectin $[L_0]$ is expressed in terms of the maximum carbohydrate binding capacity of the surface, all concentration terms can then be expressed as SPR response in RU, eliminating the need to convert from mass to molar concentration:

$$\frac{dR}{dt} = k_a C(R_{max} - R) - k_d R \tag{8}$$

where dR/dt is the rate of change in SPR signal, C the concentration of the carbohydrate in the solution (generally the analyte), R the SPR signal in RU at time t, and R_{max} the maximum analyte binding capacity in RU. R_{max} is easily calculated from the known molecular weights of both carbohydrate and lectin and the measured lectin (ligand) mass immobilized on the surface. The valence of the bound ligand is also considered:

$$R_{max} = \left(\frac{M_{wCarb}}{M_{wLec}}\right) \times \text{lectin response} \times \text{valence}$$

Rearrangement of Eq. (8) gives

$$\frac{dR}{dt} = k_a C R_{max} - (k_a C + k_d) R \tag{9}$$

which is theoretically a straight line with slope $s = k_a C + k_d$.

If R_{max} is known, then both k_a and k_d can be determined from a single sensor-gram. However, determining R_{max} is often difficult. Therefore, measuring of a range of carbohydrate concentrations is preferable. It will result in different slopes s depending on the concentration of the carbohydrate. A plot of slopes against the concentrations $s = f(C)$ is also a straight line with slope k_a. At $C = 0$, k_d can be determined. This can be used practically only when $k_a C \gg k_d$. The rate constant of the dissociation is better calculated after the pulse of carbohydrate (analyte) has passed over the lectin surface when the lectin–carbohydrate complex dissociates in a zero-order reaction and the re-association of released carbohydrate is negligible:

$$\frac{dR}{dt} = -k_d R \tag{10}$$

Separating variables and integration gives

$$R_t = R_0 e^{-k_d (t - t_0)} \tag{11}$$

or

$$\ln \frac{R_0}{R_t} = k_d (t - t_0) \tag{12}$$

where R_0 is the response at starting time t_0 (not necessarily the beginning of the dissociation phase) and R_t the response at time t.
Deriving from previous two equations:

$$\frac{dR}{dt} = -k_d R_0 e^{-k_d (t - t_0)} \tag{13}$$

or

$$\ln \frac{dR}{dt} = \ln(-k_d R_0) - k_d (t - t_0) \tag{14}$$

Plotting $\ln(R_0/R_t)$ or $\ln(dR/dt)$ against $(t-t_0)$ thus gives a straight line with slope k_d or $-k_d$ respectively.

1.1.2. Non-linear analysis

As an alternative to plotting $dR/dt = f(R)$ or $\ln(dR/dt) = f(f)$, non-linear analysis of the sensorgram data can be used to derive rate constants from an individual response curve.

For the association phase, the response R_t at time t can be described by the integration of Eq. (9):

$$R = \frac{k_a C R_{\max}}{k_a C + k_d} (1 - e^{-(k_a C + k_d)t}) \tag{15}$$

Because the SPR response depends upon the refractive index of the samples, the bulk contribution of it expressed as R_{bulk} should be considered. Also the baseline drift (R_{drift}) can be taken into consideration:

$$R_t = \frac{k_a C R_{\max}}{k_a C + k_d} (1 - e^{-(k_a C + k_d)t}) + R_{\text{bulk}} + R_{\text{drift}} t \tag{16}$$

In the dissociation phase, the bulk contribution of refractive index is not crucial. In spite of that, the term for residual analyte is introduced. The residual (non-dissociable) analyte is the analyte which remains bound to the surface bound ligand. Thus, the dissociation phase is described as

$$R_t = R_0 e^{-k_d(t-t_0)} + R_{\text{drift}}(t - t_0) + R_{\text{residue}} \tag{17}$$

1.1.3. Steady state analysis

The equilibrium association K_A and dissociation K_D constants can be calculated from determined rate constants k_a, k_d, or the steady state (equilibrium) analysis can be used. At steady state, the association of the analyte (i.e. carbohydrate) is balanced by dissociation of the lectin–carbohydrate complex from the surface:

$$\frac{dR}{dt} = k_a C(R_{\max} - R_{\text{eq}}) - k_d R_{\text{eq}} = 0 \tag{18}$$

giving

$$k_a C(R_{\max} - R_{\text{eq}}) = k_d R_{\text{eq}} \tag{19}$$

Rearranging:

$$\frac{k_a}{k_d} = K_A = \frac{R_{eq}}{C(R_{max} - R_{eq})} \tag{20}$$

or

$$\frac{R_{eq}}{C} = K_A R_{max} - K_A R_{eq} \tag{21}$$

which is an analogue to a standard Scatchard plot.

The previous pages presented the simple mathematical evaluation of the kinetics of the lectin–carbohydrate interaction (*BIAtechnology Handbook* from the Biacore company (now owned by GE Healthcare), www.biacore.com). There are also tools which allow easier evaluation of SPR measurements. Biacore™ includes all kinetic models to its BIAEvaluation software. In the case of using the software, be aware that all necessary experimental conditions were reached. Recently, many articles dealing about kinetic evaluation of the biomolecular interactions are published in various journals (for reviews see recent surveys from Biacore). Most of them utilize the linear evaluation of the SPR data. Kinetic measurements of lectin–carbohydrate interaction have appeared in the literature since the first SPR instrument was commercialized. In those papers, the glycopeptides were immobilized on the surface and various concentrations of lectins were injected [9, 21]. Haseley et al. [24] immobilized two lectins specific for sialic acid, *Sambucus nigra* and *Maackia amurensis* on the surface of the chip and tested their interaction with 10 different oligosaccharides. The kinetic analysis was performed using the BIAEvaluation software. The results show that kinetic parameters are sensitive to subtle changes in the recognized epitopes and are affected by steric hindrances.

Non-linear kinetic analysis is applied more seldom although it can be more accurate. A good example of such analysis is the interaction between an O-poly-saccharide from *Salmonella* serogroup B and an anticarbohydrate antibody [26]. The kinetic analysis could not be performed with the Fab because of trace amounts of oligomers as well as with single chain antibody variable domains (scFv). Their interaction with O-polysaccharide resulted in biphasic binding profiles when the simple interaction model could not be applied. The comparison of monomeric and dimeric antibodies showed fivefold increase in association phase of the latter. The created complex was much more stable in the case of dimeric antibody. In that case, the dissociation rate constant (described as k_{off}) was 20-fold lower. The authors described that while the association phase fitted well with the simple interaction model, the dissociation phases were biphasic for the antibody dimers. They concluded that probably carbohydrate binding proteins in general raise the possibility of engineering higher affinity by manipulating the dissociation rate constant.

Another application of non-linear analysis was the investigation of the interaction of galectin 3 with laminin [22]. Galectin 3 is a β-galactoside binding lectin. It contains a single C-terminal carbohydrate recognition domain (CRD), homologous in three-dimensional structure to other galectins, and a unique N-terminal domain. The kinetic data show comparable affinities of the intact galectin 3 at sub-micromolar concentration and its isolated CRD at all measured concentrations. At concentrations of galectin 3 higher than 1 μM, the off rate decreased 10 times. The kinetics of galectin 3 binding appeared to be closer to those of plant lectins than the binding characteristics of selectins and other C-type mammalian lectins. In the case of selectin, the on- and off-rates are extremely rapid and are thought to be necessary to mediate transient adhesions of leukocytes to cell layers during tethering and rolling. The galectins expressed at basal surfaces of polarized epithelia probably have different binding kinetics. They contribute to more stable adhesive interactions between cells and a laminin-rich basement membrane. The sensorgrams of the galectin 3–laminin interaction were biphasic. The kinetic parameters obtained for binding of intact galectin 3 to a laminin substratum fitted best with a model in which galectin 3 binds initially through its CRD interaction with carbohydrate. The second step involved the employment of additional lectin molecules from free solution. The major role of the N-terminal domain was found predominantly in the second step of the interaction. The kinetic and equilibrium data confirmed that homophilic interactions between CRD as well as N-terminal domains are implicated in galectin 3 interaction with laminin leading to the positive binding cooperation.

1.1.4. Concentration analysis

Measurement of concentration is another quantitative application of the SPR technique. The SPR response is very sensitive especially to large molecules and can detect very low concentrations of samples. Generally, two approaches can be used for concentration determinations:

- Measurement of the bound analyte after a fixed sample injection time
- Measurement of the analyte binding rate at the beginning of sample injection

Lectin-based assays may require using an additional reagent [27]. Thus, the anti- α_1-acid glycoprotein (anti-AGP) antibody was utilized for capturing the AGP protein from plasma samples [28]. The fucose-binding lectin *Aleuria aurantia* (AAL) was then injected over the surface. The AAL response was linearly dependent on the average number of fucose residues per AGP molecule (Fig. 6). While the total AGP concentration in all samples was constant and giving the same SPR response for captured AGP, the detected AAL response describes the degree of AGP fucosylation.

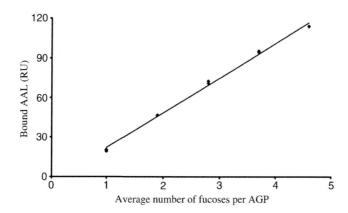

Figure 6. SPR response for bound AAL is plotted against the average number of fucoses per AGP as obtained by monosaccharide analysis of two samples. The shown linear regressions line had an $r^2 = 0.996$. Amount of captured AGP was the same for all mixtures. Copied from Ref. [28] with permission.

The authors used data simulation to determine the amount of bound AAL as a function of captured AGP. The following conditions were considered in the simulations:

- AAL binds the fucose in both monovalent and bivalent bindings
- AAL is not able to bind fucoses bivalently on the same AGP molecule
- Poisson distribution was employed to calculate the fraction of captured AGP that could be used by AAL for bivalent binding
- Only AGP molecules in spheres containing two or more AGP are available for bivalent binding

A good correlation between simulated data and experimental data was achieved (Fig. 7). After calibration, the fucose index was calculated from the SPR response for bound AAL. The advantage of the biosensor method compared with lectin-ELISA was the ability to determine both the amount of bound AGP and amount of bound lectin, and consequently calculate the fucosylation ratio independent of AGP concentration.

For low molecular weight analytes, higher sensitivities can be obtained if the binding of a ligand with higher molecular weight is measured, such as an antibody or lectin. As an example a competitive immunoassay based on SPR for the detection of the pesticide 2,4-dichloro-phenoxyacetic acid (2,4-D) can be mentioned [29]. A novel aspect was that the assay was based on the regeneration of the chip surface by the reversible interaction between monosaccharide (D-glucose) and lectin (Con A). A Con A–2,4-D conjugate was chemically synthesized, purified and used for binding to the SPR chip modified with covalently bound α-D-glucose.

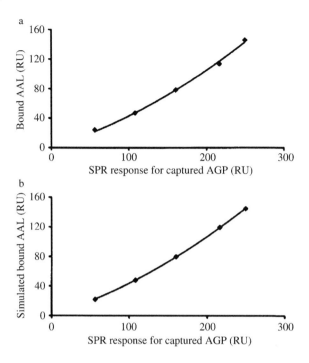

Figure 7. Experimentally obtained (a) and simulated (b) data for bound AAL plotted against SPR response for captured AGP using different dilutions of plasma.

The interaction between anti-2,4-D antibody and the surface-bound Con A–2,4-D conjugate was monitored by SPR and the response was used for the quantification of 2,4-D. The dynamic range of the calibration curve was between 3 and 100 ng/ml with a repeatability of 5–10%. This sensitivity is, however, about 100 times too low for practical measurements of e.g. pesticides in drinking water. Higher sensitivity was obtained with use of a dendrimer layer applied to a Biacore pioneer chip (uncoated gold). By coupling 2,4-D to the PAMAM dendrimer (fourth generation of polyamidoamine) sufficient binding density could be obtained in the subsequent competitive immunoassay for determinations down to 0.05 ng/ml (I. Surugiu-Wärnmark, E. S. Dey and B. Danielsson, unpublished results).

1.1.5. Qualitative analysis

Qualitative analysis based on bioaffinity interaction data can be very useful and is becoming increasingly attractive for evaluation of affinity arrays. Two examples are discussed in this section and some other are presented in Section 2.

Bacterial infection and inflammation result in massive changes in serum glycoproteins. These changes were investigated by the interaction of the saccharide moiety of the glycoprotein with lectins [18]. A panel of eight lectins (*Canavalia ensiformis, Bandeiraea simplicifolia* BS-I, *Arachis hypogaea, Phytolacca americana, Phaseolus vulgaris, Artocarpus integrifolia, Triticum vulgare* and *Pisum sativum*) was used to differentiate human serum glycoproteins obtained from patients with various bacterial infections. Lectin-functionalized sensing layers were created on gold-coated wafers using self-assembly and biotin–streptavidin chemistries and the lectin–glycoprotein interactions were monitored by SPR. The interaction of the lectin panel with serum glycoproteins produced unique patterns. Principal component analysis (PCA) was used to analyze the patterns. The actual panel of eight lectins enabled discrimination between sera obtained from patients sick with bacterial infection and healthy individuals. Extended lectin panels have the potential to distinguish between types of bacterial infection and identify specific disease state.

Another line of work in our laboratory involves optical lectin-based biosensors as tools for bacteria identification. There is a growing need for rapid, accurate and sensitive detection methods for foodborne pathogens and toxins. Conventional methods for microorganism identification are sensitive, but they usually require a long time for detection. Typically, a small sample of the analyzed food is homogenized, incubated and pathogens are identified after at least one to three days by stains, biochemical tests and/or serological reaction. Many of the new highly sensitive methods, such as mass spectrometry and PCR-based analysis require careful sample preparation and are rather time-consuming. Beside these methods biosensors employing biospecific interaction based on highly specific antibodies are coming into use but have the disadvantage that it takes at least three months to produce the antibodies after identification of a suitable antigen. An interesting alternative is to use a lectin array that can recognize the variations in the carbohydrate composition of the lipopolysaccharides (LPS) present on the cell surface. They are specific for each strain and there are a large number of molecules present on the cell surface, which corresponds to the amplification factor offered by PCR. In studies of lectin panels with the Biacore 3000 it was found that not more than eight different lectins are needed for identification of Gram-negative bacteria by using artificial neural network analysis of the binding data [30]. A major advantage is that a new strain can be discovered directly and does not require preparation of new analytical reagents such as new antibodies. This could be an important aspect with respect to needs for quickly adaptable methods for analysis in biological warfare and bio-terrorism.

2. Ellipsometry

Ellipsometry is an optical technique used to determine optical and structural parameters of the sample under investigation. It can be characterized as a measurement of the change of the state of polarization of light upon reflection from the sample surface. The term ellipsometry was coined by Rothen [31] in 1944, who developed the instrument as an evolution of the interferometric method using deposited barium stearate films. The principle of the ellipsometric technique is based on the simple fact that a polarized light beam reflected in a surface will change its state of polarization. Due to the characteristics of the surface, the characteristics of the polarization state of the reflected beam will be subsequently changed. As the optical properties of the bare surface are known the subsequent change in polarization from a deposited film can be used to calculate the thickness and refractive index of the film on the surface. The experimental technique of the ellipsometry was originally introduced by Drude [32] in 1889. The technique was first used to demonstrate the presence of surface layers and their characterization. When the ellipsometer was introduced it became possible to measure thicknesses development of thin films in a more objective way. During the years, with continuous development and increasing power of calculation devices, ellipsometry became less complicated and thus available for more applications. A very good historical review was made by Vedam [33].

The principle of ellipsometry is shown in Fig. 8.

The quantities measured by an ellipsometer are the so-called ellipsometric angles Ψ and V. The change in polarization upon reflection of light in a surface can be described by the ratio of the complex reflection coefficient:

$$\Phi = \frac{R_p}{R_s} = \tan \psi^{i\Delta}$$

Ψ describes the ratio of the amplitude change of p-polarized and s-polarized light. Δ describes the difference in phase shift. R_p and R_s are the reflection coefficients for light polarized perpendicular (R_p) and parallel (R_s, senkrecht) to the plane of the reflecting surface. They are dependent on the wavelength of the light, the angle of incidence of the light beam, the refractive index of the ambient, the complex refractive index of the surface for a bare substrate, and the thickness and refractive index of a deposited film.

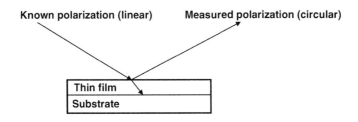

Figure 8. Principles of light reflection.

In a simple two-phase (ambient/sample) system, when the angle of incident Φ_0 and the refractive index of the ambient N_0 are known, the material refractive index N_1 can be obtained directly by the following equation:

$$N_1 = N_0 \sin\Phi_0 \sqrt{1 + \left(\frac{1-r}{1+r}\right)^2 \tan^2\Phi_0} \tag{22}$$

Since the change of reflection is different in the p- and s-directions, the reflected light usually has an elliptic polarization state. This is compensated for by introducing a quarter retardation plate in the system. To estimate the phase shift, the analyzer is rotated to give a minimum in light, as measured with the photodiode, giving the phase shift Δ. This step normally takes some time, and for some applications, e.g. when scanning a surface, a need for a faster registration of the process arise. One way to do this is the use of off-null ellipsometry. Thus, the analyzer and the polarizer are adjusted close to minimum of the light to the detector, and this reflected light is continuously measured. In a scanning mode, as the light spot from the light source moves along the investigated surface, the amount of reflected light will be proportional to the surface properties (Fig. 9).

2.1. Scanning ellipsometry

A scanning ellipsometer has been developed for making images of lectin–glycoprotein interactions at gold-coated wafers. The scanning ellipsometry is a two-dimensional imaging method used in the off-null mode previously described. The amount of adsorbed material on the surface is measured as the increase in intensity of the reflected light, and is proportional to the square of the thickness.

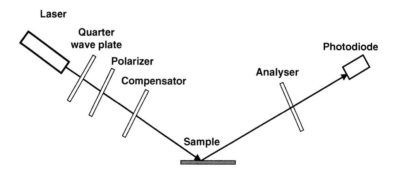

Figure 9. Schematic set up of the ellipsometer instrument.

Images of protein distributions are obtained by placing the sample on an X–Y-table. By scanning the surface in the X and Y directions, and subsequently measuring the intensity of reflected light with a photodiode, it will be possible to visualize various areas with adsorbed or immobilized material. The raw data can be treated with signal and imaging analysis. Since the scanning instrument enables measurements of a large surface, the total protein distribution and not only smaller patches of the surface can be investigated.

The scanning consisted of a laser diode equipped with an aperture and a quarter retardation plate and reflected light was passing through the polarizer and finally measured by a photodiode. The polarizer was fixed in a position giving a minimum of light passing through. The samples were placed on an X–Y table and a computer was used to control the position of the sample and to store the data obtained from the photodiode by using a common A/D converter. The maximum resolution was defined by the size of laser spot, which was normally in the order of $50 \times 50\,\mu m^2$. The equipment allowed for scan areas up to $20 \times 20\,mm^2$ with a resolution up to 200×200 pixels. The raw values obtained from the experiments were evaluated with an image analysis program, making it possible to obtain images, surface plots and line scans.

The lectin sensing surfaces were prepared using the well-known streptavidin–biotin scheme. Thus, gold surfaces were subjected to alkanethiols, forming self-assembled monolayer (SAM) and activated by use of a chemical cross-linker such as EDAC (1-ethyl-3-(3-dimethylaminopropyl)-carbodiimide). After incubation, the surface was rinsed with distilled water, blown dry and immediately incubated with a solution containing amino-biotin. The surface was again rinsed with distilled water and streptavidin was then applied to the biotinylated SAM surface followed by biotinylated lectin. The scheme is shown in Fig. 10.

An example of the technique was illustrated by a panel consisting of eight different lectins and four different sera were used: human, pig, sheep and guinea pig. In the resulting scanning ellipsometry image, Fig. 11, the different lectin–sera interactions can be seen. In the figure, the signal intensities obtained for different combinations of lectins and sera are shown.

A dendrogram was used to show the dissimilarity between the different sera and is shown in Fig. 12. The figure indicates that human and pig sera are the most related of the four different sera investigated, which is expected, since it is well known from literature that human and pig are most alike. These results open interesting possibilities for future experiments with lectin arrays. Thus, lectin panels and imaging optical methods could be a useful tool for judging the relationship between different species. Also, it would be possible to decide the degree of importance of specific lectins for glycoprotein–lectin binding in various sera.

Lectin panels have also been used for the investigation of the binding of unknown mixtures of proteins from fresh meat of cattle, chicken, pig, cod, turkey and lamb [34]. The lectin panels were made by using a contact-printing technology.

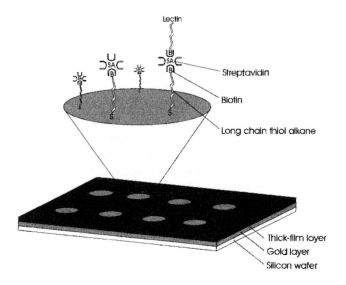

Figure 10. Schematics of the multiplayer scheme for lectin immobilization.

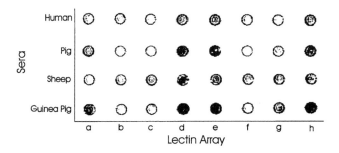

Figure 11. Scanning ellipsometry image of the eight-lectin panel with added sera from human, pig, sheep and guinea pig.

Thus, a stamp template was made of poly(dimethyl)siloxane master patterned with 4×3 circular holes, 2 mm in diameter. The stamp was inked with a COOH-terminated thiol solution, dried under a stream of nitrogen and within seconds gently pressed against a clean gold surface for 30 s. The printed surface was thereafter immediately immersed in a solution of a CH_3-terminated thiol. After an incubation time, the hydrophobic surface with its hydrophilic pattern was rinsed in ethanol, and blown dry under a stream of nitrogen.

The contact printing technique used produces SAM – patterned surfaces with well-defined regions of different chemical functionality: a hydrophobic surface with hydrophilic patches.

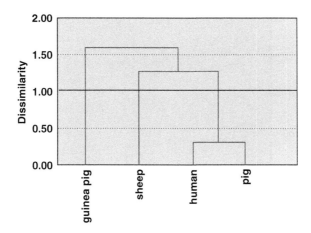

Figure 12. Dendrogram.

The well-known streptavidin-biotinylated lectin technique was used to produce lectin panels on the contact printed gold surfaces.

Lactoferrin and meat juice from six different species – cattle, chicken, pig, cod, turkey and lamb – were added to the lectin panel. Lactoferrin was used as an internal standard and was added to four patches on each lectin surface. A quadruple of each meat juice was added to one surface of each type of lectin.

Scanning imaging ellipsometry was used to produce two-dimensional matrices of the surfaces, which were treated with image analysis for visualization of binding patterns. A typical image is shown in Fig. 13.

A null ellipsometer was used to measure the thickness of bound meat juice proteins, which was related to the thickness of an internal standard of lactoferrin. The ellipsometer was aligned at an angle of incidence of 70° with respect to the surface normal. The instrument was equipped with an He–Ne laser (632.8 nm) light source with a beam area of approximately 1 mm^2. Assumptions of the refractive index (n_f) of the adsorbed proteins had to be made. Often n_f is set to ~1.5, i.e. $k_f = 0$ (extinction coefficient); it is assumed that no absorption of light takes place here. n_f was here chosen to be 1.465. Null ellipsometry is based on an instrument where the polarizing elements (polarizer and analyzer) are rotated until the signal at the detector is zero (null). The optical parameters can then be deduced from the angular positions of the polarizer and analyzer. By entering the ellipsometric angles, Δ and Ψ, into the McCrackin algorithm a value of the thickness can be calculated.

Thickness data from the null-ellipsometric measurements were treated with multivariate data analysis (MVDA) to identify possible separation or grouping of data. PCA is used to get an overview of a dataset and to identify patterns in the data. It explains the variance in the experimental data and reduces the

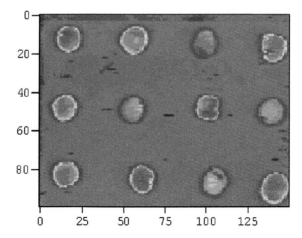

Figure 13. Scanning ellipsometry image: Con A (*Canavalia ensiformis*) with lactoferrin and meat juice from lamb and turkey.

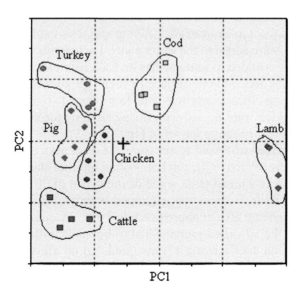

Figure 14. Score-plot showing the grouping of the different meat juices.

immense data to plots that can be easily surveyed. The PCA produces a score-plot that visualizes differences between the observations (experiments). This plot can be used to classify or group the observations. A loading plot is also produced which shows the importance of the different variables and the correlation between the variables and the observations. In Fig. 14, a PCA plot based on the null-ellipsometric data is shown. In the plot the six different meat juice groups can be distinguished from each other. From these experiments lamb seems to diverge most from the other meat juices.

References

[1] A.F. Collings and F. Caruso, Rep. Prog. Phys., 60 (1997) 1397.
[2] P. Leonard, S. Hearty, J. Brennan, L. Dunne, J. Quinn, T. Chakraborty and R. O'Kennedy, Enzyme Microb. Tech., 32 (2003) 3.
[3] E. Kress-Rogers (Ed.), Handbook of biosensors and electronic noses: Medicine, food, and the environment, CRC Press, Boca Raton, 1997.
[4] R. Narayanaswamy and O.S. Wolfbeis (Eds.), Optical sensors: Industrial, environmental and diagnostic applications. Springer series on chemical sensors and biosensors, Springer, Heidelberg, 2004.
[5] R.B. Thompson (Ed.), Fluorescence: Sensors and biosensors, CRC Press, Boca Raton, 2006.
[6] B. Liedberg, C. Nylander and I. Lundström, Biosens. Bioelectron., 10 (1995) i–ix.
[7] K. Nagata and H. Handa (Eds.), Real-time analysis of biomolecular interactions. Applications of BIACORE, Springer, Tokyo, 2000.
[8] R.L. Rich and D.G. Myszka, Curr. Opin. Biotechnol., 11 (2000) 54.
[9] I. Okazaki, Y. Hasegawa, Y. Shinohara and T. Kamasaki, J. Mol. Recognit., 8 (1995) 95.
[10] E. Dzuverger, N. Frison, A.-C. Roche and M. Monsigny, Biochimie, 85 (2003) 167.
[11] J.K. van der Wetering, A. van Remoortere, A.B. Vandrager, J.J. Batenburg, L.M.G. van Golde, A.H. Hokke and J.J. van Hellemond, Am. J. Respir. Cell. Mol. Biol., 31 (2004) 565.
[12] D. Kisiela, A. Laskowska, A. Sapeta, M. Kuczkowski, A. Wieliczko and M. Ugorski, Microbiology, 152 (2006) 1337.
[13] H.R. Juul-Madsen, T. Krogh-Meibom, M. Henryon, M. Palaniyar, P.M.H. Hegaard, S. Purup, A.C. Willis, I. Tornoe, K.L. Ingvartsen, S. Hansen and U. Holmskov, Immunogenetics, 58 (2006) 129.
[14] D.G. Myszka, J. Mol. Recognit., 12 (1999) 279.
[15] Y. Shinohara, Y. Hasegawa, H. Kaku and N. Shibuya, Glycobiology, 7 (1997) 1201.
[16] D. Mislovicová, J. Masárová, J. Svitel, R. Mendichi, L. Soltés, P. Gemeiner and B. Danielsson, Bioconjug. Chem., 13 (2002) 136.
[17] J. Carlsson, M. Mecklenburg, I. Lundström, B. Danielsson and F. Winquist, Anal. Chim. Acta, 530 (2005) 167.
[18] M. Mecklenburg, J. Svitel, F. Winquist, J. Gang, K. Ornstein, E. Dey, B. Xie, E. Hedborg, R. Norrby, H. Arwin, I. Lundström and B. Danielsson, Anal. Chim. Acta, 459 (2002) 25.
[19] J. Masárová, E. Szwajcer Dey, J. Carlsson and B. Danielsson, J. Biochem. Biophys. Methods, 60 (2004) 163.
[20] J.K. Scott, D. Loganathan, R.B. Easley, X. Gong and I.J. Goldstein, Proc. Natl. Acad. Sci. U.S.A., 89 (1992) 5398.

[21] Y. Shinohara, F. Kim, M. Shimizu, M. Tosu and Y. Hasegawa, Eur. J. Biochem., 223 (1994) 189.

[22] E.A.M. Barboni, S. Bawumia and R.C. Hughes, Glycoconj. J., 16 (1999) 365.

[23] J. Nahálková, J. Svitel, P. Gemeiner, B. Danielsson, B. Pribulová and L. Petrus, J. Biochem. Biophys. Methods, 52 (2002) 11.

[24] S.R. Haseley, P. Talaga, J.P. Kamerling and J.F.G. Vliegenhart, Anal. Biochem., 274 (1999) 203.

[25] J.L.J. Blanco, S.R. Haseley, J.P. Kamerling and J.F.G. Vliegenhart, Biochimie, 83 (2001) 653.

[26] C.R. MacKenzie, T. Hirama, S.-J. Deng, D.R. Bundle, S.A. Narang and N.M. Young, J. Biol. Chem., 271 (1996) 1527.

[27] L.S. Kelly, M. Kozak, T. Walker, M. Pierce and D. Puett, Anal. Biochem., 338 (2005) 253.

[28] M. Liljeblad, I. Rydén, S. Ohlson, A. Lundblad and P. Påhlsson, Anal. Biochem., 288 (2001) 216.

[29] J. Svitel, A. Dzgoev, K. Ramanathan and B. Danielsson, Biosens. Bioelectron., 15 (2000) 411.

[30] J. Masarova, E.S. Dey and B. Danielsson, Pol. J. Microbiol., 53 (2004) 23.

[31] A. Rothen, Rev. Sci. Instrum., 16 (1944) 26.

[32] P. Drude, Ann. Phys. Chemie, 39 (1890) 481.

[33] K. Vedam, Thin Solid Films, 313 (1998) 1.

[34] J. Carlsson, F. Winquist, B. Danielsson and I. Lundström, Anal. Chim. Acta, 547 (2005) 229.

Lectins: Analytical Technologies
C.L. Nilsson (Editor)
© 2007 Elsevier B.V. All rights reserved.

Chapter 6

Use of Natural and Synthetic Oligosaccharide, Neoglycolipid and Glycolipid Libraries in Defining Lectins from Pathogens

Krista Weikkolainen[a], Jari Helin[b], Ritva Niemelä[b], Halina Miller-Podraza[c] and Jari Natunen[b]

[a] Department of Biological and Environmental Sciences, Faculty of Biosciences, University of Helsinki, Viikinkaari 5, 00790 Helsinki, Finland
[b] Glykos Finland Ltd., Viikinkaari 6, 00790 Helsinki, Finland
[c] Institute of Biomedicine, Department of Medical Chemistry and Cell Biology, Göteborg University, P.O. Box 440, SE 405 30, Göteborg, Sweden

1. Introduction

1.1. Carbohydrate mediated pathogen binding

Pathogen–host adhesion interactions, which have been characterized on molecular level, mostly appear to involve carbohydrate binding [1–3]. These interactions are believed to affect the expression of receptor glycans on host cells thus leading to new adaptations and having evolutionary consequences [4, 5].

The present review is focused on analytical methods that may be useful in fields related to pathogen–host carbohydrate interactions. One of the basic approaches referred here is thin layer chromatography (TLC) overlay assay developed by Professor Karl-Anders Karlsson and colleagues [6]. In this technique the glycans are presented on artificial TLC surfaces as unimolecular epitopes in form of natural glycolipids or neoglycolipids. Labeled pathogens or their components are incubated with the separated glycolipids to reveal the binding structures. The TLC overlay method has been proven to be useful in defining specificities of pathogen lectins. Therefore, many glycan epitopes recognized by microbes have been identified in this way. The specific natural epitopes are usually present in tissues among a great variety of glycans. After a binding glycan has been identified, the binding can be further defined by a library of similar glycolipids. In pathogen adhesion the actual biological role of glycolipid sequences among other glycans such as protein-linked glycans has been emphasized by studies on specific inhibition of glycolipid biosynthesis [7].

During recent years several methods have been developed using libraries of synthetic and natural oligosaccharides. We present following examples about recent developments and results:

(1) Novel neolacto-oligosaccharide neoglycolipid library for additional definition of the new binding specificity of human gastric pathogen *Helicobacter pylori*.
(2) Screening of tissue glycolipid libraries for analysis of lacto-binding specificity of human gastric pathogen *H. pylori*.
(3) Novel branch specific poly-*N*-acetyllactosamine oligosaccharide library for hemagglutination inhibition studies of influenza virus.

The oligosaccharides in these libraries were produced by novel combinations of enzymatic synthesis methods. Additional chemical methods can be used to increase the chemical variety of the libraries such as coupling of reducing oligosaccharides to divalent constructs by simple divalent aldehyde-reactive reagents such as the aminooxy reagent "DADA" or by modifying the carboxylic acid of glucuronic acid introduced as a novel neolactoepitope.

1.2. Adhesins and toxins of pathogen binding tissue glycans

Carbohydrate-mediated binding has been described for all major groups of pathogenic organisms such as bacteria, viruses, yeasts/fungi and eukaryotic parasites [4]. The binding is mediated usually through specific adhesins present on the surface of the pathogen. In addition, many microbes secrete toxic lectins (toxins), which have major roles in many infective diseases [2, 8]. The adhesion mechanism is considered to be a key step in infection. It has been realized that free glycans or glycoconjugates can inhibit the adhesion and prevent infection, Fig. 1.

1.3. Possible alternative mechanism of the pathogen adhesion

1.3.1. The binding of bacterial carbohydrates to host lectins or carbohydrate–carbohydrate interactions

Beside protein-carbohydrate interactions other types of pathogen-host interactions may be discussed. E.g. polysaccharides of pathogens may bind to tissue surface lectins of the host [9]. Further, there may be carbohydrate-carbohydrate interactions between polyvalent carbohydrates or polysaccharides, but very little evidence is available about this [10].

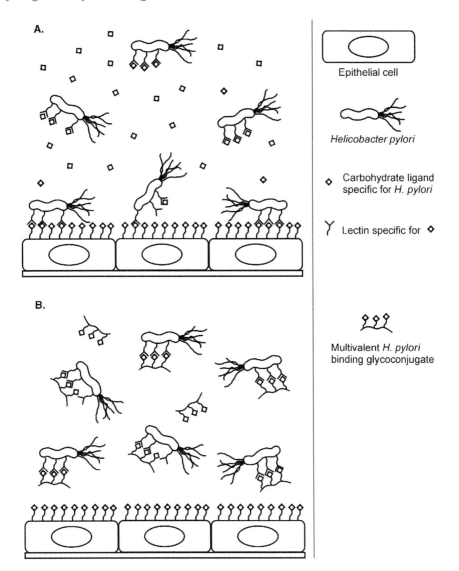

Figure 1. Schematic presentation of bacterial adhesion inhibition by (A) monovalent and (B) polyvalent carbohydrate ligands. Courtesy of K. Weikkolainen.

1.3.2. Carbohydrate binding enzymes such as glycosidases in pathogenic processes

Pathogens can, in addition to binding to the tissue glycans, also modify these by enzymatic reactions. A classic example of this is the action of the "receptor-destroying enzyme" neuraminidase (or sialidase) of influenza virus. The role of destroying the receptors on the surface of newborn viruses has been considered to be in the successful release of viruses after propagation in infected cells.

In other contexts carbohydrate modifying enzymes have also been known as cell surface adhesion molecules as demonstrated in early works of Rauvala and colleagues for a mannosidase [11] and fucosyltransferase [12], and by Shur and colleagues for cell surface galactosyltransferase [13]. As pathogens have cell surface enzymes, their roles may also be studied as adhesion receptors.

2. Analytical Technologies for Analysis of Carbohydrate Pathogen Binding

2.1. General overview of methods

This review is limited to methods that utilize whole pathogens. The overlay assays using labeled pathogens and immobilized glycans has been the key method in screening for adhesion properties of microbes. The advantage of the use of intact microbes is that under optimum conditions it should be possible to reveal all binding specificities/counter receptors of the pathogen.

However, the method has limitations. The specificity of a single receptor type may be difficult to assign, if multiple receptors are present on the pathogen. On the other hand, it has been shown that many bacteria can vary their adhesins according to growth conditions or by specific genetic regulation known as phase variation. The overlay method using whole pathogens is useful in screening of this type of changes in adhesion. After the adhesion mechanisms have been revealed and primarily characterized, the final characterization and ligand optimization (e.g., for drug development of individual receptor), may be more effective using a purified protein, if the protein can be identified from the pathogen.

The identification of individual adhesins can be performed by methods such as photo-affinity cross-linking as demonstrated by retagging methods shown elsewhere in this book (Chapter 12) by Dr. Elisabet Carlsohn and colleagues. When the individual adhesin is known, it can be produced and further analyzed by other methods developed for lectins as reviewed in this book.

2.2. Binding tests and binding inhibition assays

Various assays can be performed in two major forms, either by measuring the direct binding of a pathogen to a carbohydrate or biological material containing the carbohydrate, and/or by measuring the inhibition of the binding by soluble oligosaccharides or multi/polyvalent conjugates of the saccharides obtained thereof.

The direct binding assays are often solid phase assays or methods based on measuring direct binding in solution. Methods for direct binding have been mainly reported in non-pathogen contexts for purified proteins, using for example fluorescence polarization difference [14] or NMR technologies such as transfer-NOE methods [15]. Methods for direct visualization and quantification of binding of polyvalent glycans on surface of *H. pylori* bacteria has been demonstrated by the Borén group [16].

2.3. Screening versus quantification of the binding

Many direct binding assays are designed for screening of binding to various glycan types without quantitation of the binding. The direct binding is obviously an effective screening method. Development of glycan arrays may further improve this. Recently, glycan arrays have been used for screening of binding specificities of pandemic human influenza viruses revealing interesting differences between human and avian influenzas [17].

The quantitation of the direct binding may be performed by measuring binding to various amounts of target carbohydrate in the area/spot where the glycan is immobilized. This is referred here to as "dilution series method" [18].

2.4. Solid phase assay formats

Several assay formats have been reported for solid phase analysis of pathogen binding. The major differences are in the format of solid phase such as TLC plates, microtiter wells, membrane blots [19], and most recently microarrays as solid phase [17, 20].

The TLC overlay has been originally selected for the analysis of glycolipid mixtures as the TLC has been the key method in separation of glycolipids. TLC overlay used together with natural glycolipid mixtures is essentially screening of a library of natural carbohydrates against potential library of pathogen adhesins on the surface of the pathogen. The TLC overlay can be performed even with mixtures comprising unknown components. Several times the assays with glycolipid libraries derived from human or animal tissues have lead to successful determination of new type of binding of a pathogen to even totally novel glycan. These works include important glycan characterization efforts together with characterization of the adhesion. Examples of these include revealing novel poly-*N*-acetyllactosamine receptors for human influenza viruses [21] and revealing a quite unusual glycolipid of cat comprising repeating Galα3-structure as receptor of enterohemorrhagic *Escherichia coli* (EHEC) [22].

The membrane-based blot assays have been demonstrated for glycoproteins and *H. pylori* [19]. The use of glycoproteins may also involve protein-directed binding which should be useful in the work with multiple proteins and pure specific glycoforms of the individual glycoproteins if these are available. The first examples of the use of glycan microarrays in analysis of pathogen binding include the work with influenza virus [17] and with bacteria [20]. The microtiter well assays have been used quantitatively for characterization of bacterial toxin binding [8].

2.5. Assays using host cells and tissues

Regular assays using cells include the use of (i) cells or tissues as solid phase and (ii) precipitation of cells in precipitation assays such as hemagglutination-inhibition.

The relevance of various cells and tissues ex vivo in binding assays can be discussed. Most of the models involve the use of cultivated cell lines. This method screens binding against all glycans available on the cell surface, which is useful for revealing novel receptors in random carbohydrate libraries.

2.5.1. Consideration of glycosylation in the selection of cell/tissue model
There are following key factors to be considered in context of selecting biological material. The glycosylations are known to be specific for:

(1) Species: Living species show large differences concerning glycosylation patterns and synthesis of carbohydrate structures [23–26]. However, limited groups of glycosylations with specific functions can be mainly or partially conserved resulting in carbohydrate structures common for different groups of organisms.

(2) Tissue type within species: Tissue glycosylations are also different within an organism [23–26]. The tissue specificity is under dramatic genetic control as demonstrated by a study of Sda-antigen synthesis in mouse [27].

(3) Status of a tissue: The glycosylations are known to change e.g., during inflammatory processes. Further, there are examples of time-dependent changes in glycosylations regarding parasite pathogenesis. The pathogenesis-associated changes in carbohydrate patterns in infected species may be caused by host's attempts to get rid of glycans that bind pathogen/parasites [28].

(4) Cell line and subcell lines: Gene transfections/protein production in the cells may in general affect glycosylation. For example, it has been shown that transfection of a fucosyltransferase to CHO cells causes expression of several CHO-cell glycosyltransferases and changes glycosylation pattern [29]. Cultivated cells may experience dramatic spontaneous or induced mutations in glycosylations. Even the same cell line cultivated in different laboratories may display different glycosylations [30]. Also, cell lines may spontaneously lose major classes of saccharides such as complex N-type glycans without any visible change in behaviour and viability, while on tissue level the effects of mutations are very dramatic [31]. It is a well-known fact that cell culture conditions affect glycosylation (see, e.g., [32]).

However, this does not indicate that all animal model-based results would be non-relevant with regard to human pathogens or that cell models would be totally useless. The key message is that it could be very useful to characterize the glycosylation of the model used in comparison with natural target tissue, if natural type binding is aimed to be analyzed.

2.5.2. Human tissue models

It seems that a relevant target tissue is the most useful *ex vivo* model. The availability of tissue materials is naturally a limiting factor. One solution would be transgenic model organisms with "humanized" glycosylations. An example of this was achieved by expressing human fucosyltransferase in transgenic mouse in studies of the Lewis b binding by *H. pylori* [33].

The tissue material may be obtained directly from patients. For example, material derived from human obesity operations was used for isolation of gastric glycolipids [34], which were then screened for *H. pylori* receptors. Similar material has also been used in very sophisticated assay measuring binding of *H. pylori* to Lewis b structures present on cells in human gastric tissue culture [35]. Very recently Armstrong and colleagues have published disaccharide (*N*-acetyllactosamine)-conjugated nanoparticles for inhibition of *E. coli* colonization of the epithelium of human intestinal biopsies [36], which correlated with previous cell models and appear to be another promising approach to human ex vivo pathogenesis assays.

2.6. Inhibition assays

It should be noted that most of the binding assays are useful for inhibition assays if suitable oligosaccharides or glycoconjugates are available. A traditional form of inhibition assay is hemagglutination inhibition, where agglutination of erythrocytes is inhibited by soluble inhibitors. A variant of this technology includes the use of erythrocytes from different species or different blood groups, thus containing different patterns of glycans on their surfaces. Relevance of the glycosylation with regard to the pathogenesis is not known unless the adhesin and its target saccharides are characterized. Due to availability of the cells and effective and varying carbohydrate presentation the method is still used for screening purposes.

One variant of the hemagglutination-inhibition assay utilizes erythrocytes desialylated and specifically resialylated with specific sialyltransferase such as α3- or α6-sialyltransferase and sialic acid donor CMP-Neu5Ac [37, 38]. Hemagglutination of resialylated erythrocytes is useful in studies on viruses and bacteria, such as influenza virus [39], *Streptococcus suis* [40] or *H. pylori* [41].

3. Carbohydrate Libraries

Independent of assay techniques, structural characterization of glycans involved in adhesion is often recommended. The latter is important for detailed identification of the epitopes on binding surfaces and for explanation of chemical conditions for interactions. The availability of pure glycans has been a major problem in characterization of adhesion and lectins. This problem has been solved by isolation of oligosaccharides and glycoconjugates from natural sources and chemical and/or enzymatic synthesis. In addition to availability of the glycans, they would usually need to be immobilized or oligo/polymerized by conjugation.

3.1. Libraries of natural glycolipids and their derivatives

Glycolipids have been major sources of purified glycans and glycan mixtures for pathogen-binding assays. The glycolipids comprise practically most of the natural terminal epitopes including sialylated, sulfated, fucosylated, and AB-blood group structures present on *N*-acetyllactosamine backbones (Lacto- and Neolacto-series), and/or other major glycolipid core structures like globo- or ganglio-carbohydrate chains. These and poly-*N*-acetyllactosamine variants, especially polyglycosyl-ceramides [42], and even hybrid structures of the core structure create enormous variation in structures. Natural glycolipid structures have been reviewed in [43]. The major structures lacking from the glycolipid libraries but present in proteins are the core structures of the N- and O-linked glycans. A major benefit of glycolipids is that these can be immobilized on surfaces such as TLC plates or microtiter well, based on hydrophobic interactions.

However, some glycolipid libraries, especially those containing neoglyco-lipids, may be laborious to produce. Such libraries may, on the other hand, reveal totally new binding specificities and glycan structures among large variety of cancer and animal tissues. For example, extremely high affinity receptor epitopes were found in polyglycosylceramides containing unique variation of polylacto-samine structures. It should be noticed that it is nowadays possible to collect and synthesize a reasonable variation of structures based on commercially available sources.

Commercial sources of glycolipids include mainly bovine (or other animal) brain gangliosides and globosides. These can be obtained from companies like Calbiochem, Sigma. More specialized lipid/glycolipid companies include Avanti Polar Lipids and Matreya, but their product selection is limited. Individual special items including unique glycans can be available from companies like Wako, Japan/Germany. Best variety ever was produced by the Swedish pioneering company Biocarb AB, the remains of the stock is still being sold by Accurate Chemical and Scientific Company and the preparative work is now being continued by Frank Lindh, Isosep AB (Tullinge, Sweden).

The variety of glycan epitopes can be increased by enzymatic methods modifying existing glycolipids by, for example, desialylation (although easiest by mild acid), sialylation, fucosylation, or by specific galactosidases, and *N*-acetylhexosaminidases. Interestingly, many biologically active forms are asialo forms of common gangliosides. Effective glycosyltransferase reactions are more challenging for glycolipids than for free oligosaccharides. Such factors as donor amount, potential need of divalent cation Mn^{2+} and the type of substrate glyco-lipid as well as its concentration should be optimized for each glycosyltransferase reaction. A soluble polymer carrier may help to use glycosyltransferases in the synthesis of glycolipids [44]. Recently, glycosynthase derived from ceramide glycanase has been shown to couple oligosaccharide to ceramides [45]. Enzymes are available from companies like Calbiochem, Seikagaku, Prozyme, Fluka and Sigma. Part of the enzyme preparations is sensitive to temperature and may suffer from transport or may be suitable for only a specific glycan type. The availability of recombinant sialyltransferases and fucosyltransferases is unfortunately low but a few of the items in Calbiochem catalog are available.

3.2. Neoglycolipid libraries for analysis of pathogen binding

3.2.1. Neoglycolipids from natural oligosaccharides
Neoglycolipid means a non-natural glycolipid produced by conjugation of an oligosaccharide with a lipid carrier. Neoglycolipids can be used in solid phase binding or inhibition assays in the same way as natural glycolipids. Neoglycolipid libraries include non-glycolipid sequences, e.g., glycans prepared from polysaccharides or proteins. Many research groups have successfully used the neoglycolipid technology for investigation of various types of glycans such as O-glycans and N-glycans released from proteins and glycosaminoglycan epitopes.

3.2.2. Neoglycolipids from synthetic oligosaccharides
We show here an example of a neoglycolipid library produced from a synthetic oligosaccharide library. This technique is useful for characterization of fine speci-ficities of pathogen adhesion. By chemical and enzymatic synthesis, we produced an oligosaccharide library of analogs of the novel neolacto epitope, for studies on the human gastric pathogen *H. pylori*.

3.2.3. Use of neoglycolipids in investigation of lectins from pathogens
Neoglycolipids may be useful in many studies dealing with carbohydrate–protein interactions. Current technologies allow to create, through chemical or enzymatic synthesis, an unlimited number of carbohydrate structures, which can be coupled

to lipid tails and screened for interaction with other molecules. In classical approaches neoglycolipids are immobilized on artificial surfaces, in cluster forms to increase strength of the weak protein–carbohydrate interaction, and the surfaces are then exposed to protein ligands on whole pathogens.

As mentioned in the introduction, one simple approach, which is suitable for investigation of binding specificities of different kinds of glycolipid conjugates is TLC binding assay [6]. In this technique the lipid-linked saccharides are separated on silica gel TLC plates using organic solvents, after which the plates are overlaid with a labeled protein. The carbohydrate–protein complexes are then visualized on the silica surfaces after removal of the non-bound fraction of the protein. This assay provides a possibility to directly investigate mixtures of different glyco epitopes because oligosaccharides in form of glycolipids can easily be separated on the plates. Other artificial surfaces like nitrocellulose or PVDF membranes [46] can be used, as well as microtiter wells [47]. A sensitive oligosaccharide array strategy has been developed using neoglycolipids immobilized on nitrocellulose membranes, which permits simultaneous analysis of larger numbers of carbohydrate epitopes and their interaction with lectins, antibodies and other proteins like cytokines or chemokines [46].

Neoglycolipid probes can also be tested in form of liposomes [48–51]. These artificial vesicles allow experiments in water solutions, which is a prerequisite of many biological studies involving, for example, drug or vaccine development [48, 50]. An important advantage of the neoglycolipid technique is that both the common structures and unusual synthetic carbohydrate sequences can be analyzed as discussed here earlier.

Neoglycolipid technology is of special importance for screening for new carbohydrate–protein interactions. Carbohydrate epitopes are often coupled to proteins and it is not infrequent that one protein molecule carries several different saccharide structures. It may be difficult to distinguish biological activities of sugars in complex glycoprotein species using technologies common to proteins. However, carbohydrate chains may be released from proteins, converted to neoglycolipids [52–54] and analyzed for binding using artificial surfaces [53, 54] as mentioned above.

Tang and colleagues revealed an effective technology for synthesis of neoglycolipids where free oligosaccharides are coupled to a commercially available lipid, phosphatidylethanolamine (PE) [55]. The conjugation was performed using reductive amination, which couples the reducing end of the saccharide and the primary amine group of the PE lipid. The reaction results in the formation of a stable bond of a secondary amine (R_1–CH_2–NH–R_2) between the sugar (R_1) and the lipid (R_2). The method has the clear benefit that no blocking or protection of the oligosaccharide is needed. This neoglycolipid technology has been used by many research groups and various methods for chemical synthesis of glycolipids have been described [56–61]. An effective reductive amination involving carboxylic acid catalysts has been reported in [62].

3.2.4. Choice of carrier lipids

We present here two other alternatives for the PE as the lipid carrier. The new neo-glycolipids described here were designed not to contain negatively charged groups present in PE-based conjugates. It was considered that PE, as a natural phospho-lipid, may provoke a bacterial binding through the charged P group and that charge may create additional background problems. In addition, synthetic lipids are expected to be more resistant to cellular enzymes than natural lipids, which may possibly be cleaved by enzymatic activities of the organism.

The first lipid tail used was hexadecylaniline (HDA), containing an aromatic amine and a 16-carbon aliphatic chain. The aromatic ring provides two advan-tages: (i) the aromatic amine is effective in reductive amination, and (ii) the sub-stance has specific UV absorption, which may be used for quantitation. The other lipid used in our studies was a synthetic lipid constructed from two palmitate fatty acids (various lengths of fatty acids can be used) and one lysine molecule. The fatty acids are amidated to amines of lysine and diaminopropane is amidated to the carboxylic acid group. This was produced by a cost-effective custom synthesis by a specialized company. The aim was to design a lipid anchor with two alkyl chains similar to natural lipids like ceramide. The lipid contains 42 carbon atoms and is more hydrophobic than HDA carrier with 22-carbon atoms. This provides a bene-fit especially with larger oligosaccharides under assay conditions. The hydrophilic nature of oligosaccharides promotes the release of neoglycolipids to solution from TLC during the assay procedure, if the alkyl anchor is not hydrophobic enough.

3.2.4.1. Examples of novel carriers HDA and C42

FAB mass spectra were produced for the neoglycolipids obtained by reductive amination. Fig. 2 shows a HDA-based conjugate (example of lacto-series neogly-colipids) and Fig. 3 illustrates a neoglycolipid with the new-branched aminolipid containing 42 carbon atoms (later called C42) representing the neolacto neogly-colipid library.

3.2.4.2. Development possibilities regarding synthesis of neoglycolipids

A more hydrophobic amino–aromatic lipid such as alkyl-aniline with two alkyl chains would be useful for neoglycolipid technologies. Also ceramide analogs easily conjugatable with reducing carbohydrates would be useful. The organic synthesis of fatty acid containing materials appears to require an experienced and specialized chemist, and this may not be easily available within the custom synthesis laboratories.

A key factor concerning reductive amination is the structure of the reducing end monosaccharide residue. The secondary amine formed from the reducing end C1 of the saccharide is an open chain structure. Obviously this part of the structure can be quite different in biological recognition, when compared with oligosaccharide hav-ing a closed ring structure as the reducing end monosaccharide residue, though similar hydroxyl groups are available in a more flexible format. For this reason

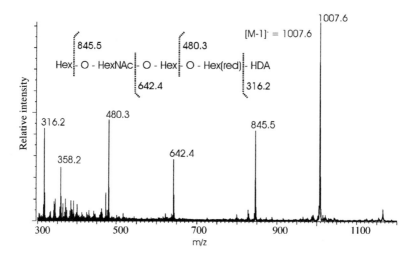

Figure 2. Example of synthesis of hexadecylaniline (HDA)-based neoglycolipids: Negative FAB MS of Galβ3GlcNAcβ3Galβ4Glc-HDA (Lacto-*N*-tetraose-HDA). The glycolipid was formed from lactotetraose and HDA using reductive amination. The spectrum shows the pseudomolecular ion at *m/z* 1007.6 and the series of Y ions [63], as indicated in the picture. The lacto epitope for *H. pylori* was described in [34].

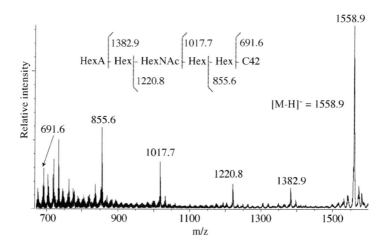

Figure 3. Example of synthesis of C-42-based neoglycolipids: Negative FAB MS of GlcAβ3Galβ4GlcNAcβ3Galβ4GlcC42. The glycolipid was built from GlcAβ3Galβ4GlcNAcβ3Gal β4Glc and a Pal-Lys branched aminolipid containing 42 carbon atoms (C42), using reductive amination. Fragment ions of the Y series of ions [63] and the molecular ion (*m/z* at 1558.9) are indicated in the picture.

oligosaccharide sequences comprising an extra monosaccharide at reducing end may be selected for this model to ensure that the whole epitope is presented. In some cases specific chemistries providing effective coupling of non-protected carbohydrates to lipid carriers and preserving the reducing end ring structures may also be applied. Synthetic methods resulting in neoglycolipids with unmodified saccharide parts have been reported by several research groups [56, 59–61].

3.3. Libraries of other glycoconjugates

3.3.1. Proteins
Natural glycoproteins have widely been used in pathogen-binding assays. The traditional reagents include desialylated variants of glycoproteins such as asialofetuin or transferrin and even agalactosyl forms produced by galactosidases. The specific glycosylation of the protein should be taken into consideration and attention should be paid to the fact that variations between batches are common due to natural sources of the proteins and differences in the production processes. It has also been shown that most proteins contain multiple glycoforms at a certain glycosylation site and some of them contain multiple glycosylation sites. Furthermore, it should be considered that the polypeptide part of proteins may display specific binding effects.

3.3.2. Neoglycoproteins
Neoglycoproteins containing glycans linked to, e.g., human or bovine serum albumin (BSA) have become standard reagents in glycobiology in various assays such as inhibition assays with soluble neoglycoproteins. For example, polyvalent albumins were effectively used for characterization of the Lewis b binding adhesin of *H. pylori* in experiments with polyvalent fluorescent conjugates, as described by Borén and colleagues [64]. Albumins are commonly used as carriers of carbohydrate epitopes because these proteins are not naturally glycosylated, although there may be exceptions and possibilities for contamination if the raw material albumin is not pure enough.

Albumins are generally used as polyvalent conjugates comprising multiple oligosaccharides. The valency of the conjugate is a key factor. This may vary from a few to about 40 and would obviously have major effects on the strength of the interaction. Thus, in quantitative comparative studies, the neoglycoproteins used should have similar valencies.

Another key factor is conjugation chemistry, which vary from reductive amination (note discussion for neoglycolipids) to elaborate spacer schemes. The length and flexibility of the spacer are key factors. It should also be noted that in derivatization of proteins with, for example biotin or a fluorescent tag, the number of available reactive groups on the surface of the protein, such as NH_2 of the lysine side chains, is decisive for the final valency.

3.3.3. Oligosaccharides for synthesis of glycoconjugates or inhibitors

A major source of natural oligosaccharides for research purposes is human milk, comprised of lactose-based oligosaccharides (HMOs) including species containing 3 to over 20–30 monosaccharide units. Major components include sialylated and fucosylated lactoses. The core structure of the larger glycans is lacto-*N*-tetraose or lacto-*N*-neotetraose, present also in lacto- and neolactoseries glycolipids. These can be elongated to larger poly-*N*-acetyllactosamines and/or decorated with fucose and sialic acid residues. The major lower size oligomers (hexa- to octamers) may easily be obtained by chromatographic separations in mg-scale and even low gram-scale for the most abundant species. The providers of HMOs include Accurate Chemical and Scientific Company, Calbiochem, Dextra, Glycorex Finechem, Glycoseparations, Calbiochem, Isosep, and Sigma. At least Isosep and the new company Glycoseparations appear to have own active purification processes.

Another source of oligosaccharides are major glycoproteins. The glycans, which can be isolated include common N-glycan structures, which are aimed mainly for standards for chromatographic methods and are available in nmol/microgram amounts, some with fluorescent labels. Presently, at least Glycoseparations is providing larger amounts of glycoprotein derived glycans and Glycotech USA has had 100 µg packages of biantennary N-glycans. A variety of standard glycans are also available from Ludger, Takara and Sigma.

Miscellaneous oligosaccharides obtained from natural sources or by enzymatic and/or chemical synthesis are available from various companies which include Accurate Chemical and Scientific Company, Calbiochem, Dextra, Glycorex Finechem, Glycoseparations, Isosep, Sigma and Toronto Research Chemicals.

3.3.4. Synthetic oligosaccharide and glycoconjugate libraries

Synthetic oligosaccharide libraries have been a key academic resource in studies involving lectins and their specificities. At the most sophisticated level, disaccharide libraries, with each hydroxyl group separately converted to the deoxy form, have been used for screening of the specificities of bacterial lectins, such as PapG adhesin of uropathogenic *E. coli* [65] and the binding specificity of *Streptococcus suis* [66a].

3.3.5. Polyacrylamide conjugates, biotin glycans and glycan arrays

Polyacrylamide (PAA) conjugates developed by Bovin and colleagues [66b] are among the most useful reagents for various assays in glycobiology. The materials commercialized through Syntesome and later through Lectinity Holdings include a vast variety of natural glycans and their synthetic analogs. The reagents are available as monovalent biotinylated, as polyvalent PAA-conjugates and even as biotinylated polyvalent PAA-conjugates. The PAA conjugates have been used, e.g., for studies of influenza virus [67].

A non-commercial US-based Consortium for Functional Glycomics (http:// glycomics.scripps.edu/CFGad.html) has also created impressive library of oligosaccharides and glycoconjugates by chemo–enzymatic synthesis. These include great variety of oligosaccharides with azidoethyl spacer, O-glycan-Thr-conjugates, biotin conjugates and PAA conjugates (www.functionalglycomics.org/static/ consortium/resources/resourcecored.shtml). The materials are also used as a printed glycan microarray with 285 glycan targets. The array concept was recently demonstrated with pandemic influenza viruses [17]. Another pioneering study demonstrated carbohydrate microarrays for studies of bacterial adhesion [20].

4. Libraries in Analysis of Novel *H. pylori* Binding Specificities

4.1. Carbohydrate binding specificities of H. pylori

H. pylori is a pathogenic bacterium, which colonizes human stomach leading to gastric diseases such as gastritis, peptic ulcer and gastric cancer [3, 68–70]. It is a common pathogen, which has been estimated to colonize the majority of people in the world. Between 10% and 20% of the infected individuals develop clinical symptoms. The bacterium causes medical problems, which have global dimensions. New preventive and therapeutic approaches are under discussion and one of the ideas is an anti-adhesion therapy based on inhibition of binding of the bacterium to target cells [3]. As many other pathogens, *H. pylori* adheres to selected carbohydrate epitopes and it has been shown that the bacterium has the ability *in vitro* to recognize a variety of carbohydrate structures including some "naked" lacto, neolacto and ganglio-based carbohydrates as well as some fucosylated, sialylated and sulfated variants of the molecules [3]. An important epitope is Lewis b structure [16], which has been proposed to function as a natural binding epitope for the bacterium in the stomach [71, 72]. It is believed that *H. pylori* contains a complex system of surface carbohydrate-binding proteins. However, these have not yet been entirely mapped and up to now only two lectin-like proteins belonging to the outer membrane protein (OMP) family of *H. pylori* have been described, the Lewis b binding adhesin [16] and the sialic acid-binding adhesin [73].

4.2. Neoglycolipid libraries for analysis of pathogen binding

Most recently two novel binding specificities of *H. pylori* have been characterized at the Göteborg University, the lacto- and the neolacto specificities. These have been shown to have some interesting characteristics including:

(1) Unusually high frequency of binding among *H. pylori* strains to about 90% of strains tested while the other specificities such as Lewis b and sialic acid binding have frequencies of about 30%.

(2) Presence in human gastric tissues. Lactotetraose was characterized from glycolipids and X2 type variant of neolacto-specificity can be detected by antibodies from gastric tissues.
(3) Possible association with a rare secretor-negative blood group status of the lacto-specificity.

4.3. Novel neolacto-binding specificity of H. pylori

4.3.1. Analysis of neolacto-binding specificities by glycolipids
We have reported a new *H. pylori*-binding epitope based on the neutral neolacto core chain [18]. The binding specificity was revealed by screening of natural glycolipid mixtures prepared from rabbit thymus and the major natural binding glycolipid of this source was identified as NeuGcα3Galβ4GlcNAcβ3 Galβ4GlcNAcβ3Galβ4GlcβCer. Chemical desialylation and successive degradations using specific glycosidases revealed later that the effective binding species also include Galβ4GlcNAcβ3Galβ4GlcβCer, GlcNAcβ3Galβ4GlcNAcβ3 Galβ4GlcβCer and Galβ4GlcNAcβ3Galβ4GlcNAcβ3Galβ4GlcβCer.

4.3.2. Quantification of binding using dilution series
The binding to tetra-, penta- and hexaglycosylceramides of the neolacto nature were analyzed by dilution series. The pentasaccharide glycolipid showed the highest binding activity and the amount of the glycolipid on the TLC line could be decreased to 20 pmol per spot without losing the activity, indicating a strong interaction.

4.3.3. Specificity analysis using pure glycolipids
The screening of natural glycolipid libraries revealed, in addition to the GlcNAcβ3Galβ4GlcNAc-binding, also binding to three other terminal structures based on type II *N*-acetyllactosamine (Neolacto-glycolipids): Galα3Galβ4GlcNAc-, GalNAcα3Galβ4GlcNAc- and GalNAcβ3Galβ4GlcNAc-. The first one is the so-called Galα-xenoantigen, not present in human or in old world monkeys but present in other mammalian species. The screening with natural library showed thus a non-human specificity variant for a human pathogen.

Interestingly, neolacto-type binding specificity is also known for animal *Helicobacteria* [74] which could indicate the role of the flexible binding specificities in spreading of the bacteria from other species to humans or *vice versa*, which is known to occur. It would be interesting to study with regard to the neolacto-binding specificity both acute zoonotic animal Helicobacter species infecting human and human *H. pylori* species isolated from infected animals.

The GalNAcβ3-structure can be considered a hybrid structure containing the terminal GalNAcβ3 of the globo-series glycolipids on the backbone of a neolacto-glycolipid. This structure has been isolated from human erythrocyte glycolipids [75] and has also been observed by antibody from human gastric tissue (discussed in Ref. [18]).

4.3.4. Molecular modeling

Comparison of the binding epitopes by molecular modeling of the binding glycans revealed similar conformations for the key neolacto structures Galα3Galβ4GlcNAc-, GalNAcα3Galβ4GlcNAc-, GalNAcβ3Galβ4GlcNAc- and GlcNAcβ3Galβ4GlcNAc-, and the conclusion was that the epitopes represent the same receptor specificity.

Intriguingly, this specificity has been shown to have similarities to the binding of *Clostridium difficile* toxin A to human glycolipids [76] and binding specificity of Galα-xenoantibodies [76]. The neolacto-binding specificity appears to be a common mechanism shared by multiple carbohydrate binding phenomena.

Based on the modeling and experimental data, we provide here new reducing end variants of the neolacto specificity with different reducing end monosaccharides and new non-reducing end residue GlcA, which can be derivatized from acid group or reduced to Glc.

4.4. New fine specificities of neolacto-binding by neoglycolipid library

The fine specificity of the neolacto binding was studied by synthesizing a series of natural type and modified neolacto oligosaccharides and these were tested against the *H. pylori* to reveal further possible modifications of the binding structure.

4.4.1. Analysis of the binding neolacto neoglycolipids

The Neolacto oligosaccharide library was conjugated to two alternative lipid carriers (HDA and C42) by reductive amination. Mass spectrometry (MS) was used to verify the conjugations, as exemplified in Fig. 3 for the pentasaccharide GlcAβ3Galβ4GlcNAcβ3Galβ4Glc. The synthesis of the novel reducing end and non-reducing end variants of the neolacto epitope are presented in Scheme 1.

The synthesis of neoglycolipids using phosphoethanolamine-related lipids is also very useful [46], however, phospholipids may be recognized by *H. pylori* [77] as epitope structures and are less useful for binding studies with this pathogen.

The neoglycolipids were tested by the TLC-overlay assay for binding to *H. pylori*. After initial screening of the binding, the bindings were quantitated with dilution series of the neoglycolipids. An example of the quantitative TLC binding studies using neoglycolipids is shown in Fig. 4.

The tested glycolipid (GlcNAcβ3Galβ4GlcNAcβ6GlcNAc-HDA) was applied on the plates in the form of dilution series and the binding by *H. pylori* of different amounts of the material was evaluated visually. As shown, the binding by two tested

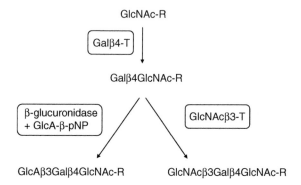

Scheme 1. Synthesis of neolacto-oligosaccharide library from various GlcNAcα/β-di/trisaccharides (R).

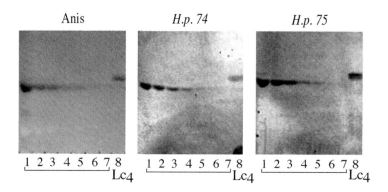

Figure 4. Example of binding studies using neoglycolipids: binding of *Helicobacter pylori* (strains CCUG 17874 and CCUG 17875, respectively) to GlcNAcβ3Galβ4GlcNAcβ6GlcNAc-HDA on silica gel TLC plates. Lanes 1–7, 2-fold dilution series of the neoglycolipid (1 nmol in the first lane); Lane 8 (Lc4), lactotetraosylceramide (positive control). The plates were developed in C/M/0.25% KCl in H₂O, 50:40:10, and visualized either by spraying with anisaldehyde (chemical staining, left plate) or by ³⁵S-labeled *H. pylori* (middle and right plates). The examined neoglycolipid was among the best binders of *H. pylori*.

strains of *H. pylori* was noticed down to lane 5 containing 30 pmols of the material, thus indicating an effective interaction with the epitope. Using similar semi-quantitative tests we could compare binding strengths of different carbohydrate epitopes. We have found that among the best binders of *H. pylori* were two neolacto-based structures, GlcNAcβ3Galβ4GlcNAcβ3Galβ4Glc-R [18] and GlcA(*N*-Me)β3Galβ4GlcNAcβ3Galβ4Glc-R (to be published in details later) respectively, where "GlcA(*N*-Me)" stands for methylamide of GlcA. We used the same binding carbohydrates in the form of glycosphingolipids and different neoglycolipids carrying different lipid parts and could in this way eliminate the possible influence of lipid parts on the binding.

On the other hand, one should remember that both carbohydrate spacers and lipid aglycons may influence the recognition leading to false negative results. We used neoglycolipid technique to investigate the importance of various structural factors of core chains and lipid parts, and some examples of these studies are shown in Table 1. Briefly, two binding trisaccharide epitopes, GlcNAcβ3Galβ4GlcNAc- and GlcAβ3Galβ4GlcNAc- were coupled to different saccharide structures and different lipids, and the molecules were tested for binding by *H. pylori* on TLC plates. The first epitope belongs to the most potent binders of *H. pylori* [18] and the GlcA-containing trisaccharide has been classified as a less active structure (to be published). As shown in the table, the interaction with neoglycolipids was influenced by both carbohydrate and lipid part and in extreme cases the result was only occasional binding (Nos. 4 and 11 in Table 1).

Structures 3 and 4 were expected to give strongly positive results, however, the hindrance in the core chain GlcNAcα) impeded the recognition. On the other hand neoglycolipids containing GlcNAcβ or Galβ6 in the same position turned out to be excellent binding molecules thus showing that a single glycosidic bond in the core chain may be a decisive factor of the interaction. The 6-linkage provides possibility of higher flexibility of the coupled group as compared with the 3-linkage

Table 1. Importance of distal sugars for the interaction of bacterial proteins with carbohydrate epitopes: Investigation of interaction of *Helicobacter pylori* with GlcNAcβ3Galβ4GlcNAc and GlcAβ3Galβ4GlcNAc, respectively, using as epitope carriers neoglycolipids with different core chains and two different lipid parts, HDA and C42 (see 4.4.2).

No.	GlcNAcβ3Galβ4GlcNAc-R	Binding
1	GlcNAcβ3Galβ4GlcNAc**β6GlcNAc**-HDA	+++
2	GlcNAcβ3Galβ4GlcNAc**β6GlcNAc**-C42	++
3	GlcNAcβ3Galβ4GlcNAc**α6GlcNAc**-HDA	(+)
4	GlcNAcβ3Galβ4GlcNAc**α6GlcNAc**-C42	(+)
5	GlcNAcβ3Galβ4GlcNAc**β3Man**-HDA	+
6	GlcNAcβ3Galβ4GlcNAc**β3Man**-C42	+
7	GlcNAcβ3Galβ4GlcNAc**β6Gal**-HDA	+++
8	GlcNAcβ3Galβ4GlcNAc**β6Gal**-C42	++
	GlcAβ3Galβ4GlcNAcβ-R	**Binding**
9	GlcAβ3Galβ4GlcNAc**β6GlcNAc**-HDA	+
10	GlcAβ3Galβ4GlcNAc**β6GlcNAc**-C42	+
11	GlcAβ3Galβ4GlcNAc**β3Galβ4Glc**-HDA	(+)
12	GlcAβ3Galβ4GlcNAc**β3Galβ4Glc**-C42	+

+++, binding with high frequency at the lower pmol level (10–50 pmol/spot); ++ and +, less strong bindings; (+), rare binding.

in oligosaccharides, which may explain the observed differences in activities. The influence of the open chain-structure created during reductive amination on the binding activity has already been discussed earlier in this chapter.

We have also noticed differences between binding of *H. pylori* to HDA and C42 glycolipid products. This is not surprising because HDA represents a non-branched lipid and C42 is a branched hydrophobic conjugate. One can speculate that such drastic differences in lipid structures should influence the final presentation of the epitope on TLC plates. The conditions on TLC surfaces mimic, to some extent, natural conditions on hydrophobic surfaces of plasma membranes because silica gel plates are covered by a layer of a hydrophobic plastic material.

In summary, neoglycolipids provide a convenient way for investigation of carbohydrate protein interactions using both natural and synthetic carbohydrate structures. However, one should bear in mind that various structural factors may influence the recognition by proteins, therefore double tests should be recommended regarding screening for new activities, possibly using neoglycolipids with different lipid parts.

4.4.2. Conjugation of oligosaccharides to neoglycolipids

Synthesis of HDA-based neoglycolipids was described before [78]. The C42 conjugates were prepared in the following way: 200–600 nmol of a reducing carbohydrate was dissolved in 300 μl of methanol/water 1:1 and mixed with 280 μl of C42 in chloroform/methanol (C/M), 5:1 (20 mg/ml) and 100 μl of NaBH$_3$CN in methanol (62 mg/ml). The sample was kept at room temperature for 1–2 days. The progress of the reaction was monitored using TLC. The plates (HPTLC silica gel 60 plates, Merck) were run in chloroform/methanol/0.25% KCl in water, 50:40:10, and the glycolipids were visualized on the plates using anisaldehyde [78]. The newly synthesized neoglycolipids were purified using small silica gel 60 columns. The material was applied on the gel in C/M, 5:1 after which different fractions were eluted from the gel using mixtures of chloroform/methanol/water (C/M/H$_2$O) with increasing polarity (C/M, 2:1; C/M, 1:1; C/M/H$_2$O, 60:35:8 and C/M/H$_2$O, 50:40:10). The eluates were analyzed by TLC (see above), and the fractions containing neoglycolipids were collected and evaporated under nitrogen.

The C42 represents a new molecule (Pal-Lys(Pal)CONH(CH$_2$)$_4$-NH$_2$), synthesized through custom synthesis by RAPP Polymere GmbH (Germany), using lysine (Lys) and palmitate (Pal) as substrates.

The *H. pylori* strains CCUG 17874 and CCUG 17875 were from culture collection of Göteborg University, Sweden.

4.4.3. Synthesis of neolacto oligosaccharides

LNβ1-6Gn: Gnβ1-6Gn was prepared by HF-reversion of N-acetylglucosamine [79] and isolated by gel filtration chromatography on a column of Superdex 30 (5 × 95 cm). The disaccharide was then β4-galactosylated by bovine milk β4-galactosyltransferase (Calbiochem) [80]. The title trisaccharide was isolated in pure form by chromatography on a column of graphitized carbon (Hypercarb, Thermo-Hypersil Ltd., UK) with gradient 1, and characterized by MALDI-TOF MS and ¹H-NMR spectroscopy.

Gnβ1-3LNβ1-6Gn: LNβ1-6Gn obtained as above was incubated with UDP-GlcNAc and human serum containing a β3-GlcNAc transferase essentially as described previously [81]. The reaction mixture was desalted and deproteinized by passing through a mixed bed column of AG 1-X8 (Ac⁻) and AG 50W-X8 (H⁺) resins (200-400 mesh, Bio-Rad). The tetrasaccharide was finally isolated by chromatography on a column of graphitized carbon (Hypercarb) using gradient 2, and characterized by MALDI-TOF MS and ¹H-NMR spectroscopy.

Gnβ1-3LNα1-6Gn: Gnα1-6Gn was prepared by HCl-reversion of *N*-acetylglucosamine [82] followed by β4-galactosylation and β3-GlcNAc transferase reactions as above. The title tetrasaccharide was isolated in pure form by Hypercarb chromatography with gradient 1, and characterized by MALDI-TOF MS and ¹H-NMR spectroscopy.

Gnβ1-3LNβ1-3Man: GlcNAcβ1-3Man (Sigma) was β4-galactosylated by bovine milk β4-galactosyltransferase as described above. A crude oligosaccharide fraction was isolated by gel filtration chromatography on a column of Superdex peptide, and subjected to β3-*N*-acetylglucosaminylation by human serum. The serum reaction mixture was desalted and deproteinized by passing through a mixed bed column of AG 1-X8 (Ac⁻) and AG 50W-X8 (H⁺) resins. The title tetrasaccharide was isolated by Hypercarb chromatography with gradient 1, and characterized by MALDI-TOF MS and ¹H-NMR spectroscopy.

Gnβ1-3LNβ1-6Gal: GlcNAcβ1-6Gal (Sigma) was β4-galactosylated by bovine milk β4-galactosyltransferase as described above. A crude oligosaccharide fraction was isolated by gel filtration chromatography on a column of Superdex peptide, and subjected to human serum β3-GlcNAc transferase as above. The serum reaction mixture was desalted and deproteinized by passing through a mixed bed column of AG 1-X8 (Ac⁻) and AG 50W-X8 (H⁺) resins. The title tetrasaccharide was isolated by Hypercarb chromatography with gradient 1, and characterized by MALDI-TOF MS and ¹H-NMR spectroscopy.

GlcAβ1-3LNβ1-6Gn: LNβ1-6Gn obtained as above was subjected to β-glucuronylation using β-glucuronidase transglycosylation reaction for the novel acceptor structure in conditions essentially as described previously [83]. Excessive nitrophenol was removed by solid-phase extraction with BondElut C-18 columns (Varian), and the oligosaccharide products were isolated by gel filtration chromatography on a column of Superdex 30 (5 × 95 cm). The title tetrasaccharide was isolated in pure form by ion-exchange chromatography with Resource™Q (Amersham Pharmacia Biotech AB, Sweden), and characterized by MALDI-TOF MS and ¹H-NMR spectroscopy.

GlcAβ1-3LNβ1-3Lac: LNβ1-3Lac (IsoSep, Lund, Sweden) was subjected to β-glucuronylation using β-glucuronidase transglycosylation reaction essentially as above. The oligosaccharide products were isolated by gel filtration chromatography on a column of Superdex 30 (5 × 95 cm). The title pentasaccharide was isolated in pure form by ion-exchange chromatography with Resource™Q, and characterized by MALDI-TOF MS and ^{1}H-NMR spectroscopy.

GlcA(NMe)β1-3LNβ1-6Gn: GlcAβ1-3LNβ1-6Gn obtained as described above was dissolved in 90% aqueous pyridine containing 3x molar excess of methylamine, HBTU (Novabiochem) and diisopropylethylamine (Fluka). Reaction was conducted at RT for three days and the crude oligosaccharide fraction was isolated by gel filtration chromatography on Superdex peptide. The methylamidated species were isolated by anion-exchange chromatography (neutral fraction), further purified by gel filtration chromatography and characterized by MALDI-TOF MS and ^{1}H-NMR spectroscopy.

GlcA(NMe)β1-3LNβ1-3Lac: GlcAβ1-3LNβ1-3Lac was amidated with methylamine in a reaction containing HBTU and diisopropylethylamine as above. The methylamidated species were isolated by anion-exchange chromatography (neutral fraction), further purified by gel filtration chromatography and characterized by MALDI-TOF MS and ^{1}H-NMR spectroscopy.

4.4.4. Chromatographic methods

Gel filtration chromatography was performed with Superdex™Peptide HR 10/30 (10 × 300 mm) (Amersham Pharmacia Biotech, Sweden) with 50 mM NH_4HCO_3 as eluent, at a flow rate of 1 ml/min or Superdex 30 (5 × 95 cm) (Amersham Pharmacia Biotech, Sweden) with 50 mM NH_4HCO_3 as eluent, at a flow rate of 5 ml/min. All experiments were monitored at 214 nm. Fractions of 10 ml were collected in Superdex 30 runs, while product peaks from the Superdex peptide column were collected manually.

Chromatography on graphitized carbon was performed using a Hypercarb column (5 µm, 250 × 4.6 mm; Thermo-Hypersil Ltd., UK). Ammonia was used in the solvents to catalyze mutarotation, as separations of the anomeric forms of oligosaccharides were often observed. Two different gradients were employed: Gradient 1; the column was equilibrated with 10 mM NH_3, and elution was performed with a linear gradient of acetonitrile (0–40% over 60 min). Gradient 2; the column was equilibrated with 10 mM NH_3, and elution was performed with a linear gradient of acetonitrile (8–40% over 80 min). Absorbance at 214 nm was recorded.

Anion-exchange chromatography was performed on a column of Resource™ Q. The column was eluted with water for 4 min, then with a linear gradient of 0–50 mM NaCl over 8 min, followed by a steeper linear gradient of 50–500 mM NaCl over 8 min. Absorbance at 214 nm was recorded.

4.4.5. Mass spectrometry
MALDI-TOF mass spectra of oligosaccharides were recorded on a Voyager-DE™ STR BioSpectrometry™ (PerSeptive Biosystems) time-of-flight instrument. Samples were analyzed in either positive ion delayed extraction mode using 2,5-dihydroxybenzoic acid (DHB) (Aldrich) matrix (10 mg/ml in H_2O) or in negative ion linear delayed extraction mode using 2,4,6-trihydroxyacetophenone (Fluka) (3 mg/ml in acetonitrile/20 mM aqueous diammonium citrate, 1:1, by volume).

4.5. Lacto-binding specificity of H. pylori

4.5.1. Analysis by binding to natural glycolipid library
The lacto epitope for *H. pylori* was described in [34]. It was revealed by studying natural glycolipid libraries and purified components thereof, Table 2 and glycolipids isolated from human gastric tissues. A tetrasaccharide glycolipid was revealed by the TLC overlay assay and analyzed to by MS and NMR to be Lactotetraosylceramide.

This *H. pylori* binding glycolipid was present in only one of the seven samples from different individuals. This person had a secretor negative blood group, which does not express the α2-fucosyltransferase, modifying type 1 lactosamines to blood group glycans. It is considered that the secretor negative blood group may reveal the Lactotetra-glycolipid receptor and thus be a risk factor for *H. pylori* binding to gastric epithelial tissue surface.

4.5.2. Inhibition by lacto-N-tetraose oligosaccharide
The lactotetraosyl-binding specificity was tested by inhibition assay using oligosaccharide lacto-*N*-tetraose, containing the same oligosaccharide sequence as the glycolipid as inhibitor against the binding of *H. pylori* to the lactotetraosyl-ceramide. The free oligosaccharide inhibited the binding effectively at a concentration as low as 0.1 mg/ml, while the control saccharide lactose had no effect with the same concentration. Interestingly, the concentration of lacto-*N*-tetraose in human milk may be as high as 1 mg/ml and would thus be most likely high enough for in vivo inhibition of *H. pylori* binding. An infant formula comprising the oligosaccharide at the same level could work in similar fashion.

5. Poly-*N*-Acetyllactosamine Receptors for Influenza Virus

5.1. Analysis of the receptor from glycolipids

Using TLC binding methods, a high affinity receptor for human influenza A virus was found among glycolipids prepared from human neutrophils. Selected fractions containing complex glycolipids were shown to bind the virus particles with high

Table 2. Lacto-binding specificity.

No.	Trivial Name	Glycolipid Structure[a] (Adapted PCT SE0002567)	Binding[b]	Source	Ref.
1	Lacto-tri	GlcNAcβ3Galβ4Glcβ1Cer	−	Malignant melanoma	c
2	Lacto-tetra	Galβ3GlcNAcβ3Galβ4Glcβ1Cer	+	Human meconium	[84]
3		Galβ3GlcNH$_2$β3Galβ4Glcβ1Cer	−	Human meconium	d
4	H5-1	Fucα2Galβ3GlcNAcβ3Galβ4Glcβ1Cer	−	Human meconium	[85]
5	Le[a]-5	Galβ3(Fucα4)GlcNAcβ3Galβ4Glcβ1Cer	−	Human small intestine	[86]
6	Le[b]-6	Fucα2Galβ3(Fucα4)GlcNAcβ3Galβ4Glcβ1Cer	−	Human small intestine	[87]
7	B6-1	Galα3(Fucα2)Galβ3GlcNAcβ3Galβ4Glcβ1Cer	−	Monkey intestine	e
8		Galα3Galβ3GlcNAcβ3Galβ4Glcβ1Cer	−	Monkey intestine	f
9	B7-1	Galα3(Fucα2)Galβ(Fucα4)3GlcNAcβ3Galβ4Glcβ1Cer	−	Monkey intestine	e
10	A6-1	GalNAcα3(Fucα2)Galβ3GlcNAcβ3Galβ4Glcβ1Cer	−	Human meconium	[85]
11		NeuGcα3Galβ3GlcNAcβ3Galβ4Glcβ1Cer	−	Rabbit thymus	[88]

[a] The Galβ3GlcNAc parts have been underlined.
[b] +marks a significant darkening on the autoradiogram when 2 µg was applied on the thin-layer plate, while − marks no darkening.
[c] B.E. Samuelsson and K.A. Karlsson, unpublished results.
[d] Glycosphingolipid no. 3 was produced from Galβ3GlcNAcβ3Galβ4Glcβ1Cer from human meconium (no. 2) by treatment with anhydrous hydrazine (to be reported separately).
[e] N. Strömberg and K.A. Karlsson, unpublished results.
[f] Glycosphingolipid no. 8 was generated from Galα3(Fucα2)Galβ3GlcNAcβ3Galβ4Glcβ1Cer from monkey intestine (no. 7) by incubation in 0.05 M HCl at 80°C for 2h.

strength. The smallest binding glycolipids corresponded to sialylated hexaglycosyl-ceramides. The binding of the human virus to less complex glycolipids was not detected. Analysis of the binding species by sialic acid-binding lectins, *Sambucus nigra* agglutinin (SNA) and *Maackia amuriensis* agglutinin (MAA), revealed binding to α6-linked sialic acid (as identified by SNA) in the material, but not α3-linked sialic acid (as identified by MAA). Further analysis by MS revealed sialylated lactosamine type structures in the fraction of the bound glycolipids [21].

A control experiment with a regular avian (duck) influenza strain confirmed the expected α3-sialic acid binding specificity and a strong binding to less complex sialylated glycolipid with tetrasaccharide core chains [21].

These results were somewhat in contrast to previous results, which suggested binding of human influenza strains to α3-sialylated and possible fucosylated structures of human neutrophils [89]. These differences could be due to strain specificity variety or even to differences in the resolution of the TLC method.

It is also of interest that the recent microarray analysis also indicated that biantennary α6-sialylated N-glycans are especially effective receptors for pandemic human and avian influenza viruses [17].

5.2. Poly-N-acetyllactosamine library to reveal effective receptors

Poly-*N*-acetyllactosamines are effective molecules for the presentation of bioactive oligosaccharide structures. This has been demonstrated, for example, by binding of mammalian lectins involved in lymphocyte homing [90] and by studies on mammalian fertilization [91] using branched synthetic polylactosamines produced at the laboratory of Prof. O. Renkonen at the Institute of Biotechnology at the University of Helsinki [92]. On pathogen adhesion, the natural polyglycosylceramides were shown to have good activity regarding binding by various pathogens, as reviewed in [42].

To search for the most effective poly-*N*-acetyllactosamines binding to influenza viruses, a library of the complex glycans was synthesized. Schemes 2–8 reveal examples of the synthetic strategies applied for the production of the library. The major aim of the work was to reveal potential differences in binding depending on the asymmetry of the biantennary lactosamines. The asymmetry was achieved by

(i) using HMO lacto-*N*-hexaose as an asymmetric precursor, and
(ii) utilizing fine specificities of glycosyltransferases capable of branch specific reactions on poly-*N*-acetyllactosamines.

The branch specificity has been studied for several enzymes and branch specific enzymes may provide additional alternatives in synthesis as mentioned above [93–95].

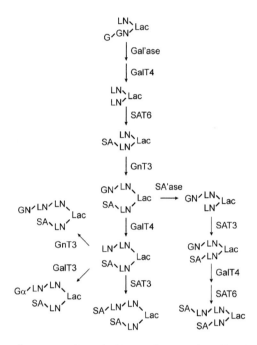

Scheme 2. Synthesis of asymmetric poly-*N*-acetyllactosamine libraries. Abbreviations: LN, Galβ4GlcNAc; Lac, Galβ4Glc; G, Gal; GN, GlcNAc; Gal'ase, β-galactosidase; GalT4, β4-galactosyltransferase; GalT3, α3-galactosyltransferase; GNT3, β3-*N*-acetylglucosaminyltransferase; SAT3/6, α3/6sialyltransferase; SA, Neu5Ac; / indicates 1–3-linkage; \ indicates 1–6 linkage.

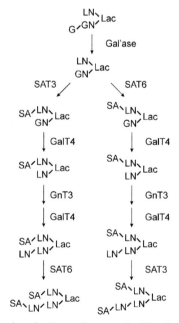

Scheme 3. Synthesis of asymmetric poly-*N*-acetyllactosamine libraries. Abbreviations as in Scheme 2.

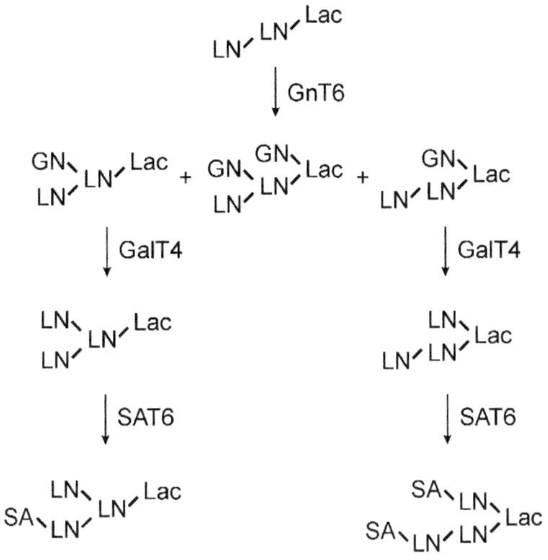

Scheme 4. Synthesis of asymmetric poly-*N*-acetyllactosamine libraries. Abbreviations as in Scheme 2.

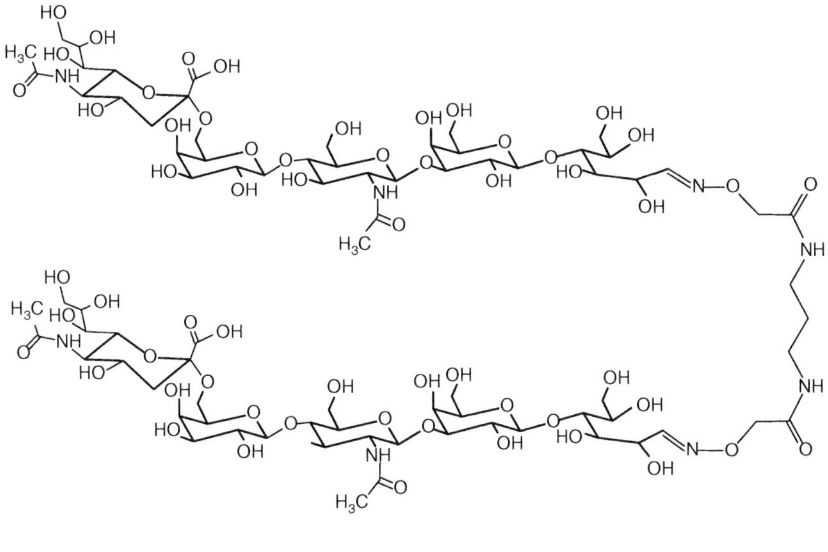

Scheme 5. Divalent oligosaccharide conjugate no. 25 (Table 3).

5.2.1. Synthesis of the branched oligosaccharide library

The oligosaccharide library represented in Table 3 was synthesized using the preferred methods according to the invention and as described in Schemes 2–8.

The oligosaccharides were purified using chromatographic methods and the products were characterized by MALDI-TOF MS and NMR spectroscopy.

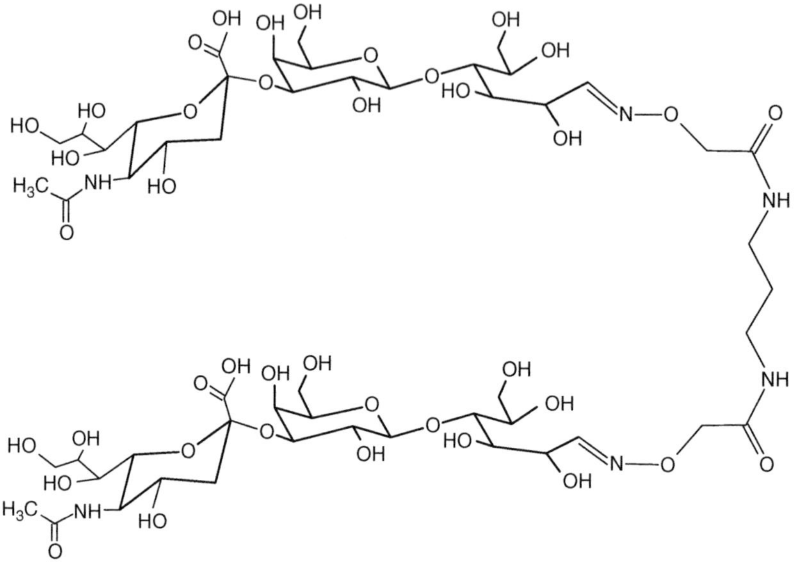

Scheme 6. Divalent oligosaccharide conjugate no. 26 (Table 3).

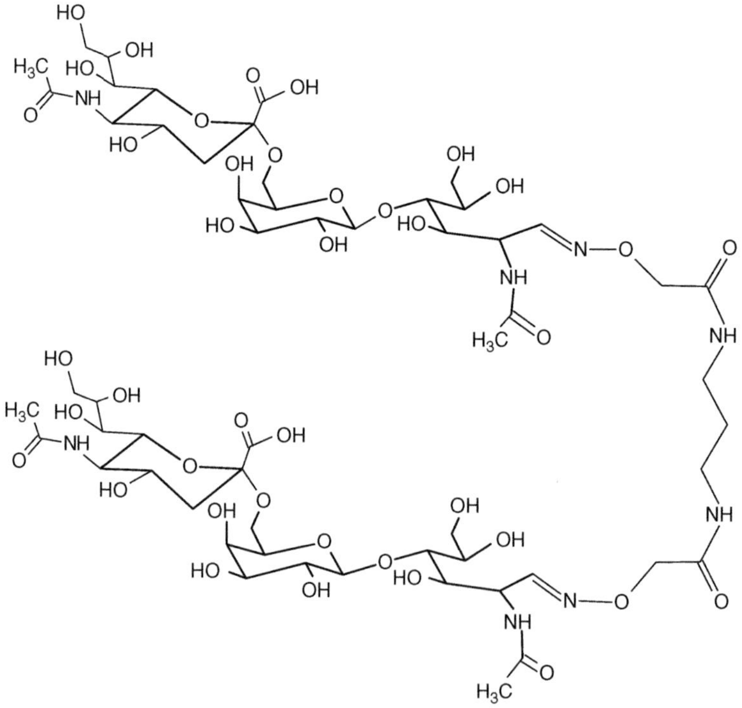

Scheme 7. Divalent oligosaccharide conjugate no. 27 (Table 3).

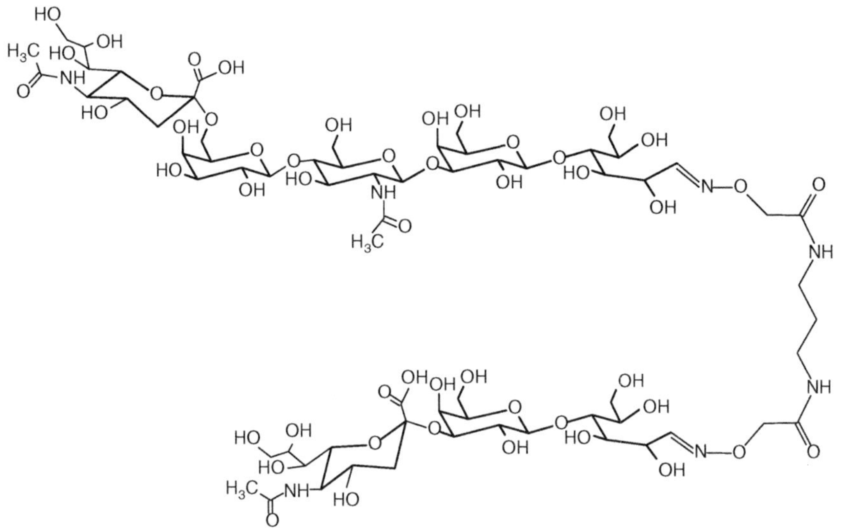

Scheme 8. Divalent oligosaccharide conjugate no. 28 (Table 3).

5.2.2. Preparation of divalent conjugates by conjugating oligosaccharides from reducing ends

A divalent aminooxy reagent (*N,N'*-diaminooxyacetic acid amide of 1,3-diaminopropane, DADA) was used to produce divalent carbohydrate molecules through oxime formation. One micromole of DADA was incubated with 5 μmol of reducing carbohydrate in 0.2 M sodium acetate buffer, pH 4.0, for 42 h at 37°C. The divalent carbohydrate oxime was purified with gel-permeation chromatography, and subjected to NMR spectroscopic analysis.

Four different oligosaccharide-DADA conjugates were prepared using three sialylated oligosaccharides (Schemes 5–8, Table 3, compounds 25–28). The ^{1}H-NMR spectra of the DADA conjugates revealed all characteristic signals for the glycan residues and the DADA unit. In addition, the NMR data confirmed the formation of oxime bond, but it is also clear that about 50% of the reducing monosaccharide unit exists in pyranose form.

5.2.3. Hemagglutination inhibition studies

The glycans were analyzed in a hemagglutination inhibition assay (Table 4). In the hemagglutination inhibition studies the influenza viruses were incubated at room temperature for 1 h in a mixture containing 25 μl influenza viruses (about eight hemagglutination units), 10 μl buffered inhibitor solution in various concentrations, and 25 μl erythrocytes. Hemagglutination inhibition was determined as lowest microscopically detectable inhibitory concentration. Results revealed an

Table 3. Library of branched poly-*N*-acetyllactosamines including simple monosialylated structures.

No.	Structure
1	Neu5Acα2-6Galβ1-4GlcNAcβ1-3Galβ1-4Glc
2	Neu5Acα2-3Galβ1-4GlcNAcβ1-6Galβ1-4Glc
3	Neu5Acα2-6Galβ1-4GlcNAcβ1-6Galβ1-4Glc
4	Neu5Acα2-3Galβ1-4GlcNAcβ1-3Galβ1-4Glc
5	Neu5Acα2-6Galβ1-4GlcNAcβ1-3(Galβ1-4GlcNAcβ1-6)Galβ1-4Glc
6	Neu5Acα2-3Galβ1-4GlcNAcβ1-3(Neu5Acα2-3Galβ1-4GlcNAcβ1-6)Galβ1-4Glc
7	Neu5Acα2-6Galβ1-4GlcNAcβ1-3(Neu5Acα2-6Galβ1-4GlcNAcβ1-6)Galβ1-4Glc
8	Neu5Acα2-6Galβ1-4GlcNAcβ1-3(Neu5Acα2-6Galβ1-4GlcNAcβ1-6)Galβ1-4Glc
9	Neu5Acα2-3Galβ1-4GlcNAcβ1-3(Neu5Acα2-3Galβ1-4GlcNAcβ1-6)Galβ1-4Glc
10	Neu5Acα2-6Galβ1-4GlcNAcβ1-3Galβ1-4[Fucα1-3]GlcNAcβ1-3Galβ1-4Glc
11	Neu5Acα2-6Galβ1-4GlcNAcβ1-3Galβ1-4GlcNAcβ1-3Galβ1-4Glc
12	Neu5Acα2-3[Galβ1-4GlcNAcβ1-3(Galβ1-4GlcNAcβ1-6)Galβ1-4Glc]
13	Neu5Acα2-6Galβ1-4GlcNAcβ1-3(Galβ1-4[Fucα1-3]GlcNAcβ1-6)Galβ1-4Glc
14	Neu5Acα2-6Galβ1-4GlcNAcβ1-3(Galβ1-4GlcNAcβ1-6)LNβ1-3Galβ1-4Glc
15	Neu5Acα2-6[Galβ1-4GlcNAcβ1-3(Galβ1-4GlcNAcβ1-6)Galβ1-4Glc]
16	Neu5Acα2-6Galβ1-4GlcNAcβ1-3Galβ1-4GlcNAcβ1-3 (Neu5Acα2-6Galβ1-4GlcNAcβ1-6)Galβ1-4Glc
17	Neu5Acα2-6Galβ1-4GlcNAcβ1-3(Neu5Acα2-3Galβ1-4GlcNAcα1-3Galβ1-4GlcNAcβ1-6)Galβ1-4Glc
18	Neu5Acα2-3Galβ1-4GlcNAcβ1-3(Neu5Acα2-6Galβ1-4GlcNAcβ1-3Galβ1-4GlcNAcβ1-6)Galβ1-4Glc
19	Neu5Acα2-6Galβ1-4GlcNAcβ1-3(Neu5Acα2-3Galβ1-4GlcNAcβ1-3Galβ1-4GlcNAcβ1-6)Galβ1-4Glc
20	Neu5Acα2-3Galβ1-4GlcNAcβ1-3(Neu5Acα2-3Galβ1-4GlcNAcβ1-3Galβ1-4GlcNAcβ1-6)Galβ1-4Glc
21	Neu5Acα2-6Galβ1-4GlcNAcβ1-3(Neu5Acα2-6Galβ1-4GlcNAcβ1-3Galβ1-4GlcNAcβ1-6)Galβ1-4Glc
22	Neu5Acα2-3Galβ1-4GlcNAcβ1-3(Neu5Acα2-6Galβ1-4GlcNAcβ1-3Galβ1-4GlcNAcβ1-6)Galβ1-4Glc
23	Neu5Acα2-6Galβ1-4GlcNAcβ1-3(Galα1-3Galβ1-4GlcNAcβ1-3Galβ1-4GlcNAcβ1-6)Galβ1-4Glc
24	Neu5Acα2-6Galβ1-4GlcNAcβ1-3(GlcNAcβ1-3Galβ1-4GlcNAcβ1-3Galβ1-4GlcNAcβ1-6)Galβ1-4Glc
25	[Neu5Acα2-6Galβ1-4GlcNAcβ1-3Galβ1-4Glc]2-DADA- oxime
26	[Neu5Acα2-3Galβ1-4GlcNAc]2-DADA-oxime
27	[Neu5Acα2-6Galβ1-4GlcNAc]2-DADA-oxime
28	Neu5Acα2-6Galβ1-4GlcNAcβ1-3Galβ1-4Glc-(Neu5Acα2-6Galβ1-4GlcNAc-)DADA-oxime

Table 4. The hemagglutination inhibition activity of large oligosaccharides as measured with influenza virus A/Victoria/3/75. Martina Pantzar is acknowledged for the experiment.

Oligosaccharide (No.)	Relative effectivity (% of inhibition)
7	66
16	26
17	15
18	8
19	<5
20	7.5
21	100
20	19
1−4	<2

interesting selectivity for structures containing two long poly-N-acetyllactosamine arms, structure no. 21 in Table 3, or longer β6-linked arm (structure 7 in Table 3). This correlated with a molecular model revealing that this structure may bridge between the known primary and secondary sialic acid binding regions of influenza virus hemagglutinin, while the opposite isomer (structure 16) with longer β3-branch cannot.

These data indicate that saccharide 7 is the most active isomer among the branched sialylated isomeric decasaccharide structures. The data also shows that the double α6-sialylated 12-meric saccharide 21 was even more effective. The data further indicate that even certain α3-sialylated branched polylactosamine structures have reasonable activity in inhibiting hemagglutination. The divalent DADA constructs produced from reducing oligosaccharides with larger distance between the terminal sialic acid epitopes (25, 28) had similar activities (not shown).

5.2.4. Synthesis and analysis methods for the branched oligosaccharides

5.2.4.1. Starting materials for enzymatic synthesis

LNnT (LNβ3L) and Gnβ3LNβ3L were from commercial sources or enzymatically synthesized. Gnβ6L was purchased from Sigma (USA). LNH [LNβ3(Gβ3Gnβ6)L] was purchased from Dextra Laboratories (UK) or IsoSep (Sweden).

5.2.4.2. Enzymatic synthesis

α1,3-Galactosyltransferase (GalT3): 5 mM acceptor and 10 mM UDP-galactose were incubated with GalT3 (0.02 mU of enzyme/nmol of acceptor site was used; recombinant enzyme from bovine, Calbiochem, cat. no. 345647) in 0.1 M

2-(*N*-morpholino)ethanesulfonic acid (MES), pH 6.5 and 20 mM MgCl$_2$ for 24 h at 37°C. The reaction was terminated by incubation in a boiling water bath for 3 min.

β1,6-N-acetylglucosaminetransferase (GnT6): 8 mM acceptor and molar excess of UDP-*N*-acetylglucosamine were incubated with GnT6 (fresh, concentrated rat serum was used as the enzyme source) in 20 mM EDTA, 0.5 mM ATP, 200 mM Gal, 60 mM γ-galactonolactone, 8 mM NaN$_3$ for 7 days at 37°C. The reaction was terminated by incubation in a boiling water bath for 3 min.

β1,4-Galactosyltransferase (GalT4): 4 mM acceptor and 8 mM UDP-galactose were incubated with GalT4 (0.05 mU of enzyme/nmol of acceptor was used; enzyme from bovine milk, Calbiochem, cat.no. 345649) in 50 mM MOPS pH 7.4 and 20 mM MnCl$_2$ for 6 h at 37°C. The reaction was terminated by incubation in a boiling water bath for 3 min.

β1,3-N-acetylglucosaminetransferase (GnT3): 2 mM acceptor and molar excess of UDP-*N*-acetylglucosamine were incubated with GnT3-preparate (see below) in 0.1 mM ATP, 0.04% NaN$_3$ and 8 mM MnCl$_2$ for five days at 37°C. Reaction was terminated by incubation on a boiling water bath for 3 min.

Concentrated human plasma was used as GnT3-preparate: Human plasma was purchased from Finnish Red Cross, a protein concentrate from ammonium sulphate precipitation of 25–50% was obtained, dissolved to 50 mM Tris-HCL, pH 7.5 and 0.5 M NaCl, dialyzed against highly purified H$_2$O and lyophilized. Prior to use, enzyme preparate was dissolved to 50 mM Tris-HCL, pH 7.5.

α2,6-Sialyltransferase (SAT6):
(1) Enzyme from ICN: 8 mM acceptor and molar excess of CMP-*N*-acetylneuraminic acid were incubated with SAT6 (3 μU of enzyme/nmol of acceptor site was used) in 50 mM MES, pH 6.0 and 0.2% BSA for 48 h at 37°C. The reaction was terminated by incubation in a boiling water bath for 3 min.
(2) Enzyme from bovine colostrums: 10 mM acceptor and molar excess of CMP-*N*-acetylneuraminic acid were incubated with SAT6-preparate (see below) and 0.02% NaN$_3$ for 72 h at 37°C. The reaction was terminated by incubation in a boiling water bath for 3 min.

Fat was removed by centrifugation of fresh or at −20°C stored bovine colostrum. Casein was removed by acid precipitation after which the preparate was neutralized. A protein concentrate from ammoniumsulphate precipitation of 40–60% was obtained, dissolved to 10 mM MES, pH 6.0, dialyzed against the same buffer and lyophilized. Prior to use, the enzyme preparate was dissolved to highly purified H$_2$O.
(3) Recombinant enzyme from rat, Calbiochem, cat.no. 566222
8 mM acceptor and molar excess of CMP-*N*-acetylneuraminic acid were incubated with SAT6 (4 μU of enzyme/nmol of acceptor site was used) in

50 mM MES, pH 6.0, 0.1% Triton-X-100, 0.5 mg/ml BSA and 0.02% NaN$_3$ for 48 h at 37°C. The reaction was terminated by incubation in a boiling water bath for 3 min.

α2,3-Sialyltransferase (SAT3): 4 mM acceptor and 8 mM CMP-*N*-acetylneuraminic acid were incubated with SAT3 (0.1 U of enzyme/μmol of acceptor site was used; enzyme from rat, recombinant, Calbiochem, cat.no. 566218) in 50 mM MOPS, pH 7.4 and 0.2 mg/ml BSA for 48 h at 37°C. The reaction was terminated by incubation in a boiling water bath for 3 min.

α1,3-Fucosyltransferase (FucT3): 1 mM acceptor and 5 mM GDP-fucose were incubated with FucT3 (0.02 mU of enzyme/nmol of acceptor was used; enzyme human FucT VI, recombinant, Calbiochem, cat.no. 344323) in 50 mM 3-(*N*-morpholino)propanesulfonic acid (MOPS), pH 7.2 and 20 mM MnCl$_2$ for 24 h at 37°C. The reaction was terminated by incubation in a boiling water bath for 3 min.

5.2.4.3. Enzymatic degradation reactions
α2-3,6,8-Sialidase (SA'ase): 8 mM substrate was incubated with sialidase (13 mU/μmol of substrate; *Vibrio cholerae* sialidase was from Calbiochem, cat. no. 480717) in 50 mM Na-acetate buffer, pH 5.5 and 1 mM CaCl$_2$ for 24 h at 37°C. The reaction was terminated by incubation in a boiling water bath for 3 min.

β1,3-Galactosidase (Gal'ase): 8 mM substrate was incubated with galactosidase (13 mU/μmol of substrate; recombinant enzyme from Calbiochem, cat. no. 345795) in 50 mM Na-citrate buffer, pH 4.5 for 24 h at 37°C. The reaction was terminated by incubation in a boiling water bath for 3 min.

5.2.4.4. MALDI-TOF MS
MALDI-TOF mass spectra were acquired as described for the neolacto oligosaccharide syntheses.

5.2.4.5. NMR-spectroscopy of the branched oligosaccharide library
NMR spectroscopy was performed in D$_2$O at 23°C using a 500 MHz NMR-spectrometer.

The primary structure of lactosaminoglycans can be determined from one-dimensional NMR spectra. Structural elements are identified from signals having characteristic chemical shifts. The integration of these signals gives the relative amount of different types of monosaccharides within the glycan. Typical structure reporting signals are the anomeric H1 protons and other protons at or near the site of glycosidic linkage. The anomericity of the monosaccharides is obtained from the H1–H2 coupling constant. Typical values are 3–4 Hz for α anomer and 7–8 Hz for β anomer. In branched lactosaminoglycans H1 protons of Gal and GlcNAc have distinct chemical shift depending on whether Gal is 3- or 6-substituted of if

Gal is both 3- and 6-substituted. In repeating 4GlcNAcβ1-3Galβ1 structures Gal H1 and GlcNAc H1 resonate at 4.46 ppm and 4.70 ppm, respectively. Another characteristic feature of GlcNAcβ1-3Gal structures is the Gal H4 signal, which resonates at 4.16 ppm. In structures having Galβ1-4GlcNAcβ1-6Gal the chemical shift of GlcNAc H1 is ~4.63 ppm. The terminal Gal H1 has different chemical shifts depending on whether it is in the 3- or 6-branch. The chemical shifts are 4.48 and 4.46 ppm, respectively. A GlcNAc in all structures is also identified from the methyl signal of the *N*-acetyl group between 2.02–2.07 ppm. Sialylated lactosaminoglycans also have easily recognizable signals. The linkage isomers, e.g., Neu5Acα2-3/6Gal can be distinguished. In Neu5Acα2-3Gal the equatorial and axial H3 of Neu5Ac resonate at 2.76 and 1.80 ppm, respectively. The Gal H3 signal is observed at 4.12 ppm and Gal H1 resonates at 4.56 ppm. In Neu5Acα2-6Gal the equatorial and axial H3 of Neu5Ac resonate at 2.67 ppm and 1.72 ppm, respectively. The Gal H3 in buried under the signals of other skeletal protons and cannot be assigned from the one-dimensional spectrum. In both isomeric forms the methyl signal of the *N*-acetyl group is observed at approximately 2.06 ppm. More specific shifts were used in defining the key branched precursor structures.

References

[1]   D. Ilver, P. Johansson, H. Miller-Podraza, P.G. Nyholm, S. Teneberg and K.A. Karlsson, Methods Enzymol., 363 (2003) 134–157.

[2]   N. Sharon, Biochim Biophys Acta, 1760 (2006) 527–537.

[3]   K.A. Karlsson, Glycobiology, 10 (2000) 761–771.

[4]   A. Varki, Glycobiology, 3 (1993) 97–130.

[5]   A. Varki, Cell, 126 (2006) 841–845.

[6]   K.A. Karlsson and N. Stromberg, Methods Enzymol., 138 (1987) 220–232.

[7]   M. Svensson, F.M. Platt and C. Svanborg, FEMS Microbiol. Lett., 258 (2006) 1–8.

[8]   J. Ångstrom, S. Teneberg and K.A. Karlsson, Proc. Natl. Acad. Sci. USA, 91 (1994) 11859–11863.

[9]   B.J. Appelmelk, M.A. Monteiro, S.L. Martin, A.P. Moran and C.M. Vandenbroucke-Grauls, Trends Microbiol., 8 (2000) 565–570.

[10] I. Bucior and M.M. Burger, Curr. Opin. Struct. Biol., 14 (2004) 631–637.

[11] H. Rauvala and S.I. Hakomori, J. Cell. Biol., 88 (1981) 149–159.

[12] H. Rauvala, J.P. Prieels and J. Finne, Proc. Natl. Acad. Sci. USA, 80 (1983) 3991–3995.

[13] B.D. Shur, S. Evans and Q. Lu, Glycoconj. J., 15 (1998) 537–548.

[14] P. Sörme, B. Kahl-Knutson, U. Wellmar, U.J. Nilsson and H. Leffler, Methods Enzymol., 362 (2003) 504–512.

[15] H. Maaheimo, P. Kosma, L. Brade, H. Brade and T. Peters, Biochem., 39 (2000) 12778–12788.

[16]   D. Ilver, A. Arnqvist, J. Ogren, I.M. Frick, D. Kersulyte, E.T. Incecik, D.E. Berg, A. Covacci, L. Engstrand and T. Borén, Science, 279 (1998) 373–377.

[17]   J. Stevens, O. Blixt, T.M. Tumpey, J.K. Taubenberger, J.C. Paulson and I.A. Wilson, Science, 312 (2006) 404–410.

[18]   H. Miller-Podraza, B. Lanne, J. Ångstrom, S. Teneberg, M.A. Milh, P.A. Jovall, H. Karlsson and K.A. Karlsson, J. Biol. Chem., 280 (2005) 19695–19703.

[19]   A. Walz, S. Odenbreit, J. Mahdavi, T. Borén and S. Ruhl, Glycobiology, 15 (2005) 700–708.

[20]   M.D. Disney and P.H. Seeberger, Chem. Biol., 11 (2004) 1701–1707.

[21]   H. Miller-Podraza, L. Johansson, P. Johansson, T. Larsson, M. Matrosovich and K.A. Karlsson, Glycobiology, 10 (2000) 975–982.

[22]   S. Teneberg, J. Ångstrom and A. Ljungh, Glycobiology, 14 (2004) 187–196.

[23]   A. Kobata, Eur. J. Biochem. 209 (1992) 483–501.

[24]   K. Furukawa and A. Kobata, Curr. Opin. Biotechnol., 3 (1992) 554–559.

[25]   P. Gagneux and A. Varki, Glycobiology, 9 (1999) 747–755.

[26]   T. Angata, N.M. Varki and A. Varki, J. Biol. Chem., 276 (2001) 40282–40287.

[27]   K.L. Mohlke, A.A. Purkayastha, R.J. Westrick, P.L. Smith, B. Petryniak, J.B. Lowe and D. Ginsburg, Cell., 96 (1999) 111–120.

[28]   F.J. Olson, M.E. Johansson, K. Klinga-Levan, D. Bouhours, L. Enerback, G.C. Hansson and N.G. Karlsson, J. Biol. Chem., 277 (2002) 15044–15052.

[29]   B. Potvin, R. Kumar, D.R. Howard and P. Stanley, J. Biol. Chem., 265 (1990) 1615–1622.

[30]   S. Goelz, R. Kumar, B. Potvin, S. Sundaram, M. Brickelmaier and P. Stanley, J. Biol. Chem., 269 (1994) 1033–1040.

[31]   P. Stanley and E. Ioffe, Faseb. J., 9 (1995) 1436–1444.

[32]   M. Gawlitzek, U. Valley, M. Nimtz, R. Wagner and H.S. Conradt, J. Biotechnol., 42 (1995) 117–131.

[33]   P.G. Falk, L. Bry, J. Holgersson and J.I. Gordon, Proc. Natl. Acad. Sci. USA, 92 (1995) 1515–1519.

[34]   S. Teneberg, I. Leonardsson, H. Karlsson, P.A. Jovall, J. Ångstrom, D. Danielsson, I. Naslund, A. Ljungh, T. Wadstrom and K.A. Karlsson, J. Biol. Chem., 277 (2002) 19709–19719.

[35]   F.O. Olfat, E. Naslund, J. Freedman, T. Borén and L. Engstrand, J. Infect. Dis., 186 (2002) 423–427.

[36]   R.M. Hyland, T.P. Griener, G.L. Mulvey, P.I. Kitov, O.P. Srivastava, P. Marcato and G.D. Armstrong, J. Med. Microbiol., 55 (2006) 669–675.

[37]   S. Kelm, A.K. Shukla, J.C. Paulson and R. Schauer, Carbohydr. Res., 149 (1986) 59–64.

[38]   J.C. Paulson and G.N. Rogers, Methods Enzymol., 138 (1987) 162–168.

[39]   G.N. Rogers, G. Herrler, J.C. Paulson and H.D. Klenk, J. Biol. Chem., 261 (1986) 5947–5951.

[40]   J. Liukkonen, S. Haataja, K. Tikkanen, S. Kelm and J. Finne, J. Biol. Chem., 267 (1992) 21105–21111.

[41] S. Hirmo, S. Kelm, R. Schauer, B. Nilsson and T. Wadstrom, Glycoconj. J., 13 (1996) 1005–1011.

[42] H. Miller-Podraza, Chem. Rev., 100 (2000) 4663–4682.

[43] C.L. Stults, C.C. Sweeley and B.A. Macher, Methods Enzymol., 179 (1989) 167–214.

[44] S.I. Nishimura and K. Yamada, J. Am. Chem. Soc., 119 (1997) 10555–10556.

[45] M.D. Vaughan, K. Johnson, S. DeFrees, X. Tang, R.A. Warren and S.G. Withers, J. Am. Chem. Soc., 128 (2006) 6300–6301.

[46] S. Fukui, T. Feizi, C. Galustian, A.M. Lawson and W. Chai, Nat. Biotechnol., 20 (2002) 1011–1017.

[47] M.J. Martin, T. Feizi, C. Leteux, D. Pavlovic, V.E. Piskarev and W. Chai, Glycobiology., 12 (2002) 829–835.

[48] N. Murahashi, H. Ishihara, M. Sakagami and A. Sasaki, Biol. Pharm. Bull., 20 (1997) 704–707.

[49] Y. Harada, T. Murata, K. Totani, T. Kajimoto, S.M. Masum, Y. Tamba, M. Yamazaki and T. Usui, Biosci. Biotechnol. Biochem., 69 (2005) 166–178.

[50] M. Fukasawa, Y. Shimizu, K. Shikata, M. Nakata, R. Sakakibara, N. Yamamoto, M. Hatanaka and T. Mizuochi, FEBS Lett., 441 (1998) 353–356.

[51] A.S. Palma, T. Feizi, Y. Zhang, M.S. Stoll, A.M. Lawson, E. Diaz-Rodriguez, M.A. Campanero-Rhodes, J. Costa, S. Gordon, G.D. Brown and W. Chai, J. Biol. Chem., 281 (2006) 5771–5779.

[52] W. Chai, C.T. Yuen, T. Feizi and A.M. Lawson, Anal. Biochem., 270 (1999) 314–322.

[53] Y. Ozaki, K. Omichi and S. Hase, Anal. Biochem., 241 (1996) 151–155.

[54] R. Horie and K. Nakano, Carbohydr. Res., 264 (1994) 209–226.

[55] P.W. Tang, H.C. Gool, M. Hardy, Y.C. Lee and T. Feizi, Biochem. Biophys. Res. Commun., 132 (1985) 474–480.

[56] G. Magnusson, A.Y. Chernyak, J. Kihlberg and L.O. Kononov, Synthesis of neoglycoconjugates. In: Neoglycoconjugates: Preparation and applications, Academic Press, San Diego, 1994.

[57] M.S. Stoll, T. Feizi, R.W. Loveless, W. Chai, A.M. Lawson and C.T. Yuen, Eur. J. Biochem. 267 (2000) 1795–1804.

[58] D. Lafont, B. Gross, R. Kleinegesse, F. Dumoulin and P. Boullanger, Carbohydr. Res., 331 (2001) 107–117.

[59] I.D. Manger, T.W. Rademacher and R.A. Dwek, Biochemistry, 31 (1992) 10724–10732.

[60] I.D. Manger, S.Y. Wong, T.W. Rademacher and R.A. Dwek, Biochemistry, 31 (1992) 10733–10740.

[61] T. Feizi. In: XXIII International Carbohydrate Symposium, Whistler, Canada 2006.

[62] R.A. Evangelista, F.T.A. Chen and A. Guttman, J. Chromatogr. A, 745 (1995) 273–280.

[63]   B. Domon and C.E. Costello, Glycoconj. J., 5 (1988) 397–409.
[64]   T. Borén, P. Falk, K.A. Roth, G. Larson and S. Normark, Science, 262 (1993) 1892–1895.
[65]   R. Striker, U. Nilsson, A. Stonecipher, G. Magnusson and S.J. Hultgren, Mol. Microbiol., 16 (1995) 1021–1029.
[66a] S. Haataja, K. Tikkanen, U. Nilsson, G. Magnusson, K.A. Karlsson and J. Finne, J. Biol. Chem., 269 (1994) 27466–27472.
[66b] N.V. Bovin and Glycoconj J., 15 (1998) 431– 446.
[67]   A.S. Gambaryan, E.Y. Boravleva, T.Y. Matrosovich, M.N. Matrosovich, H.D. Klenk, E.V. Moiseeva, A.B. Tuzikov, A.A. Chinarev, G.V. Pazynina and N.V. Bovin, Antiviral Res., 68 (2005) 116–123.
[68]   B.E. Dunn, H. Cohen and M.J. Blaser, Clin. Microbiol. Rev., 10 (1997) 720–741.
[69]   R.M. Peek Jr. and M.J. Blaser, Nat. Rev. Cancer, 2 (2002) 28–37.
[70]   M. Gerhard, S. Hirmo, T. Wadström, H. Miller-Podraza, S. Teneberg, K.A. Karlsson, B. Appelmelk, S. Odenbreit, R. Haas, A. Arnqvist and T. Borén, Helicobacter pylori, and adherent pain in the stomach, Horizon Scientific Press, Wymondham, UK, 2001.
[71]   S. Linden, H. Nordman, J. Hedenbro, M. Hurtig, T. Borén and I. Carlstedt, Gastroenterology, 123 (2002) 1923–1930.
[72]   J.H. Van de Bovenkamp, J. Mahdavi, A.M. Korteland-Van Male, H.A. Buller, A.W. Einerhand, T. Borén and J. Dekker, Helicobacter, 8 (2003) 521–532.
[73]   J. Mahdavi, B. Sonden, M. Hurtig, F.O. Olfat, L. Forsberg, N. Roche, J. Ångstrom, T. Larsson, S. Teneberg, K.A. Karlsson, S. Altraja, T. Wadstrom, D. Kersulyte, D.E. Berg, A. Dubois, C. Petersson, K.E. Magnusson, T. Norberg, F. Lindh, B.B. Lundskog, A. Arnqvist, L. Hammarstrom and T. Borén, Science, 297 (2002) 573–578.
[74]   S.O. Hynes, S. Teneberg, N. Roche and T. Wadstrom, Infect. Immun., 71 (2003) 2976–2980.
[75]   J.J. Thorn, S.B. Levery, M.E. Salyan, M.R. Stroud, B. Cedergren, B. Nilsson, S. Hakomori and H. Clausen, Biochemistry, 31 (1992) 6509–6517.
[76]   S. Teneberg, I. Lonnroth, J.F. Torres Lopez, U. Galili, M.O. Halvarsson, J. Ångström and K.A. Karlsson, Glycobiology, 6 (1996) 599–609.
[77]   C.A. Lingwood, M. Huesca and A. Kuksis, Infect. Immun., 60 (1992) 2470–2474.
[78]   H. Miller-Podraza, P. Johansson, J. Ångström, T. Larsson, M. Longard and K.A. Karlsson, Glycobiology, 14 (2004) 205–217.
[79]   J. Defaye, A. Gadelle and C. Pedersen, Carbohydr. Res., 186 (1989) 177–188.
[80]   K. Brew, T.C. Vanaman and R.L. Hill, Proc. Natl. Acad. Sci. USA, 59 (1968) 491–497.
[81]   F. Piller and J.P. Cartron, J. Biol. Chem., 258 (1983) 12293–12299.
[82]   K. Blumberg, F. Liniere, L. Pustilnik and C.A. Bush, Anal. Biochem., 119 (1982) 407–412.
[83]   T. Yasukochi, K. Fukase, Y. Suda, K. Takagaki, M. Endo and S. Kusumoto, Bull. Chem. Soc. Jpn., 70 (1997) 2719–2725.
[84]   K.A. Karlsson and G. Larson, J. Biol. Chem., 254 (1979) 9311–9316.
[85]   K.A. Karlsson and G. Larson, J. Biol. Chem., 256 (1981) 3512–3524.

[86] E.L. Smith and J.M. McKibbin, J. Biol. Chem., 250 (1975) 6059–6064.

[87] J.M. McKibbin, W.A. Spencer, E.L. Smith, J.E. Mansson, K.A. Karlsson, B.E. Samuelsson, Y.T. Li and S.C. Li, J. Biol. Chem., 257 (1982) 755–760.

[88] M. Iwamori, M. Iwamoto, K. Hayashi, K. Kiguchi and Y. Nagai, J. Biochem. (Tokyo), 98 (1985) 1561–1569.

[89] J. Muthing, Carbohydr. Res., 290 (1996) 217–224.

[90] J.P. Turunen, M.L. Majuri, A. Seppo, S. Tiisala, T. Paavonen, M. Miyasaka, K. Lemstrom, L. Penttila, O. Renkonen and R. Renkonen, J. Exp. Med., 182 (1995) 1133–1141.

[91] E.S. Litscher, K. Juntunen, A. Seppo, L. Penttila, R. Niemela, O. Renkonen and P.M. Wassarman, Biochemistry, 34 (1995) 4662–4669.

[92] O. Renkonen, Cell. Mol. Life Sci., 57 (2000) 1423–1439.

[93] D.H. Joziasse, W.E. Schiphorst, D.H. Van den Eijnden, J.A. Van Kuik, H. Van Halbeek and J.F. Vliegenthart, J. Biol. Chem., 262 (1987) 2025–2033.

[94] O. Renkonen, A. Leppanen, R. Niemela, A. Vilkman, J. Helin, L. Penttila, H. Maaheimo, A. Seppo and J. Suopanki, Biochem. Cell. Biol., 70 (1992) 86–89.

[95] A. Seppo, L. Penttila, A. Leppanen, H. Maaheimo, R. Niemela, J. Helin, J.M. Wieruszeski and O. Renkonen, Glycoconj. J., 11 (1994) 217–225.

Lectins: Analytical Technologies
C.L. Nilsson (Editor)
© 2007 Elsevier B.V. All rights reserved.

Chapter 7

Carbohydrate Microarrays for Lectin Characterization and Glyco-Epitope Identification

Denong Wang[a] and Albert M. Wu[b]

[a]Carbohydrate Microarray Laboratory, Department of Genetics, and Neurology and Neurological Sciences, Stanford University School of Medicine, Beckman Center B007, Stanford, CA 94305-5318, USA
[b]Glyco-Immunochemistry Research Laboratory, Institute of Molecular and Cellular Biology, College of Medicine, Chang-Gung University, Kwei-san, Tao-yuan, 333, Taiwan

1. Introduction

Lectins are an important class of proteins or glycoproteins of non-immune origin that bind non-covalently to characteristic carbohydrate structures with specificity or selectivity. There are many ways to classify lectins. Given the fact that a wide range of living organisms, from microbes to mammals, produce lectins, they can be divided according to species origin, such as microbial lectins, plant lectins, invertebrate lectins and vertebrate lectins. Based on the similarities in sequence of homology and activity, lectins can also be grouped into subfamilies. There are at least five groups of vertebrate lectins, such as C-type (calcium dependent), P-type, S-type, I-type, and pentraxins [1]. Cellular lectin molecules or receptors with lectin-like carbohydrate-binding domains play important roles in cell signaling, protein intracellular trafficking, and cell–cell communication [2–5].

Lectins can also be classified according to their binding specificity or selectivity with carbohydrates. This type of classification is helpful for selection of lectins as structural probes in biomedical applications. By the early seventies, the carbohydrate-binding activity of a lectin was commonly described in terms of monosaccharide specificity. Experimentally, a type of inhibition assay, such as agglutination inhibition, was performed to identify the monosaccharide that most effectively inhibited the binding reaction. Many lectins are able to cross-react with a panel of sugar chains with a common terminal sugar residue. Lately, disaccharides are considered to be a better way to define a lectin's specificity.

However, some lectin-binding sites accommodate glyco-epitopes larger than mono- and disaccharides and recognize conformational epitopes. The latter may consist of much longer oligosaccharide or polysaccharide chains. There is also evidence that lectins recognize certain cluster configurations of complex carbohydrates with high selectivity and avidity.

X-ray crystallographic studies identified some common characteristics in the binding pockets of lectins and antibodies [2, 6, 7]. Aromatic amino acid residues appear frequently in carbohydrate-binding sites of various fine specificities. Such residues are large and participate in a wide variety of van der Waals and electrostatic interactions. The majority of crystal structures so far resolved are associated with a network of H-bonds, frequently with water molecules, in their combining sites. Such bound water molecules play important roles at the contacting interface between carbohydrate ligands and receptors. In addition, polyamphiphilic surfaces were identified in the interfaces of a number of carbohydrate ligands and binding sites. Presence of relatively rigid structures is required to produce an amphiphilic surface in solution. Carbohydrate molecules with branched termini, such as many blood group substances, which are formed by oligosaccharide chains, are favorable for the generation of such contact surfaces.

Much remains to be learned regarding the specificity and cross-reactivity of the carbohydrate–lectin interactions and their functional outcome in biological systems. Challenging issues in lectin characterization may include but are not limited to (a) the spectrum or repertoire of carbohydrate structures that are reactive with given lectins, (b) conformational properties of glyco-epitopes that are selectively recognized by lectins, and (c) the cluster effect in lectin–carbohydrate interactions, especially those that take place on cell surfaces and trigger the events of intracellular cell signaling, cell differentiation and proliferation. Recent development of microarray-based broad-range binding assays offers new tools to facilitate these investigations.

In this chapter, we summarize the concept and the updated information regarding the specificity-based lectin classification (Section 2). Then, we introduce a practical platform of carbohydrate microarrays that is likely useful for lectin characterization and classification (Section 3). Lastly, we discuss a few examples to illustrate the application of this technology in lectin-related experimental investigations (Section 4).

2. The Specificity-Based Lectin Classification

Lectins are functionally classified based on their relative binding reactivities with the structural units of carbohydrate or glyco-epitopes. They are grouped according to their monosaccharide specificities and then further sub-grouped based on their reactivities with more complex structures. As listed below, carbohydrate specificities of biomedically important lectins are classified into six groups according to their specificities to monosaccharides. They are further sub-grouped based on the binding affinities to (a) GalNAcα1→*O* to Ser(Thr) of the peptide chain; (b) disaccharides; (c) trisaccharides; and (d) the number and the location of LFucα1→linked to oligosaccharides. These structures are frequently found in soluble glycoproteins and as cell surface glycoconjugates in mammals [8–11]. This scheme of lectin classification is, thus, practically useful in bio-medical application of lectins.

2.1. Group I. GalNAc-specific agglutinins

(1) **F/A**, GalNAcα1→3GalNAc (**Forssman**) and GalNAcα1→3Gal (blood group **A** determinant disaccharide) specific agglutinins – *Dolichos biflorus* (DBA) and *Helix pomatia* (HPA).

(2) **A**, GalNAcα1→3Gal (blood group **A** determinant disaccharide) – specific agglutinins – Soybean (SBA) and Lima bean (LBA).

(3) **Tn**, GalNAcα1→Ser/Thr-specific agglutinins – *Vicia villosa* B₄ (VVL-B₄) and *Salvia sclarea* (SSA).

2.2. Group II. Gal-specific agglutinins

(1) **T**, Galβ1→3GalNAc (**T$_\alpha$**, Galβ1→3GalNAcα1→ the mucin-type sugar sequence on human erythrocyte membrane or **T$_\beta$**, Galβ1→3GalNAcβ1→ at the terminal non-reducing end of the gangliosides) specific agglutinins – Peanut (PNA), and *Bauhinia purpurea alba* (BPA).

(2) **I/II**, Galβ1→3(4)GlcNAcβ1→(Lacto-*N*-biose/*N*-acetyllactosamine) spe-cific agglutinins (Human blood group type **I** (Galβ1→3GlcNAc) and type **II** (Galβ1→4GlcNAc) carbohydrate sequences [16–18], the disaccharide residues at the non-reducing end of the carbohydrate chains derived from either *N*-glycosidic or *O*-glycosidic linkages.) – *Ricinus communis* agglu-tinin (RCA₁), *Datura stramonium* (thorn apple, TAA) and wheat germ agglutinin (WGA).

(3) **B**, Galα1→3Gal (Human blood group **B** disaccharide-*Griffonia (Bandeiraea) simplicifolia* B₄ (GSI-B₄).

(4) **E**, Galα1→4Gal, blood group P^k active disaccharide – Abrin-a and Mistletoe lectin-I (ML-I).

2.3. Group III. Man and/or Glc-specific agglutinins

Man-linked oligosaccharide-specific agglutinins – *Concanavalin ensiformis* (Jack bean, ConA) and *Lens culinaris* (LCA).

2.4. Group IV. GlcNAc, and/or Galβ1→4GlcNAcβ1→linked specific agglutinins

Chitin oligosaccharide-specific agglutinins – WGA and *Griffonia (Bandeiraea) simplicifolia* II (GSA-II).

2.5. Group V. LFuc-specific agglutinins

(1) Monofucosyl-specific agglutinins – *Ulex europaeus* I (UEA-I) and UEA-II.
(2) Difucosyl-specific agglutinins – *Griffonia (Bandeiraea) simplicifolia* IV (GSA-IV).

2.6. Group VI. Sialic acid specific agglutinins

2.6.1. Limulus polyphemus agglutinin (LPA).

Glycoproteins and glycosphingolipids contain many carbohydrate epitopes or crypto-glyco-epitopes of biomedical importance. They are present on the cell surface and function as receptors in various life processes. Many known glyco-epitopes exist in soluble or gel form and serve as biological lubricants or as barriers against microbial invasion. During the past decade, 11 mammalian structural units have been used to express the binding domain of Gal- and GalNAc-specific lectins. They are **F**, GalNAcα1→3GalNAc; **A**, GalNAcα1→3Gal; **T**, Galβ1→3GalNAc; **I**, Galβ1→3GlcNAc; **II**, Galβ1→4GlcNAc; **B**, Galα1→3Gal; **E**, Galα1→4Gal; **L**, Galβ1→4Glc; **P**, GalNAcβ1→3Gal; **S**, GalNAcβ1→4Gal and **Tn**, GalNAcα1→Ser(Thr) (Fig. 1 and Table 1). Except **L** and **P**, all of the units can be found in glycoproteins. **Tn**, which is an important marker for breast/colon cancer and vaccine development, exists only in *O*-glycans. Natural **Tn** glycoprotein, the simplest mammalian *O*-glycan, is exclusively expressed in the armadillo salivary gland. Antifreeze glycoprotein is composed of repeating units of **T**. *Pneumococcus* type 14 capsular polysaccharide, which has uniform **II** disaccharide as carbohydrate side chains. Asialo human α_1-acid glycoprotein and asialo fetuin provide multi-antennary **II** structures. Human ovarian cyst glycoproteins, which belong to the complex type of glycoform, comprise most of the structural units. To facilitate the selection of lectins that could serve as structural probes, the carbohydrate-binding properties of Gal/GalNAc-reactive lectins have been classified according to their highest affinity for structural units and their binding profiles are expressed in decreasing order of reactivity. The source and the structural relationship of the proposed mammalian structural units (lectin determinants) for Gal- and GalNAc-specific lectins, which are shown in Fig.1 as part of carbohydrate structures, are shown in Fig. 2 and Table 1. Most of the lectin F_β and T_β determinants are found in glycosphingolipids [12–14], and the other determinants are present in mammalian glycoproteins [9, 15–19], especially in the human blood group A, B, H, Le[a], Le[b], and Ii active glycoproteins prepared from human ovarian cyst fluid [15, 18] and salivary glycoproteins [20]. The reactivities of lectin determinants represent a combined result of the binding of individual sugars. The contribution of each sugar to the binding is not necessarily equal, and is different among lectins, for example, both *Amaranthus caudatus* (ACL) and PNA are Galβ1→3GalNAc specific [8, 21]. However, the inhibitory profile of the monosaccharides with these two lectins is quite different;

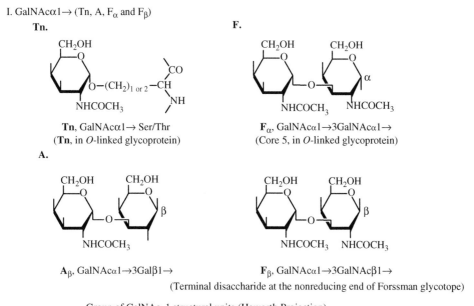

I. GalNAcα1→ (Tn, A, F$_\alpha$ and F$_\beta$)

Tn.

F.

Tn, GalNAcα1→ Ser/Thr
(**Tn**, in *O*-linked glycoprotein)

F$_\alpha$, GalNAcα1→3GalNAcα1→
(Core 5, in *O*-linked glycoprotein)

A.

A$_\beta$, GalNAcα1→3Galβ1→

F$_\beta$, GalNAcα1→3GalNAcβ1→
(Terminal disaccharide at the nonreducing end of Forssman glycotope)

Group of GalNAcα1 structural units (Haworth Projection)

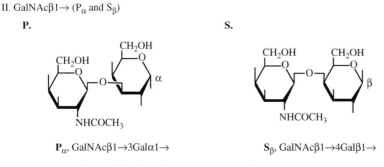

II. GalNAcβ1→ (P$_\alpha$ and S$_\beta$)

P.

S.

P$_\alpha$, GalNAcβ1→3Galα1→

S$_\beta$, GalNAcβ1→4Galβ1→

Group of GalNAcβ1→ structural units (Haworth Projection)

Figure 1. Mammalian glycoconjugates structural units used to express and classify the carbohydrate specificity of lectins.

III. Galα1→ (B and E)

B.   **E.**

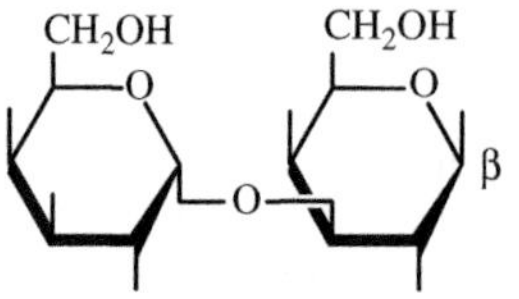

B$_\beta$, Galα1→3Galβ1→   **E$_\beta$**, Galα1→4Galβ1→

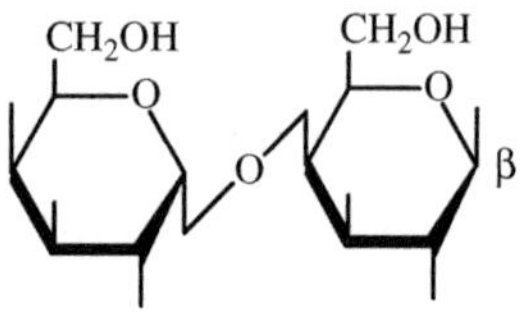

Group of Galα1→ structural units (Haworth Projection)

IV. Galβ1→ (Tα, T$_\beta$, L, I and II)

T.   **I.**

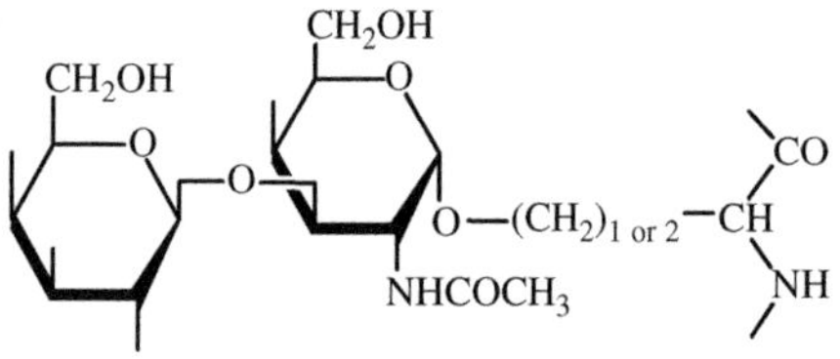

T$_\alpha$, Galβ1→3GalNAcα1→Ser/Thr   **I$_\beta$**, Galβ1→3GlcNAcβ1
(**T**, in *O*-linked glycoprotein)

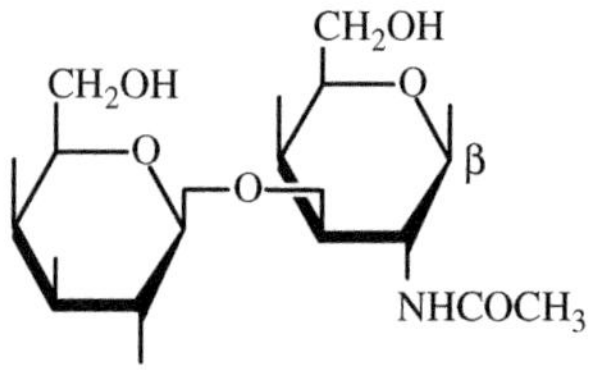

T$_\beta$, Galβ1→3GalNAcβ1→   **II$_\beta$**, Galβ1→4GlcNAcβ1→
(Terminal disaccharide at the asialo GM1; GSL)

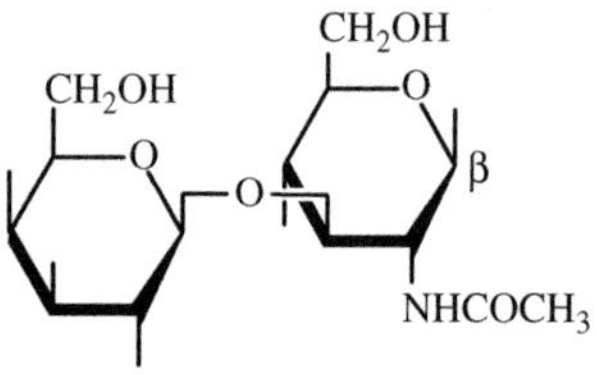

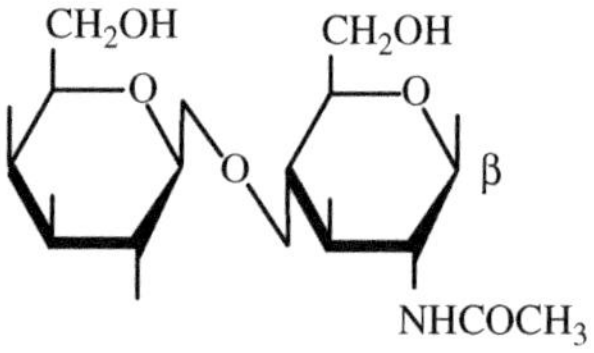

L.

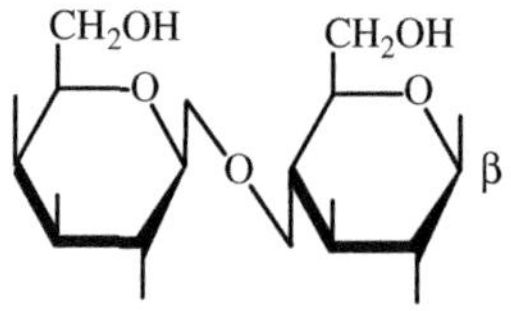

L$_\beta$, Galβ1→4Glcβ1→(GSL)

Group of Galβ1→ structural units (Haworth Projection)

Figure 1. (Continued).

Glycosphingolipids

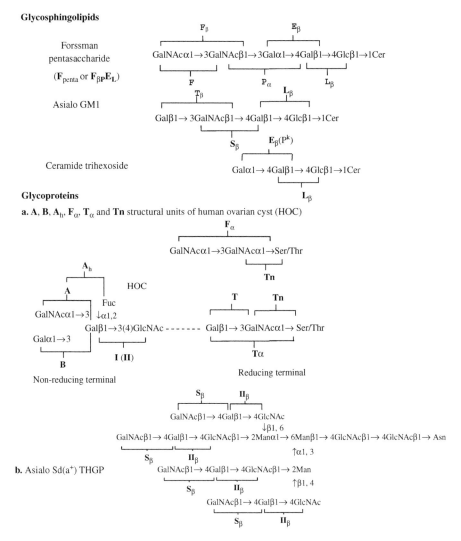

Figure 2. Mammalian structural units (lectin determinants) present in glycoconjugates. Lectin **F** determinant is found in glycosphingolipid (the principal glycolipid of mammalian tissues) of the tissues of the guinea pig, horse, cat, and chicken. It can also be found on the surface of some bacteria, viruses, human gastric carcinoma, and colon tumors [14, 15, 18, 38]. All (**F**, **A/A_h**, **B**, **I/II**, **T**, **Tn**) determinants are found in human blood groups A, B, H, Le^a, Le^b, and Ii active glycoconjugates prepared from human ovarian cyst fluid (HOC) [16, 18, 22, 42]. The Gal β1→3GalNAc group at the non-reducing end of asialo-GM$_1$ is also considered as a lectin **T**_β determinant. Determinants of lectins **I** and **II** discussed in this article are equivalent to human blood group type **I** (Lacto-*N*-biose, Galβ1→3GlcNAc) and type **II** (*N*-acetyllactosamine) carbohydrate sequences [39]. **S** determinant is found in human Tamm-Horsfall glycoprotein (THGP) and GM2. (a) **A**, **B**, **A_h**, **F**_α, **T**_α and **Tn** structural units of human ovarian cyst (HOC); (b) Asialo Sd(a$^+$) THGP; (c) Tri-antennary **II** and **T** units of asialo fetuin. It has three oligosaccharide side chains with two different structures *O*-glycosidically linked to Ser or Thr of the protein core as well as three carbohydrate side chains per molecule *N*-glycosidically linked to asparagine.

(Continued)

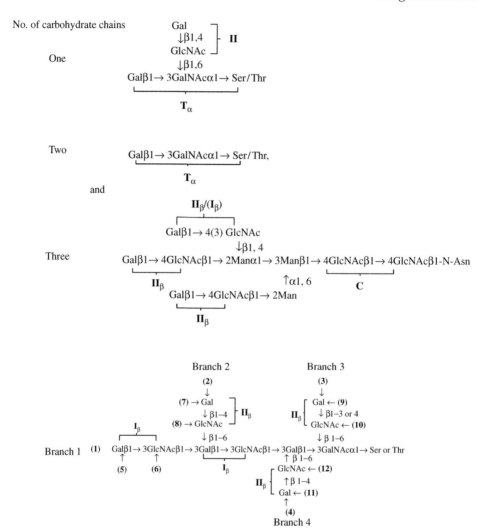

No. of carbohydrate chains

One

Two

and

Three

Branch 1

Branch 2

Branch 3

Branch 4

Figure 2. (Continued) Asialo-carbohydrate chains of fetuin are shown as above [40, 41]. C: Chitin disaccharide. (d) Representative carbohydrate chains of blood group glycoprotein from human ovarian cyst fluids. The four-branched internal structure (**I–IV**) shown is the representative internal portion of the carbohydrate moiety of blood group substances to which the residues responsible for A, B, H, Le[a], and Le[b] activities are attached. This structure represents precursor blood group active glycoproteins [16] and can be prepared by Smith degradation of A, B, H active glycoproteins, purified from human ovarian cyst fluids. Numbers in parentheses indicate the site of attachment for the human blood group A, B, H, Le[a], and Le[b] determinants. These determinants as well as the structural units at the non-reducing end are the sources of lectin **A/A$_h$**, **B/B$_h$**, **I/II**, **T**, and **Tn** determinants in (a) [9]. A megalosaccharide of 24 sugars has not been isolated. However, most of the carbohydrate chains isolated are parts of this structure.

Blood group active glyco-protein purified from human ovarian cyst fluid	Human blood group or antigenic determinant	Sugar added	Site of addition
MSS 1st Smith[2] Beach P-1[2]	Ii *Pneumococcus* type 14 polysaccharide	None	
Tighe P-1[2]			
Cyst N-1 Le[a] 20% 2x	Le[a] Le[x]	ʟFucα1→4 ʟFucα1→3	(**6**) and/or (**10**) (**8**), (**10**) and/or (**12**)
Cyst JS phenol insoluble	H	ʟFucα1→2	(**5**), (**7**), (**9**), (**11**)
Tighe phenol insoluble	H, Le[y] H, Le[x] > Le[b]	ʟFucα1→2 and others as in Le[a]	(**5**), (**7**), (**9**) and/or (**11**) (**6**), (**8**), (**12**)
MSS, native MSM Cyst 9 Cyst 14	A$_1$ or A$_2$	GalNAcα1→3 and others as in H, Le[b] and Le[y]	(**1**), (**2**), (**3**) and/or (**4**) in addition to (**5**) to (**12**)
Cyst Beach phenol insoluble	B	Galα1→3 and Same as H Le[b] and Le[y]	(**1**), (**2**), (**3**) and/or (**4**), in addition to (**5**) to (**12**)

Figure 2. (Continued).

GalNAc>>>Gal (inactive) in ACL and Gal>>GalNAc (poor) in PNA. Both RCA$_{120}$ and WGAs are Galβ1→4GlcNAcβ1→6 specific, and the inhibitory profile of the monosaccharide residues with these two lectins is also very different: Gal>>>GlcNAc in RCA$_{120}$ and GlcNAcβ1→ >> Gal in wheat germ [8, 9]. From the data available, expressions of the Gal/GalNAc-specific lectins based on mammalian structural units [8–11, 22–24] are shown in Table 2.

During the past two decades, it has been established that many multi-branched oligosaccharides exhibit a significant increase in lectin-binding reactivities as compared to their linear counterparts [25, 26], "the glycoside cluster effect" in mammalian hepatic asialoglycoprotein receptors [25] is one of the important findings in this field. Effect of polyvalencies of glyco-epitopes on the binding of a lectin from the edible mushroom, *Agaricus bisporus* [27] is another important example, in which the disaccharide **II** monomer and tri-antennary Galβ1-4GlcNAc (Tri-**II**) glycopeptides, the major carbohydrate side chains of *Pneumococcus* type 14 ps, asialo fetuin and asialo human α$_1$-acid glycoprotein, were poor inhibitors, while their polyvalent carriers were very active, implying that there are other more complicated structural factors involved in binding besides the multi-antennary **II** sequences. To explain this phenomenon, the present concept of glycoside cluster effect has to be further defined and classified into two groups – (a) the "*Multi-antennary* or simple glycoside cluster effect" as in galactosides with hepatic lectin

Table 1. Carbohydrate structural units in mammalian glycoproteins and glycosphingolipids.

	Codes[a]	Structural Units	Sources
1	**F**	GalNAcα1 →3GalNAc	Forssman pentasaccharide. Animal tissue antigens and human oncofetal glyco-epitopes, mainly in glycosphingolipids.
	F$_{penta-}$	GalNAcα1 →3GalNAcβ1 →3Galα1 →4Galβ1 →4Glc	
	F$_\alpha$	GalNAcα1 →3GalNAcα1 →Ser/Thr of protein core	In *O*-linked glycoproteins core.
	F$_\beta$	GalNAcα1 →3GalNAcβ1 →	Glycotope at the non-reducing end of **F**$_{penta-}$
2	**A**	GalNAcα1 →3Gal	Human blood group A related **disaccharide**.
	A$_h$	GalNAcα1 →3[LFucα1 →2]Gal	Human blood group A related **trisaccharide**.
3	**Tn**	GalNAcα1 →Ser/Thr of protein core	Tn antigen, only in *O*-linked glycoproteins.
4	**T**$_\alpha$	Galβ1 →3GalNAcα1 →Ser/Thr of protein core	The mucin-type sugar sequence on the human erythrocyte membrane.
	T$_\beta$	Galβ1 →3GalNAcβ1 →····ceramide	Brain glycoconjugates and gangliosides, GM$_1$.
5	**I**	Galβ1 →3GlcNAc	Human blood group precursor type I and II carbohydrate sequences.
	I$_\beta$	Galβ1 →3GlcNAcβ1	
6	**II**	Galβ1 →4GlcNAc	Branched or linear repeated **II** sequence is part of blood group I and i epitopes. **I** and **II** are precursors of ABH and Le^a, Le^b, Le^x, Le^y blood group active antigens. Most of the
	II$_\beta$	Galβ1 →4GlcNAcβ1 →	
	Tri-**II**	Tri-antennary Galβ1 →4GlcNAc	
	m**II**	Multivalent Galβ1 →4GlcNAc	

lectins reactive with **II** are also reactive with **I**. Lectin Tri-**II** and m**II** determinants are present at the non-reducing end of the carbohydrate chains derived from *N*- and *O*-glycans.

7	**B**	Galα1→3Gal	Human blood group B related **disaccharide**.
	B$_h$	Galα1→3[LFucα1→2]Gal	Human blood group B related **trisaccharide**.
8	**E**	Galα1→4Gal	Blood group P^k and P$_1$ active disaccharide. Sheep hydatid cyst glycoproteins, salivary glycoproteins of the Chinese swiftlet, glycosphingolipids in human erythrocytes, and small intestine.
9	**L**	Galβ1→4Glc	Constituent of mammalian milk.
	L$_\beta$	Galβ1→4Glcβ1→	Lactosyl ceramides in brain and part of carbohydrate structures in gangliosides.
10	**P**	GalNAcβ1→3Gal	Blood group P related disaccharide; glycotope at
	P$_\alpha$	GalNAcβ1→3Galα1→	the non-reducing end of globoside.
11	**S**	GalNAcβ1→4Gal	Brain asialo-GM$_2$ disaccharide; human blood
	S$_\beta$	GalNAcβ1→4Galβ1→	group Sd(a+) related disaccharide in most human urine secretions, Tamm-Horsfall glycoprotein.

[a]α, β: anomer of sugars; m: multivalent, and Tri: tri-antennary.

Table 2. Expression of binding properties of Gal/GalNAc-reactive lectins by carbohydrate structural units.

Codes	Lectins [2, 12, 13, 46, 47]	Carbohydrate Specificity
F/A	*Codium fragile* subspecies *tomentosoides* (CFA)	F_{penta-} and T_α $>A_h$ $>>$ **I/II** and **L**
	Dolichos biflorus (DBA)	F_{penta-} $>A_h^a$ $>A>$ **Tn** $>>$ **P**
	Helix pomatia (HPA)	F_{penta-} $>A$ $(>A_h^a)$ $=$ **Tn, T** $>>$ **P**
	Hog peanut (*Agaricus bisporus* agglutinin (ABA), *Amphicarpaea bracteata*)	$F_\beta > A >>$ **L**
	Wistaria floribunda (WFA)	A $(>A_h^b)$, $F_{penta-}>$ **F/P** $>$ **Tn, I (II)** $>$ **L**
	Geodia cydonium agglutinin (GCA)	F_{penta-}, $A_h >$ **mII** $>$**L**$>$ **II, T** $>$**I** $>>$ **E**
	Griffonia (*Bandeiraea*) *simplicifolia* A_4 (GSI-A_4,)	F_{penta-} $> A_h >$ **GalNAc** $>$ **E** $>$ **B** $>$ **I, T** $>>$ **L, II**
F/II	*Caragana arborescens* agglutinin (CAA)	F_{penta-} $>$ **II, mTn** $>$ **Sialyl Tn**
	Wistaria sinensis agglutinin (WSA)	F_{penta-} $\geq$ **mII** $>$**P**$>$**II, Tn, I** and $A_h \geq$ **L/E**
A	Lima bean (LBA)	Hexa-$A_h^a > A_h^b >>$ **B**
	Soybean (SBA, *Glycine max*)	A $(>A_h^b)$, **Tn** and **I (II)**
	Vicia villosa (VVA)(a mixture of A_4, A_2B_2 and B_4)	A $(>A_h^b)$ and **Tn** mainly
Tn	CFA	F_{penta-} and T_α $>$ **Tn** clusters $> A_h$ $>>$ **I/II** and **L**
	Vicia villosa B_4 (VVL-B_4)	Two **Tn** $>>$ one **Tn** $>>$ one or two **T**
	Salvia sclarea (SSA)	Two **Tn** $>$ single or three sequential **Tn** structures
Tn, I/II	*Glechoma hederacea* agglutinin (GHA)	**Tn** clusters $>$ **Tn** $>$ A_L, $A > A_h >$ **F, S, P, B** $>$ **E**
T	Peanut (PNA, *Arachis hypogaea*)	**T** $>>$ **I (II)** $>>$ **Tn**

T/Tn	CFA	F_α, **A**, T_α and **Tn**
	Maclura pomifera (MPA)	**T** > **Tn** >>> **I, II,** and **L**
	Artocarpus integrifolia (jacalin, AIA)	T_α > **P** > **T, Tn** >>> **I(II)** > T_β
T, Tn/II	*Bauhinia purpurea alba* (BPA)	**T** > **I(II), L,** and **Tn**
	ABA	T_α and **Tn** > **I** >> GalNAc >>>**II, L**
	Morus nigra galactose-specific lectin (Morniga G)	T_α >> **Tn** clusters > **T** > **Tn, P,** Tri-**II**
T/II	*Ricinus communis* toxin (ricin, RCA$_2$)	**T** > **I(II)** and **Tn**
	Abrus precatorius agglutinin (APA)	**T** > **I/II** > **E** > **B** > **Tn**
	Sophora japonica (SJA)[c]	**T** > **I** ≥ **II** > **L**
I(II)	*Ricinus communis* agglutinin (RCA$_1$)	**II** ≥ **I** > **E, B** > **T**
	Datura stramonium (thorn apple, TAA)	Bi-antennary **I(II)** (penta-2,6)>>**C**[c]
	Erythrina cristagalli (coral tree, ECA)	Multi-antennary **I/II**
	Phaseolus vulgaris-L	Tri-**II**, mII clusters > Penta-2,6
		> Tri-2,6 > Hepta-3,6 > **II**$_\beta$ >GlcNAcβ1,2Man
	Geodia cydonium (GCL)	**A** and mII[d] clusters, **F** > **A** > m**II** > **L** > **II, T** > **I** >> **E**
B	*Griffonia (Bandeiraea) simplicifolia* B$_4$ (GSI-B$_4$)	**B** > **E** > **A**
E	*Abrus precatorius* toxin-a (Abrin-a)	**E, B** > **T, L, I,** and **II**
	Mistletoe lectin-I (ML-I)	**E** > **II, L** > **T** and **I**

[a] Substitution of Fucα1 →2 to subterminal Gal is important for binding; [b] Substitution of Fucα1 →2 to subterminal Gal blocks binding; [c] C, chitin disaccharide; [d] mII, multivalent **II**.

[25, 26] and tri-antennary **II** with a galectin from chicken liver (CG-16) [28] as well as (b) the "*High-density polyvalent* or complex glycoside cluster effect" as in macromolecular interaction of high-density **II** cluster containing glycoproteins (*Pneumococcus* type 14 ps and asialo human α_1-acid glycoproteins) with ABA. The clustered glyco-epitopes present on macromolecules generate a great enhancement in affinity with ABA up to 4.7×10^6 times as compared to their monomeric counterparts, and demonstrate the structural importance of complex carbohydrates.

To obtain a complete description of the carbohydrate specificities of the applied lectins, the following information is suggested to be necessary – (1) monosaccharide specificity (Gal, GalNAc, GlcNAc, and/or Man); (2) expression of reactivities toward mammalian disaccharide structural units or their derivatives (Table 1) in decreasing order (Table 2); (3) the most active ligand; (4) simple multivalent or cluster effect, such as glycopeptides and multi-antennary glyco-epitopes to inhibit binding; (5) complex multivalent or cluster effects present in macromolecules with known glyco-epitopes. In this report, we illustrate only, as shown in Table 2, the abilities of mammalian disaccharide structural units or their derivatives to inhibit lectin–glycan binding and express it in decreasing order. During the past years, many cases of the effect of polyvalency on binding have been found [28–35]. However, the available data are insufficient to make a solid conclusion, therefore, it is not included in this review.

Due to the multiple reactivity of lectins toward mammalian glyco-epitopes (Table 2), the possible existence of different combining sites or subsites in the same molecule has to be examined, and the differential binding properties of these combining sites (if any) have to be characterized. Recent advances in characterization, cloning, and structural analysis have allowed us to classify plant lectins into seven families of structurally and evolutionarily related proteins. Within each lectin family the overall profile and structure of the carbohydrate-binding site(s) are conserved. A closer examination of the carbohydrate specificity further indicates that most plant lectins are not targeted against plant carbohydrates but preferentially bind foreign glycans [36, 37]. Establishing the relationship between the amino acid sequences of the combining sites of plant lectins and mammalian glyco-epitopes should be an important direction to be addressed in lectinology.

3. Carbohydrate Microarrays

Our laboratories have established a high-throughput biochip platform for constructing carbohydrate microarrays [42–44]. Using this technology, carbohydrate-containing macromolecules of diverse structures, including polysaccharides, natural glycoconjugates, and mono- and oligosaccharides coupled to carrier molecules, can be stably immobilized on a glass chip without chemical modification.

This technology takes advantage of existing cDNA microarray system, including spotter and scanner, for an efficient production and application of carbohydrate microarrays.

We have demonstrated that this current platform is able to overcome a number of expected technical difficulties, by proving that (1) carbohydrate molecules can be immobilized on a nitrocellulose-coated glass slide without chemical conjugation, (2) the immobilized carbohydrates are able to preserve their immunological properties and solvent accessibility, (3) the system reaches the sensitivity, specificity, and capacity to detect a broad range of antibody specificities in clinical specimens, and (4) this technology allows highly sensitive detection, as compared to other existing technologies, of the broad range of carbohydrate–lectin/antibody interactions.

In this section, we provide a practical protocol for this high-throughput carbohydrate microarray system. We focus on an eight-chamber sub-array system that is in active use in our laboratory's routine carbohydrate research. It is our wish that the readers who have access to a standard cDNA microarray facility would be able to explore this technology for their own carbohydrate research.

3.1. Materials

3.1.1. Apparatus

Microspotting: Cartesian Technologies' PIXSYS 5500C (Irvine, CA); or GMS 417 Arrayer, Genetic Microsystems, Inc. (Woburn, MA)
Supporting substrate: FAST Slides (Schleicher & Schuell, Keene, NH)
Microarray scanning: ScanArray 5000 Standard Biochip Scanning System (Packard Biochip Technologies, Inc., Billerica, MA)

3.1.2. Softwares

Array design: CloneTracker (Biodiscovery, Inc., Marina del Rey, CA)
Array printing: AxSys™ (Cartesian Technologies, Inc., Irvine, CA)
Array scanning and data capturing: ScanArray Express (PerkinElmer, Torrance, CA)

3.1.3. Antibodies and lectins

PHA-L and GSI-1 (*Phaseolus vulgaris*-L, EY Laboratories, Inc., San Mateo, CA) Streptavidin-FITC, -PE, -APC, and -Cy5 conjugates (Amersham Pharmacia, Piscataway, NJ).
Species-specific anti-immunoglobulin antibodies and their fluorescent conjugates, APC, PE, Cy5, Alexa647, or FITC (Sigma, St. Luis, MO; BD-PharMingen, San Diego, CA; Invitogen, San Diego, CA).

3.1.4. Reagents and buffers

Dilution buffer: Saline (0.9% NaCl)

Rinsing solution: 1× PBS, pH 7.4 w/0.05% (v/v) Tween 20

Blocking solution: 1% (w/v) BSA in PBS w/0.05% (w/v) NaN$_3$

3.2. Methods

The methods described below outline (1) design and construction of an eight-chamber sub-array system for defined purposes, (2) microspotting carbohydrate-containing molecules onto nitrocellulose-coated glass slides, (3) immunostaining and scanning of microarrays, (4) analysis of microarray data, and (5) validation of microarray findings by conventional immunological assays.

3.2.1. Design and construction of carbohydrate arrays

We have been applying an eight-chamber sub-array system to construct customized carbohydrate microarrays for defined purposes. As illustrated in Fig. 3, each microglass slide contains eight separated sub-arrays. The microarray capacity is ~500 microspots per sub-array. A single slide is, thus, designed to enable eight microarray assays. Similar sub-array designs with various array capacities are commercially available (Schleicher & Schuell, Keene, NH; ArrayIt, Sunnyvale, CA).

- Each microglass slide contains eight sub-arrays of identical content. There is chip space for 500 microspots per sub-array, with spot sizes of ~200 μm and at 300 μm intervals, center to center. A single slide is, therefore, designed to enable eight detections.
- Repeats and dilutions: We usually print carbohydrate antigens at the initial concentration of 0.1–0.5 μg/μl. The absolute amount of antigens printed on chip substrate is in the range of 0.1–0.25 ng per microspot for the highest concentration. They are further diluted at 1:3, 1:9, and 1:27. A given concentration of each preparation is repeated at least three times to allow statistic analysis of detection of identical preparation at given antigen concentration.
- Standard curves: For serological study, we include antibodies of IgG, IgA, and IgM isotype of corresponding species to serve as standard curves in microarray format. This design allows quantifying antibody signals that are captured by spotted carbohydrate antigens. In addition, such standard curves are useful for microarray data normalization and cross-chip scaling of microarray detection. For lectin study, we spot fluorescent conjugates, such as BSA conjugates of FITC, Cy5, or other dyes, to generate standard curves.

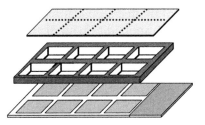

Figure 3. Illustration of the eight-chamber sub-arrays.

3.2.2. Microspotting of carbohydrates onto nitrocellulose-based substrate

Using Cartesian Technologies' PIXSYS 5500C (Irvine, CA), a high-precision robot designed for cDNA microarrays, carbohydrate antigens of various complexities are picked up by dipping quill pins into antigen/antibody solutions and printed onto nitrocellulose-coated slides in consistent amount (Schleicher & Schuell, Keene, NH; ArrayIt, Sunnyvale, CA). The complementary AxSysTM software (Cartesian Technologies, Inc., Irvine, CA) is used to instruct movement of pins about the dispense platform and the printing process.

- Prepare samples of carbohydrate antigens in 0.9% NaCl and transfer them in 96-well plates.
- Place the 96-well plates containing samples on Cartesian arrayer robot.
- Adjust program so that carbohydrate antigens are printed at spot sizes of ~150 μm and at 300 μm intervals, center to center.
- Each antigen or antibody is spotted as triplet replicates in parallel.
- The printed carbohydrate microarrays are air-dried and stored at room temperature (RT) without desiccant before application.

3.2.3. Immunostaining of carbohydrate microarrays

The staining procedure for carbohydrate microarrays is basically identical to the routine procedure for immunohistology. Immunostaining steps of carbohydrate arrays are listed below.

- Rinse printed microarray slides with 1 × PBS, pH 7.4 with 0.05% Tween 20 at RT for 5 min.
- Block slides with 1% BSA in PBS containing 0.05% NaN_3 at RT for 30 min.
- Stain each sub-array with 50 μl of test sample, which is diluted in 1% BSA PBS containing 0.05% NaN_3 and 0.05% Tween 20.
- Incubate the slide in a humidified chamber at RT for 60 min.
- Wash slides by dipping them in washing buffer (1 × PBS, pH 7.4 with 0.05% Tween 20) five times with at least one-time incubation at RT for 5 min.

- Stain slides with 50 μl of titrated secondary antibodies. Anti-human (or other species) IgG, IgM, or IgA antibodies with distinct fluorescent tags, Cy5, PE or FITC, are mixed and then applied on the chips.
- Incubate the slide in a humidified chamber, protected from light at RT for 30 min.
- Wash slides five times as specified above.
- Place slide in a 50 ml falcon centrifuge tube and spin at 1,000 rpm for 5 min to remove washing buffer.
- Cover slides in a histology slide box to prevent fluorescent quenching of signal by light.

3.2.4. Microarray scanning, data processing, and statistical analysis

- Scan microarray with ScanArray Microarray Scanner (PerkinElmer Life Science) following the manufacturer's instructions.
- Fluorescence intensity values for each array spot and its background were calculated using Packard Bioscience's QuantArray software analysis packages or the updated ScanArray Express software. A staining result is considered positive if the mean fluorescent intensity value of microspot is significantly higher than the mean background of an identically stained microarray with the same fluorescent color.
- Microarray data processing and statistical analysis: a number of advanced software packages are available for microarray data normalization, statistical analysis, and pattern-recognition-based advanced data-processing (http://genome-www5.stanford.edu/resources/restech.shtml). We have been using Stanford University's Significance Analysis of Microarrays (SAM, http://www-stat.stanford.edu/~tibs/SAM/) and SAS Institute's JMP-Genomics, Proteomics, and Microarrays (http://www.jmp.com/). It is important to conceptually understand the functions of these bioinformatics tools in order to correctly interpret the results.

3.2.5. Validation of microarray observations

It is important to verify microarray findings by other experimental approaches. We usually confirm our results by at least one of the alternative immunoassays, such as ELISA, dot blot, western blot, flow cytometry, or immunohistology. Detailed examples of such investigations have been described in our recent publications [42, 44]. However, given the fact that microarray assays are highly sensitive, a positive detection in a microarray may not necessarily be reproduced by other assays that are less sensitive than microarray assays. Under such circumstances, we usually repeat the detection by microarray assays to confirm the reproducibility of a positive result. The most important validation is, however, to further characterize the molecular targets that are discovered or suggested by microarray assays in a relevant biological system. This is discussed in the subsequent section.

4. Exploring Lectin-Binding Profiles Using Carbohydrate Microarrays

Carbohydrate microarrays share the technical advantages of miniaturized multiplex binding assays [45–51]. It is suitable for displaying a diverse panel of carbohydrates in a limited amount of chip space. It is "economical" in using carbohydrates since each saccharide is spotted on a microarray substrate in an amount (approximately sub-nanograms) that is drastically smaller than that which is required for conventional molecular or immunological assays (approximately sub-micrograms or more). Its detection sensitivity is also higher than many conventional molecular and immunological assays [43, 45, 48], which is an intrinsic property of the microarray-based miniaturized assays.

Theoretically, the improved detection sensitivity can be attributed to the fact that the binding in a microarray assay meets the so-called "ambient analyte condition" [45]. A key factor for achieving this assay condition is the use of "tiny" amounts of carbohydrate ligand in microarrays. For the microarray platform we discuss here, only sub-nanogram amounts of carbohydrates are immobilized in the microarray substrate. Binding to these miniaturized carbohydrate spots has no or minimum reduction of the concentration of lectins in the solution phase. Under such conditions, the carbohydrate–lectin interactions reach equilibrium rapidly and result in highly sensitive detection of lectins or anti-carbohydrate antibodies.

Thus, carbohydrate microarray is, in principle, well suited for measuring the relative binding reactivities of lectins with a spectrum of diverse carbohydrate structures. A few specific examples are discussed below to illustrate its application in lectin characterization.

4.1. Visualizing carbohydrate-binding profiles using carbohydrate arrays

As summarized in Section 2, the binding property of a lectin is expressed as its specific or selective reactivity with a number of glyco-epitopes that share various degrees of structural similarities or antigenic cross-reactivities. For examples, PHA-L and GS-I are Gal/GalNAc-specific lectins but have significant differences in their fine specificities (Table 2). The binding property of PHA-L is expressed as "Tri-**II** and m**II** clusters > Penta-2,6 > Tri-2,6 > Hepta-3,6 > **II**$_\beta$ > GlcNAcβ1,2Man". The specificity of GS-IA$_4$ is defined as "$F_{penta-} > A_h$ (GalNAcα1 → 2 (LFucα1 → 3)Gal) > GalNAc > **E** > **B** > **I, T** > **L, II**; GSI-B$_4$ isolectin defined as "**B** > **E** > **A**" (Table 2). These lectins may serve as good models for testing the potential of microarray technology in lectin characterization.

Therefore, we constructed a carbohydrate array to display a panel of 22 Gal/GalNAc-containing glycoconjugates for a "proof-of-concept" study. The eight-chamber sub-array system described in the Section 3 above was applied to produce this carbohydrate array. Each of carbohydrate preparations were spotted in four dilutions and with triplicate microspots for each dilution. This design is helpful for the quantitation and statistical analysis of microarray results. It also facilitates the graphical presentation or visualization of the binding profiles of lectins.

We assumed that if this microarray platform reached the sensitivity for measuring the lectin binding and if it reflected well the results of conventional binding assays, we would expect to visualize the lectin-binding profiles by glycan arrays. We applied lectin PHA-L and GS-I at 0.5 µg/ml for microarray staining, which is a titrated condition for cell or tissue section staining using these reagents [44]. The images of the stained microarrays are shown in Fig. 4 (I and II) for lectin PHA-L and GS-I, respectively. The content and location of carbohydrates are listed in the table under the microarray images. A dye marker (Cy5 conjugate) was spotted in triplicates at the bottom right corner of each array.

This assay shows clearly that PHA-L is highly specific for asialoorosomucoid (ASOR) (array location C3) that expresses glyco-epitopes Tri-II and mII. Given that PHA-L has no staining of AGOR (array location A4), which is a D-galactose derivative of ASOR, the terminal non-reducing end Gal-residues are likely key elements for the specificities. In striking contrast to the highly selective binding of PHA-L, lectin GS-I shows a broad spectrum of carbohydrate-binding activities. It is reactive with seven different Gal/GalNAc-containing glycoconjugates. They are Tn-HAS, Asialo-PSM, B-dimer-BSA, Asialo-OSM, Beach ϕOH, Tij20%fr. 2nd 10%, and Beach P1 ϕOH insoluble. These carbohydrate preparations display a number of well-characterized glyco-epitopes, including Tn, T, I, i, B, and Bh.

It is interesting to compare the microarray-binding profile of GS-I with the specificities of its isolectins. GS-I is known as a tetramer consisting of subunit A and subunit B in different ratios. GS-IA_4 and GS-IB_4 have equal binding affinity for α-galactose end groups. However, the A_4 isolectin has a greater affinity for α-GalNAc than the B_4 isoform. The natural product of GS-I is a mixture of the five isolectins, A_4, A_1B_3, A_2B_2, A_3B_1, and B_4. Such molecular diversification may extend the spectrum of carbohydrate-binding profile of GS-I.

Some but not all the GS-I reactivities detected by microarray assay can be directly attributed to its subunit A and B specificities that were determined by previous studies. For example, its binding to Tn-containing carbohydrates (location C1 and C2) reflects the affinity of A subunit for α-GalNAc. Both A and B subunits are likely responsible for the binding to the terminal Gal in B, T, I, and i glyco-epitopes. However, this preparation of GS-I has no binding to cyst 9 (blood group A/location B_4) and Hog$_4$ (blood group A/location D_4). This is unexpected given that its A subunit is specific for the A-glyco-epitopes. It is also

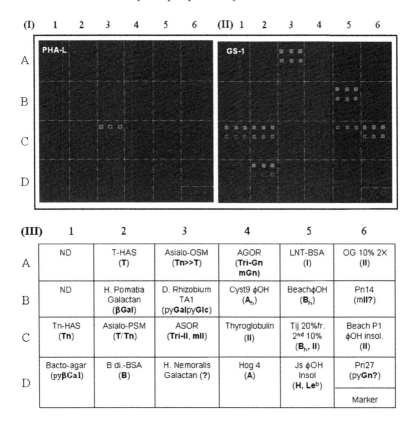

Figure 4. Carbohydrate microarrays for characterization of the carbohydrate-binding profiles of lectins. (I) An image of the microarray stained with PHA-L; (II) an image of the microarray stained with GS-I; and (III) Content and location of carbohydrate antigens in carbohydrate arrays. The carbohydrate units displayed by each carbohydrate preparation are indicated either using the codes listed in Table 1 or with its structural key elements.

not clear why this lectin is strongly reactive with asialo-OSM that displays T-glyco-epitopes but is completely negative to T-HAS. A possible interpretation of this result is that asialo-OSM displays the T-glyco-epitopes in a specific cluster configuration that substantially facilitates the binding to GS-I. By contrast, the T-glyco-epitopes of T-HAS are in the unfavorable configuration for GS-I. It is important to further characterize the specificities of individual GS-I isolectins (A_4, A_1B_3, A_2B_2, A_3B_1, and B_4). Using carbohydrate microarrays of extended structural diversity, in combination with saccharide inhibition or blocking assays, may provide more information to better understand the selective binding to specific cluster of glyco-epitopes.

4.2. Lectins as structural probes for glyco-epitope identification

Lectins and anti-carbohydrate antibodies of defined specificities are useful probes for complex carbohydrates expressed by living organisms. We present, here, an integrated strategy using these probes to facilitate the identification of immunogenic sugar moieties that are expressed by microbial pathogens. This involves three steps of experimental investigation. The first step is to perform carbohydrate microarray characterization of the antibody responses to an infectious agent or antigen preparation in order to recognize the disease- or pathogen-associated anti-carbohydrate antibody specificities. The second step focuses on identification of lectins and/or antibodies that are specific for the glyco-epitopes that are recognized by the pathogen-elicited antibodies. This provides specific structural probes to enable the third step of investigation, that is, to identify the glyco-epitopes in the candidate pathogens using specific lectins or antibodies identified by Steps 1 and 2.

We applied this strategy to explore the glyco-epitopes expressed by a previously unrecognized viral pathogen, SARS–CoV [52–54]. Our rationale was that if SARS–CoV expressed antigenic carbohydrate structures, then immunizing animals using the whole virus-based vaccines would elicit antibodies specific for these structures. In addition, if SARS–CoV displayed a carbohydrate structure that mimics host cellular glycans, then vaccinated animals may develop antibodies with autoimmune reactivity to their corresponding cellular glycans.

By characterizing the SARS–CoV neutralizing antibodies elicited by an inactivated SARS–CoV vaccine, we detected autoantibody reactivity specific for the carbohydrate moieties of an abundant human serum glycoprotein ASOR [44]. This surveillance provides important clues for the selection of specific immunologic probes to further examine whether SARS–CoV expresses antigenic structures that imitate the host glycan. As shown in the Fig. 4 above, lectin PHA-L is specific for glyco-epitopes Tri-**II** or m**II** of ASOR. Using PHA-L as a structural probe, we confirmed that SARS–CoV expresses the PHA-L reactive antigenic structure (see Ref. [44] for details). We, thus obtained, immunological evidence that a carbohydrate structure of SARS–CoV shares antigenic similarity with host glycan complex carbohydrates. This viral component is probably responsible for the stimulation of the autoantibodies directed at a cellular glycan complex carbohydrate.

These observations raise important questions about whether autoimmune responses are in fact elicited by SARS–CoV infection and whether such autoimmunity contributes to SARS pathogenesis. ASOR is an abundant human serum glycoprotein and the ASOR-type complex carbohydrates are also expressed by other host glycoproteins [55, 56]. Thus, the human immune system is generally non-responsive to these "self" carbohydrate structures. However, when similar sugar moieties were expressed by a viral glycoprotein, their cluster configuration could differ significantly from those displayed by a cellular glycan, and in this manner generate a novel "non-self" antigenic structure.

A documented example of such antigenic structure is a broad-range HIV-1 neutralization epitope recognized by a monoclonal antibody 2G12. This antibody is specific for a unique cluster of sugar chains displayed by the gp120 glycoprotein of HIV-1 [57]. It is, therefore, important to examine whether naturally occurring SARS–CoV expresses the Tri-**II** or m**II**-type autoimmune reactive sugar moieties. During SARS epidemic, the viruses replicate in human cells. Their sugar chain expression may differ from the monkey cell produced viral particles. Scanning of the serum antibodies of SARS patients using glycan arrays or other specific immunologic tools may provide information that can shed light on this question.

In summary, the recent establishment of carbohydrate-based microarrays, and especially the availability of different technological platforms to meet the multiple needs of carbohydrate research [42, 58–64], marks an important developmental stage of postgenomic research. It is our prediction that microarray-based broad-range binding assays may substantially extend the scope of current carbohydrate research, including lectin characterization and classification. We have described here a practical platform of carbohydrate microarrays and discussed a few examples to illustrate the application of this technology in lectin characterization. We also present an integrated experimental strategy to facilitate the identification of the glyco-epitopes in biological systems, such as a newly emerged viral pathogen, SARS–CoV. This research strategy is likely applicable for the exploration of complex carbohydrates that are differentially expressed by host cells, including human cancers and stem cells at various stages of differentiation.

References

[1] S.A. Brooks, M.V. Dwek and U. Schumacher, Functional & Molecular Glycobiology, BIOS Scientific Publishers Ltd, Oxford, 2002.

[2] H. Feinberg, D.A. Mitchell, K. Drickamer and W.I. Weis, Science, 294 (2001) 2163–2166.

[3] T. Feizi, Adv. Exp. Med. Biol., 152 (1982) 167–177.

[4] T. Feizi, Trends. Biochem. Sci., 19 (1994) 233–234.

[5] P.R. Crocker and T. Feizi, Curr. Opin. Struct. Biol., 6 (1996) 679–691.

[6] S.I. Patenaude, S.M. Vijay, Q.L. Yang, H.J. Jennings and S.V. Evans, Acta Crystallogr. D. Biol. Crystallogr., 54 (1998) 1005–1007.

[7] S.I. Patenaude, C.R. MacKenzie, D. Bilous, R.J. To, S.E. Ryan, N.M. Young and S.V. Evans, Acta Crystallogr D Biol. Crystallogr., 54 (1998) 1456–1459.

[8] A.M. Wu, S.C. Song, M.S. Tsai and A. Herp, Adv. Exp. Med. Biol., 491 (2001) 551–585.

[9] A.M. Wu and S.J. Sugii, Adv. Exp. Med. Biol., 228 (1988) 205–263.

[10] A.M. Wu, S.C. Song, S. Sugii and A. Herp, Indian J. Biochem. Biophys., 34 (1997) 61–71.

[11] A.M. Wu, Mol. Cell. Biochem., 61 (1984) 131–141.

[12] G.F. Springer, Prog. Allergy, 15 (1971) 9–77.

[13] A. Makita, C. Suzuki, Z. Yosizawa and T. Konno, Tohoku J. Exp. Med., 88 (1966) 277–288.

[14] S. Hakomori and R. Kannagi, J. Natl. Cancer Inst., 71 (1983) 231–251.

[15] A.M. Wu, Adv. Exp. Med. Biol., 228 (1988) 351–394.

[16] W.M. Watkins. In: A. Gottschalk, (Ed.), Glycoproteins, Elsevier, Amsterdam, 1972, pp. 830–890.

[17] W.M. Watkins, Adv. Hum. Genet., 10 (1980) 1–136, 379–385.

[18] F. Maisonrouge-McAuliffe and E.A. Kabat, Arch. Biochem. Biophys., 175 (1976) 71–80.

[19] M. Fukuda and M.N. Fukuda, Biology of glycoproteins, Plenum Press, New York, 1984.

[20] J.M. Wieruszeski, J.C. Michalski, J. Montreuil, G. Strecker, J. Peter-Katalinic, H. Egge, H. Van Halbeek, J.H. Mutsaers and J.F. Vliegenthart, J. Biol. Chem., 262 (1987) 6650–6657.

[21] A.M. Wu, Adv. Exp. Med. Biol., 491 (2001) 55–64.

[22] J.T. Gallagher, Biosci. Rep., 4 (1984) 621–632.

[23] O. Mäkelä, Ann. Med. Exp. Biol. Fenn. Suppl., 1135 (1957) 1–156.

[24] I. Goldstein and R. Poretz, In: I. Liener, N. Sharon and I. Goldstein, (Eds.), The lectins, properties, functions and applications in biology and medicine, Academic Press, New York, 1986, pp. 33–247.

[25] R.T. Lee and Y.C. Lee, Glycoconj. J., 17 (2000) 543–551.

[26] Y.C. Lee, FASEB J., 6 (1992) 3193–3200.

[27] A.M. Wu, J.H. Wu, A. Herp and J.H. Liu, Biochem. J., 371 (2003) 311–320.

[28] A.M. Wu, J.H. Wu, M.S. Tsai, H. Kaltner and H.J. Gabius, Biochem. J., 358 (2001) 529–538.

[29] A. Wu, J. Wu, L. Lin, S. Lin and J. Liu, Life Sci., 72 (2003) 2285–2302.

[30] A.M. Wu, J.H. Wu, J.H. Liu and T. Singh, Life Sci., 74 (2004) 1763–1779.

[31] A.M. Wu, FEBS Lett., 562 (2004) 51–58.

[32] A.M. Wu, J.H. Wu, J.H. Liu, T. Singh, S. Andre, H. Kaltner and H.J. Gabius, Biochimie, 86 (2004) 317–326.

[33] T. Singh, U. Chatterjee, J.H. Wu, B.P. Chatterjee and A.M. Wu, Glycobiology, 15 (2005) 67–78.

[34] T. Singh, J.H. Wu, W.J. Peumans, P. Rouge, E.J. Van Damme, R.A. Alvarez, O. Blixt and A.M. Wu, Biochem. J., 393 (2006) 331–341.

[35] A.M. Wu, T. Singh, J.H. Wu, M. Lensch, S. Andre and H.J. Gabius, Glycobiology, 16 (2006) 524–537.

[36] W.J. Peumans, E.J. Van Damme, A. Barre and P. Rouge, Adv. Exp. Med. Biol., 491 (2001) 27–54.

[37] W. Peumans, A. Barre, G. Hao, P. Rouge and E. Van Damme, Trends Glycosci. Glycotechnol., 12 (2000) 83–101.

[38] A. Makita, C. Suzuki and Z. Yosizawa, J. Biochem., 60 (1966) 502–513.

[39] A.M. Wu, A. Herp, S.C. Song, J.H. Wu and K.S. Chang, Life Sci., 57 (1995) 1841–1852.

[40] B. Nilsson, N.E. Norden and S. Svensson, J. Biol. Chem., 254 (1979) 4545–4553.

[41] R.G. Spiro and V.D. Bhoyroo, J. Biol. Chem., 249 (1974) 5704–5717.

[42] D. Wang, S. Liu, B.J. Trummer, C. Deng and A. Wang, Nat. Biotechnol., 20 (2002) 275–281.

[43] D. Wang, Proteomics, 3 (2003) 2167–2175.

[44] D. Wang and J. Lu, Physiol. Genomics, 18 (2004) 245–248.

[45] R. Ekins, F. Chu and E. Biggart, Ann. Biol. Clin., 48 (1990) 655–666.

[46] A. Lueking, M. Horn, H. Eickhoff, K. Bussow, H. Lehrach and G. Walter, Anal. Biochem., 270 (1999) 103–111.

[47] G. MacBeath and S.L. Schreiber, Science, 289 (2000) 1760–1763.

[48] D. Stoll, M.F. Templin, M. Schrenk, P.C. Traub, C.F. Vohringer and T.O. Joos, Front. Biosci., 7 (2002) C13–C32.

[49] P.J. Utz, Immunol. Rev., 204 (2005) 264–282.

[50] W.H. Robinson, C. DiGennaro, W. Hueber, B.B. Haab, M. Kamachi, E.J. Dean, S. Fournel, D. Fong, M.C. Genovese, H.E. de Vegvar, K. Skriner, D.L. Hirschberg, R.I. Morris, S. Muller, G.J. Pruijn, W.J. Van Venrooij, J.S. Smolen, P.O. Brown, L. Steinman and P.J. Utz, Nat. Med., 8 (2002) 295–301.

[51] W.H. Robinson, P. Fontoura, B.J. Lee, H.E. de Vegvar, J. Tom, R. Pedotti, C.D. DiGennaro, D.J. Mitchell, D. Fong, P.P. Ho, P.J. Ruiz, E. Maverakis, D.B. Stevens, C.C. Bernard, R. Martin, V.K. Kuchroo, J.M. Van Noort, C.P. Genain, S. Amor, T. Olsson, P.J. Utz, H. Garren and L. Steinman, Nat. Biotechnol., 21 (2003) 1033–1039.

[52] R.A. Fouchier, T. Kuiken, M. Schutten, G. Van Amerongen, G.J. Van Doornum, B.G. Van Den Hoogen, M. Peiris, W. Lim, K. Stohr and A.D. Osterhaus, Nature, 423 (2003) 240.

[53] T.G. Ksiazek, D. Erdman, C.S. Goldsmith, S.R. Zaki, T. Peret, S. Emery, S. Tong, C. Urbani, J.A. Comer, W. Lim, P.E. Rollin, S.F. Dowell, A.E. Ling, C.D. Humphrey, W.J. Shieh, J. Guarner, C.D. Paddock, P. Rota, B. Fields, J. DeRisi, J.Y. Yang, N. Cox, J.M. Hughes, J.W. LeDuc, W.J. Bellini and L.J. Anderson, N. Engl. J. Med., 348 (2003) 1953–1966.

[54] P.A. Rota, M.S. Oberste, S.S. Monroe, W.A. Nix, R. Campagnoli, J.P. Icenogle, S. Penaranda, B. Bankamp, K. Maher, M.H. Chen, S. Tong, A. Tamin, L. Lowe, M. Frace, J.L. DeRisi, Q. Chen, D. Wang, D.D. Erdman, T.C. Peret, C. Burns, T.G. Ksiazek, P.E. Rollin, A. Sanchez, S. Liffick, B. Holloway, J. Limor, K. McCaustland, M. Olsen-Rassmussen, R. Fouchier, S. Gunther, A.D. Osterhaus, C. Drosten, M.A. Pallansch, L.J. Anderson and W.J. Bellini, Science, 300(5624) (2003) 1394–1399.

[55] R.D. Cummings and S. Kornfeld, J. Biol. Chem., 259 (1984) 6253–6260.

[56] F. Pacifico, N. Montuori, S. Mellone, D. Liguoro, L. Ulianich, A. Caleo, G. Troncone, L.D. Kohn, B. Di Jeso and E. Consiglio, Mol. Cell. Endocrinol., 208 (2003) 51–59.

[57] D.A. Calarese, C.N. Scanlan, M.B. Zwick, S. Deechongkit, Y. Mimura, R. Kunert, P. Zhu, M.R. Wormald, R.L. Stanfield, K.H. Roux, J.W. Kelly, P.M. Rudd, R.A. Dwek, H. Katinger, D.R. Burton and I.A. Wilson, Science, 300 (2003) 2065–2071.

[58] W.G. Willats, S.E. Rasmussen, T. Kristensen, J.D. Mikkelsen and J.P. Knox, Proteomics, 2 (2002) 1666–1671.

[59] F. Fazio, M.C. Bryan, O. Blixt, J.C. Paulson and C.H. Wong, J. Am. Chem. Soc., 124 (2002) 14397–14402.

[60] S. Fukui, T. Feizi, C. Galustian, A.M. Lawson and W. Chai, Nat. Biotechnol., 20 (2002) 1011–1017.

[61] B.T. Houseman and M. Mrksich, Chem. Biol., 9 (2002) 443–454.

[62] S. Park and I. Shin, Angew. Chem. Int. Ed. Engl., 41 (2002) 3180–3182.

[63] E.W. Adams, D.M. Ratner, H.R. Bokesch, J.B. McMahon, B.R. O'Keefe and P.H. Seeberger, Chem. Biol., 11 (2004) 875–881.

[64] G.T. Carroll, D. Wang, N.J. Turro and J.T. Koberstein, Langmuir, 22 (2006) 2899–2905.

Lectins: Analytical Technologies
C.L. Nilsson (Editor)

Chapter 8

Glycoproteomics Based on Lectin Affinity Chromatographic Selection of Glycoforms

Fred E. Regnier[a], Kwanyoung Jung[a], Stephen B. Hooser[b, c] and Christina R. Wilson[b, c]

[a]*Department of Chemistry, Purdue University, West Lafayette, IN, USA*
[b]*Department of Comparative Pathobiology, Purdue University, West Lafayette, IN, USA*
[c]*Animal Disease Diagnostic Laboratory, Purdue University, West Lafayette, IN, USA*

1. Proteomic Strategies

Proteome complexity generally exceeds the analytical capacity of instrumentation by an order of magnitude or more. How to deal with this has been the focus of a great deal of the proteomics literature. One school of thought is that the solution is higher resolution mass spectrometers, i.e. to "put everything in the mass spectrometer and let it and the data system sort it out." This would be wonderful but it is both very costly and ignores the fact that as sample complexity increases so does the difficulty of getting peptides into the gas phase. Relatively non-polar peptides in mixtures suppress the ionization of polar peptides in electrospray ionization (ESI) mass spectrometry [1], while in matrix-assisted laser desorption ionization (MALDI), mass spectrometry positively charged species suppress the ionization of neutral and negatively charged peptides [2]. Even in the tryptic digest of a single protein, differences in the ionization efficiency of peptides are sufficiently large that one seldom sees all the components. Ultrahigh resolution instruments are not useful if the requisite proteins and peptides needed to study a problem do not get into the mass spectrometer. At the opposite extreme is the concept that we need much higher resolution of components before they enter the mass spectrometer [3]. While it is true this approach would diminish matrix effects, it too has negative features. Among them are the need for sophisticated separation systems with huge numbers of theoretical plates, the accompanying long analysis times, and the need to sort through spectral data from 10^5 to 10^6 components to find a small number of relevant peptides.

The fallacy in both of these approaches is the idea that one needs to examine an entire proteome to answer biological questions. In many cases this will not be necessary. This chapter explores the use of structure-specific affinity selection as a way to look at a portion of a proteome. This can be done in several ways. One is to derivatize a specific functional group in proteins with an affinity selectable reagent that can be targeted with an affinity chromatography column. Avidin selection of proteins or peptides in which sulfhydryl or carbonyl groups have been derivatized with biotin is an example [4, 5]. A second approach is to directly select proteins or peptides based on shared structural features. For instance, direct selection of histidine containing peptides with copper-loaded immobilized metal affinity chromatography (Cu^{2+}-IMAC) columns [6]. Selection of phosphorylated peptides with either Fe^{3+}-IMAC or Ga^{3+}-IMAC columns is another [7]. An additional powerful approach is to select a subset of proteins which have a common post-translational modification (PTM) related to the disease being studied. For example, lectins can be used to select for specific protein glycosylation associated with inflammation or cancer. Taken together there are more than a dozen ways to target structural features in proteomics [8].

2. Recognizing Post-Translational Modifications

Most proteins undergo some type of PTM during their biological life. Among some of the important roles of PTMs are in cell signaling, generation of active forms of proteins, coding proteins for transport to specific compartments, and in marking polypeptides for degradation. For all of these reasons, the formation, fate, and role of post-translationally modified proteins in cellular regulation are major issues in proteomics. There are several ways to carry out global studies of all PTMs of a particular class. Peptides or proteins bearing a particular modification often have a unique mass spectral fragmentation signature that is relatively easily recognized, such as in the case with phosphorylation. The disadvantage of this approach is the need for MS/MS analysis of every peptide in the sample to recognize a small number that are phosphorylated, unless the phosphorylated peptides have been selected and enriched as mentioned above.

A second, much more difficult type of PTM to recognize is one that varies in structure as is the case with protein oxidation or glycosylation. There are more than a dozen different ways proteins can be oxidatively modified [9]. Lectin arrays indicate that glycosylation is even more complex [10]. It is generally the case that both the type and number of monosaccharide residues appearing in the glycan appended to a particular site on a polypeptide are variable. Commonly one finds 10–50 glycoforms at a single site on a protein [11]. Whereas polypeptide structure is determined by a single mRNA template, glycan structure is determined by the sequential addition and trimming of sugars from the glycan by a series of glycosyltransferases and glycosidases. Consequently, glycans can vary substantially in structure [12].

3. Lectin Selectors

The study of glycans and glycoproteins has been greatly facilitated by the fact that a number of protein lectins exist in nature that recognize and bind to both mono- and oligosaccharides. When immobilized and used in affinity chromatography columns, lectins can be a valuable asset in glycoprotein purification. Although much of the seminal work on lectin based affinity chromatography occurred before the advent of proteomics, the power of lectin affinity chromatography in glycoproteomics is beginning to emerge. The discussion below will therefore focus first on work from the pre-proteomics era and then on the use of lectins in modern glycoproteomics.

The name lectin was derived from the Latin term "lectus" – meaning to select. Although lectins bind to monosaccharides, they generally bind with much higher affinity to complex oligosaccharides with specific structural determinants. Lectins that bind preferentially to fucose, galactose, and mannose containing glycans are shown in Tables 1–3, respectively. The monosaccharide structure code for these tables is shown at the bottom of Table 1. Naturally, different glycans appear with varying frequency in body fluids. For instance, recent studies in our laboratory, using single lectin affinity selection, have shown that concanavalin A (Con A), *Lotus tetragonolobus* agglutinin (LTA) and Jacalin select for 11%, 4%, and 0.5% of the total protein from human serum, respectively.

Although lectin affinity columns are highly specific, it is important to point out that non-glycosylated proteins can also bind to columns during lectin affinity chromatography [13, 14]. Non-specific binding is generally of a weak

Table 1. Lectins that bind to glycans containing fucose.

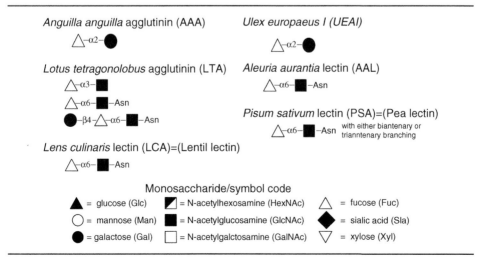

Anguilla anguilla agglutinin (AAA)

Ulex europaeus I (UEAI)

Lotus tetragonolobus agglutinin (LTA)

Aleuria aurantia lectin (AAL)

Pisum sativum lectin (PSA)=(Pea lectin)
with either biantenary or trianntenary branching

Lens culinaris lectin (LCA)=(Lentil lectin)

Monosaccharide/symbol code

▲ = glucose (Glc)   ▨ = N-acetylhexosamine (HexNAc)   △ = fucose (Fuc)

◯ = mannose (Man)   ■ = N-acetylglucosamine (GlcNAc)   ◆ = sialic acid (Sla)

● = galactose (Gal)   □ = N-acetylgalctosamine (GalNAc)   ▽ = xylose (Xyl)

Table 2. Lectins that bind to glycans containing galactose.

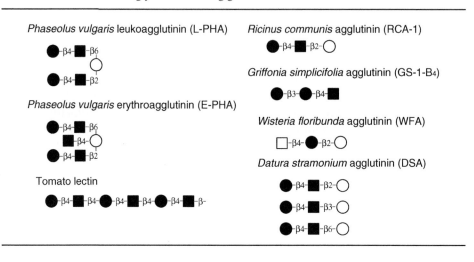

Table 3. Lectins that bind to glycans containing mannose.

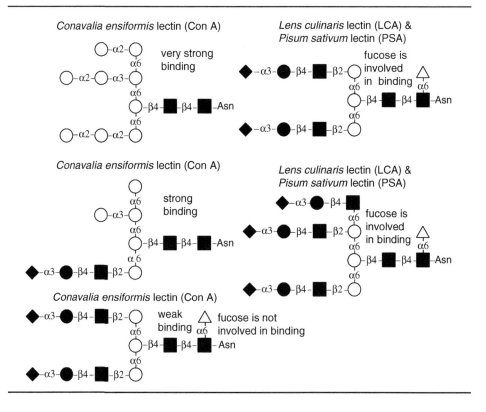

hydrophobic nature and can occur on either the chromatographic support matrix or the immobilized lectin. Because glycoproteins and glycopeptides can be eluted from affinity columns with small hydrophilic sugar displacers without disturbing the binding of non-glycosylated proteins, contamination of eluted fractions with non-glycosylated proteins tends not to be a problem. However, non-specifically bound proteins should be removed from columns before another sample is applied because they reduce the binding capacity of affinity columns. While it is known that serum albumin is generally not enzymatically glycosylated, non-specific binding of albumin to the chromatographic support matrix during lectin affinity selection of serum has previously been a concern. However, our investigations using Con A, *L. tetragonolobus* agglutinin, and Jacalin affinity selection of glycoproteins from human serum indicate that only trace amounts of albumin are retained (unpublished data).

4. Serial Lectin Affinity Chromatography

The specificity with which lectins recognize glycans can be both a blessing and a curse. The blessing is that specific structural types can often be targeted for selection. The curse is that no single lectin has sufficient breath of selectivity to cover all the features one would like to examine in a glycoproteome. Cummings and Kornfeld [15] recognized that a solution to this dilemma was to apply lectin chromatography columns in series in what has come to be known as serial lectin affinity chromatography (SLAC). They used Con A to capture mannose rich *N*-glycosylated peptides derived from the proteome of mouse lymphoma BW5147 cells and then further selected with *Datura stramonium* agglutinin and *Phaseolus vulgaris* leukoagglutinin affinity columns. These two lectins bind to the repeating *N*-acetyllactosamine sequence at the periphery of bi-, tri-, and tetra-antennary Asn-linked oligosaccharides. Up to two thirds of the tetra-antennary Asn-linked glycans selected contained repeating *N*-acetyllactosamine sequences, whereas 50% and 10% of tri- and bianterrary glycans, respectively, were thus glycosylated. Being able to select glycoproteins and glycopeptides based on sets of unique structural features contained in their glycans is a very distinct advantage in proteomics, particularly when those structural features are associated with a particular biological phenomenon. This is the great value of this SLAC.

Using SLAC with broad selectivity lectins has made it possible to select general classes of glycoproteins, such as those in which glycans are attached to asparagine (*N*-glycosylated) or serine/threonine (*O*-glycosylated). Con A, for example, is a broad selectivity lectin that selects glycans with a mannose rich core as would be the case in many *N*-glycosylated proteins [16, 17]. When Con A selection is coupled to selection with either a lentil lectin (LCA), *Ricinus communis* agglutinin (RCA-1) or wheat-germ agglutinin (WGA) affinity column, a still broader set of glycans is captured. LCA binds to mannose rich glycans that also contain fucose (Table 3) while RCA-1 binds to Gal-β4-GlcNAc-β2- Man rich glycans (Table 2), and WGA binds to N-acetylglucosamine rich glycans. Predominantly N-glycosylated proteins and peptides are selected by these lectins although WGA does select some O-glycosylated (alkali-labile) species. Using SLAC with broad selectivity lectins has made it possible to select general classes of glycoproteins, such as those in which glycans are attached to asparagine (N-glycosylated) or serine/threonine (O-glycosylated).

Glycoproteins and glycopeptides that pass through a Con A [18] and E-PHA SLAC column set are largely O-glycosylated. Species in this unbound, eluted fraction carrying the Gal-β3-GalNAc- sequence can be selected with Jacalin [19]. The Gal-β3-GalNAc- sequence occurs widely in O-glycosylated proteins and peptides, but not in all. It is important to note that Jacalin also binds to some N-glycosylated polypeptides and should only be used in the selection of O-glycosylated species after the removal of N-glycosylated proteins and peptides with a Con A and E-PHA column set. Knowing this, SLAC has been less widely used in the study of O-glycosylated proteins.

An important use of SLAC has been in the recognition of aberrations in glycosylation with cancer. Changes in the glycosylation of asparagine-linked oligosaccharides have been shown in various tumor cells, including human colon cancer [20]. SLAC employing immobilized Con A, D. stramonium agglutinin, and tomato lectin showed that highly metastatic carcinoma cells express more poly-N-acetyllactosaminyl side chains with branched galactose residues than cells with low metastatic potential. In the case of prostate carcinoma, asparagine (N)-linked sugar-chain structures of prostate-specific antigen (PSA) were investigated using Con A and E-PHA in tandem. It was seen that protein(s) with branched N-acetylglucosamine-β4-mannose glycans increased in concentration in association with disease progression [21]. Sialylation can vary in cancer patients as well. Using WGA, Elderberry lectin (SNA), and *Maackia amurensis* lectin (MAL) to differentiate between α(2,3)- and α(2,6)-sialylations, it has been shown that both the type of sialic acid linkage and amount of sialylation can vary with different types of cancer [22]. Aberrations in fucosylation have also been associated with certain types of cancer [23]. The literature [24] indicates that α(1→6) fucosylation has been detected in cultured neoplastic cells [25], human tumors [26], and secreted host glycoproteins [27]. Fucosylation sites on the surface of cancer cells have been shown to interact with the adhesion molecule selectin on the vascular endothelium [28]. In breast and ovarian cancer patients, an isoform of α-1-protease inhibitor with increased fucosylation was found to be a marker of unresponsiveness to chemotherapy [29]. This is also the case in human prostate carcinoma [30] and lymphosarcoma in canines [31]. Roughly 50 fucosylated proteins were elevated in concentration during disease progression in the lymphosarcoma studies, most of which returned to normal following chemotherapy. Two of the fucosylated proteins identified, CD44 and E-selectin, are known to be involved in cell adhesion and cancer cell migration.

5. Proteomic Strategies Based on Lectin Selection

The strategy chosen for a proteomics analysis is generally driven by the objective of the study. Proteome cataloging, identification of glycosylation sites, glycoform analysis, expression analysis, monitoring temporally dependent changes in glycosylation, and glycan sequencing have each been the focus of recent studies, each using different analytical methods. One of the key differentiators in lectin-based glycoproteomics is whether lectin selection is being achieved at the glycoprotein (Fig. 1) or glycopeptide (Fig. 2) level. The rationale for these two selection strategies will be developed below.

Concern has been expressed at recent scientific conferences that polypeptide structure might contribute to lectin selection. If true, some glycoproteins could be missed when glycoprotein identification is based on lectin selection of glycopeptides alone. It is important to note however that there is no evidence in the affinity chromatography literature to support this fear. In fact, the "glycol-catch" method pioneered by Hirabayashi et al. [32] for the study of the *C. elegans* glycome uses lectin selection of glycopeptides to identify glycans. As more studies are published the question of the involvement of polypeptide structure in lectin binding will be answered.

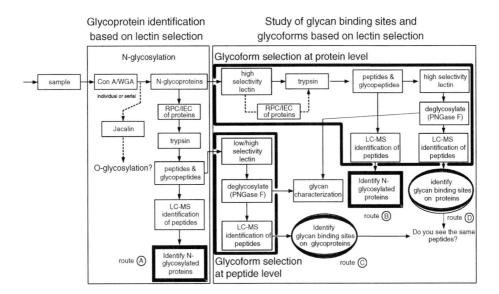

Figure 1. Glycoproteomics based on affinity chromatography at the protein level with immobilized lectin columns.

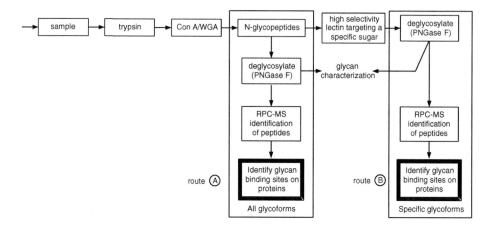

Figure 2. Glycoproteomics based on affinity chromatography at the glycopeptide level with immobilized lectin columns.

5.1. Proteome cataloging

The objective in cataloging is generally to obtain a wide definition of a glycoproteome. One way to do this is through the use of broad selectivity lectins in the first separation dimension as seen in Fig. 1, route A. Con A and WGA are suggested in the figure for the selection of *N*-glycosylated proteins, but E-PHA, RCA-1, or LCA could have been used instead of WGA [33]. Jacalin can then be used to select *O*-glycosylated proteins carrying the Gal-α6- or Gal-β3-GalNAc- determinant from the Con A/WGA flow through fraction [34]. As noted above, Gal-β3-GalNAc- occurs widely in *O*-glycosylated proteins, but not in all *O*-glycans.

N-glycoproteins selected by broad selectivity lectins can be identified in a number of ways. One of the better ways is to further resolve the glycoprotein mixture through either reversed phase chromatography (RPC) or ion exchange chromatography (IEC) as illustrated in the first panel entitled "glycoprotein identification based on lectin selection." It is critical in the case of protein RPC that one uses either a wide pore (>30 nm) or non-porous silica-based packing material to assure high recovery [35]. The advantage of non-porous (NP) particles is that separations can be achieved faster and higher resolution because they have no stagnant mobile mass transfer limitations. But care must be taken not to overload NP sorbents. They are of much lower surface area and loading capacity can be 100 times lower than porous sorbents. Protein peak capacities with state-of-the-art RPC columns are roughly 100.

Assuming that 1,000 glycoproteins were selected from a sample by the broad selectivity Con A/WGA lectin column in Fig. 1 (route A) and that these proteins were further resolved on an RPC column with a peak capacity of 100, roughly 10 glycoproteins would be eluting in each RPC peak. Collection and tryptic digestion of these RPC fractions would produce approximately 300 peptides per fraction assuming the average glycoprotein produces 30 tryptic peptides on proteolysis. The assumptions being made can be adjusted to fit any particular system. One has the option of either trying to identify glycoproteins directly from these tryptic digests by MALDI-MS or further resolving them by another dimension of RPC-MS. The later is the best, but most time consuming.

The need to use macro-porous or non-porous columns is similar with IEC. Because glycoproteins often vary in sialic acid content, IEC columns may fractionate glycoforms of a protein. Although this is an advantage in looking at glyco-forms, it is undesirable in cataloging.

Glycoprotein identification in this approach will be achieved most easily through the identification of non-glycosylated peptides. Tryptic digests of protein fractions from the RPC or IEC column contain a mixture of both non-glycosylated and glycosylated peptides in a ratio of roughly 30 to 1 based on the assumption that the average glycoprotein yields 30 tryptic peptides and has one glycosylation site. Obviously there are exceptions that deviate widely from this rule. In both ESI and MALDI, non-glycosylated peptides are much easier to ionize than glycopep-tides. Also, the mass of glycopeptides will in general be greater than 2,000 atomic mass units (amu). This, along with the fact that the mass and structure of the gly-can are generally unknown, makes it difficult to identify glycoproteins through their glycopeptides.

Glycopeptides and glycosylation sites are most easily identified after deglyco-sylation [31]. In this approach (1) all glycopeptides are globally selected from tryptic digests of glycoproteins with a broad selectivity lectin set, (2) glycans are removed from peptides with PNGase F, and (3) glycosylation site peptides are identified by MALDI-MS or ESI-MS as illustrated in route C of Fig. 1. In this process, PNGase F converts the asparagine residue at the glycosylation site to an aspartate residue. Executing the PNGase F hydrolysis in $H_2^{18}O$ specifically labels new aspartate residues formed during hydrolysis, facilitating identification of the amino acid residue to which the glycan was attached [36]. This scheme for identi-fication of glycosylation sites is similar to the one described in Fig. 2, but has the great advantage that the glycoprotein mixture was fractionated into a hundred frac-tions before site analysis. This means that on average glycopeptide mixtures will be 100-fold simpler than the one in Fig. 2. Unfortunately, it will also take much longer to examine samples. Although probably the most definitive approach, it will also take the longest.

The basic premise in the SLAC approach to proteomics is that a broad selectivity lectin set is used to select all glycoforms while a narrow selectivity lectin is used in tandem to select glycoforms. Based on the concern that polypeptide structure could play a role in lectin selection, glycoform selection can be a carried out as illustrated in route B of Fig. 1. In the case of a complex glycoproteome, it is best to further frac-tionate the protein mixture from the affinity columns by either RPC or IEC. From this point on glycoprotein identification in route B is identical to route A.

Glycosylation site identification is achieved in this method by again using the high selectivity lectin (route D in Fig. 1), but in this case to select glycopeptides instead of glycoproteins. Glycosylation site identification from this point on is identical to that in route C. However, it is important to note that the sites identified will depend on the selectivity of the lectin(s) in the selection process. Use of a narrow selectivity lectin will result in fewer glycosylation sites being identified in route D.

An alternative approach to protein cataloging is to affinity select at the peptide level as is shown in Fig. 2. The first step in this approach is to trypsin digest the proteome. Glycopeptides are then selected with the same broad selectivity lectin set used in Fig. 1. This produces a peptide mixture that has at least 30-fold fewer peptides than the one in Fig. 1. Because of the problems noted above in directly identifying glycopeptides, the mixture of glycopeptides selected by the SLAC column(s) must be deglycosylated with PNGase F and the resulting peptides identified by RPC/MS. Again, the reaction can be carried out in $H_2^{18}O$ to label the aspartic acid residue formed in the hydrolysis of the asparagine residue to which the glycan is attached. The analytical protocol for broadly identifying all *N*-glycosylated proteins and their glycosylation sites is illustrated in route A of Fig. 2. Route B is used in the identification of proteins and glycosylation sites selected by high specificity lectins that target a particular glycoform.

Again assuming a mixture of 1,000 glycoproteins as above, each with a single glycosylation site, 1,000 glycopeptides would be produced on proteolysis and selected by the broad selectivity lectin column in route A of Fig. 2. After deglycosylation the resulting mixture of 1,000 peptides would be further fractionated by RPC column before MS identification. But RPC columns generally have a peak capacity of 300 or more with peptides as opposed to 100 with proteins. This means that each RPC peak would contain 3 peptides on average, 100 times fewer than in route A of Fig. 1.

This approach has many advantages. One is the much greater simplicity of the peptide mixtures being identified. Another is that both the glycoprotein and sites of glycosylation are being identified simultaneously with a single peptide. Still another is that abundant proteins are represented by one or two peaks and have less chance to mask low abundance proteins. The disadvantage of this method is that proteins are being identified based on one, or at most two peptides. Due to matrix suppression of ionization (MSI), it is generally the case that some peptides in a mixture are not seen. This means that proteins represented by single peptide are more likely to be missed. Although much simpler than the method illustrated in Fig. 1, this method is probably more likely to give false negatives.

5.2. Identification of glycosylation sites

Protocols for the identification of glycosylation sites and their relative merits have been outlined above. The procedure illustrated in Fig. 2 is the simplest, but most prone to give false negatives as noted.

5.3. Glycoform analysis

It has been illustrated in Figs. 1 and 2 that the best way to identify and differentiate between glycoforms with lectins is through the use of lectins with narrow selectivity that select glycans with a specific structural determinant. Although Figs. 1 and 2 suggest the use of broad and narrow selectivity lectins in series, there is nothing to preclude a very specific lectin being used alone in the first step of the analysis if it is the objective to find all proteins that carry a very specific glycan structure. Single lectin selection has been used in the search for fucosylated proteins in cancer patient blood samples [31].

5.4. Quantification of glycoproteins and glycopeptides

Strategies used in the quantification of glycoproteins are the same as those used to determine other proteins, as will be described below. Because most of what is being discussed here focuses on LC/MS-based proteomics using ESI, methods relating to this approach will be most extensively addressed. These methods can be grouped into two classes: (1) direct comparison of peak intensities in the mass spectrometer and (2) stable isotope labeling methods.

It would seem that a good way to determine peptide concentration in samples would be through the ion current they produce during mass spectral analysis. But unfortunately, peptides vary in the degree to which they ionize due to MSI [37]. It is generally agreed that MSI is the result of competition between peptides for space at the surface of electrospray droplets [38]. When the concentration of peptides in a droplet is larger than the available surface area, the droplet surface is dominated by more hydrophobic peptides. This means that hydrophobic peptides are more likely to be ionized and they can suppress the ionization of other peptides by 10- to 1,000-fold [39]. Put another way, the ionization efficiency of any peptide depends on (1) the presence of other, unknown peptides in samples and (2) the total concentration of peptides in the electrospray droplet. Arguments in favor of using peptide peak intensities to estimate relative concentration between sequential analyses are that (1) tryptic digests of a proteome are so complicated they are in fact constant in complexity and MSI is not really a problem, (2) the number of peptides that vary between disease state and control samples is so small they will not alter MSI, and (3) the differences in peptide concentration needed to recognize biological differences are far, far larger than variations caused by MSI. Software for direct quantification from MS ion current is available on most new mass spectrometers. In light of the literature in MSI [37–39], the accuracy of this method remains to be determined.

A better way to deal with quantification errors resulting from MSI is to use some type of stable isotope labeling or coding as it is also known [40]. The concept in stable isotope coding is to synthesize a heavy isotope version of the molecule one wishes to quantify and add it to mixture being examined at a known concentration.

During mass spectral analysis the relative concentration of standard and analyte are easily recognized. Because the molecule being determined and the isotopically coded version of the molecule show identical ionization efficiency, MSI effects are negated. Relative standard deviation in quantification with isotope coding methods seems not to exceed 6–8%, regardless of the number of steps in the analysis [41]. This is an important issue in protocols involving a lot of steps as in Fig. 1. Although the procedure described above is for a single analyte, this simple procedure has evolved in proteomics to the point that large numbers of internal standards are being generated simultaneously [42]. Proteins and peptides in control samples and experimental samples are being differentially labeled on a global scale either in vivo or in vitro, then mixed, and compared in a single, or small number of LC/MS analyses. The great advantage of this approach is that you don't have to know the structure of species being labeled and compared in order to quantify changes in concentration among hundreds of proteins. In fact, large changes in the concentration of a protein as a result of experimental stimuli are often used to target a protein for identification.

It is easy to see that one simple way to globally label all analytes would be by biosynthetic labeling during the course of growth and development. This in vivo metabolic labeling of an experimental sample is achieved either by metabolic incorporation of a stable isotope tagged amino acid or by feeding an animal a dietary protein source, such as a microorganism in which all the proteins were uniformly stable isotope labeled during their growth and development [43]. Control samples in contrast are not heavy isotope labeled. In this example, control and experimental samples are then mixed and subjected to any proteomic analysis in which the stable isotope ratio can be determined in a final step by mass spectrometry. The advantage of this approach is that labeling is done in vivo under physiological conditions. The disadvantage is that it cannot be applied to humans or to the analysis of samples that have been previously collected.

In vitro labeling, in contrast, is carried out after the biological phase of an experiment is finished. Control samples would be labeled with one isotopomer while experimental samples would be labeled with another. Among the three different in vitro stable isotope coding approaches, the AQUA method is the nearest mimic to the old stable isotope coding methods described above. In the AQUA approach, stable isotope labeled peptides, structurally identical to what one expects to find in a proteome digest, are synthesized and added to samples at a known concentration [44]. When the peptide isotopomers in the sample coelute into the mass spectrometer the relative molar response of the instrument to the peptide isoforms is identical, irrespective of the amount of MSI. The concentration ratio between the differentially coded internal standards and sample peptides determined by the mass spectrometer allows the absolute concentration of sample peptides to be calculated.

In vitro labeling can also be carried out by derivatization as opposed to peptide synthesis. In this case, the requisite stable isotope coding agent is added to either proteins or proteolytic fragments of proteins by derivatization of a reactive functional group. For example, sulfhydryl groups are derivatized in the isotopically coded affinity tagging (ICAT) method [45]. The derivatizing agent is this case is an iodoacetamide derivative of biotin that reacts with sulfhydryl groups on either sulfhydryl-containing peptides or proteins. Control samples are derivatized with one biotinylated isotopomer while experimental samples are derivatized with a second, heavier isoform. After mixing the control and experimental samples, they would be lectin selected as shown in Fig. 1 (route A) and then further fractionated by RPC. After trypsin digestion, biotinylated peptides would be selected from the trypsin digest by avidin affinity chromatography. Quantification would be achieved by RPC-MS as in route A (Fig. 1). The great advantage of the ICAT approach here is that the complexity of the peptide mixture would be reduced at least five- to sixfold. In the 1,000 glycopeptide example, one would expect 50–60 peptides per fraction instead of 300. The isotope ratio is directly proportional to the relative concentration of peptides and proteins in samples. This approach could be used in most protein based selection methods described in Fig. 1. Because glycopeptides might not contain cysteine, the ICAT method would not be as useful in the glycopeptide based selection methods described in Fig. 2.

The limitation of labeling rare amino acids as a route to protein quantification is that only a portion of all peptides will be isotopically coded. There are global methods such as GIST [46], iTRAQ [47], and ^{18}O labeling [48] that derivatize all proteolytic fragments derived from a proteome. These methods exploit the fact that during proteolysis amino and carboxyl groups are globally generated on all peptide cleavage fragments. After proteolysis these functional groups can be labeled. In the ^{18}O tagging method the proteolytic digest is generally dried and $H_2^{18}O$ is added along with the enzyme originally used in the proteolysis. Because serine proteases covalently bind their peptide cleavage fragments reversibly, they will incorporate two moles of ^{18}O into the carboxyl group of all peptides in the digest except the C-terminus of the protein. This allows differential coding with a mass difference of 4 amu [49].

GIST [46] and iTRAQ [47] are amine coding methods. These reagents are isotopically coded acids with *N*-hydroxysuccinimide (NHS) activated carboxyl groups. Carboxyl groups activated in this way readily react with the α-amino group of peptides and the ε-amino group of lysine to form amide bonds [50]. The iTRAQ reagent is unusual in that it allows quantification in up to four samples at a time (Fig. 3). This is achieved by having four different forms of the reagent that have the same molecular weight but fragment in an isotopically distinct (sample specific) pattern during tandem mass spectrometery (MS/MS). This means that each sample source can have an isotopically different and recognizable set of ions in the MS/MS dimension of analysis. But there are now a variety of coding agents that exploit the derivatization of primary amino groups. Any of these global coding agents can be used for quantification with the methods described in Figs. 1 and 2.

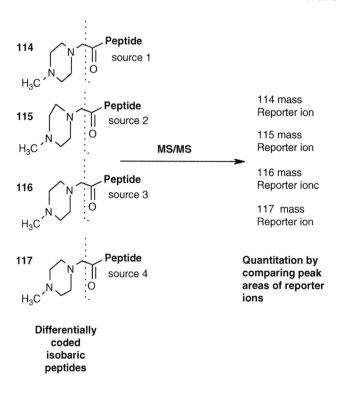

Figure 3. Peptides from different sources differentially coded with four different isotopomers of the iTRAQ isotopic labeling agent.

5.5. Expression analysis

Protein glycosylation and glycan processing is a complex process that is poorly understood. The critical role glycoproteins play in the cell cycle along with the fact that protein glycosylation seems to become aberrant in a broad range of diseases is stimulating the need to know more about how expression and PTM of glycoproteins is regulated. Based on the fact that expression and PTM are regulated in different ways, it is necessary to have analytical methods that differentiate between them.

Quantification of expression requires determination of the total amount of a protein synthesized, irrespective of whether it is post-translationally modified. Because *N*-glycosylation of proteins is a co-translational process, there will be no non-glycosylated isoform of the protein. (*Mutations that preclude N-glycosylation are generally lethal.*) Targeting the mannose-rich glycan core and other structural features of *N*-glycosylated proteins with broad selectivity lectins allows selection of most *N*-glycosylated proteins as was discussed in Section 5.1 (Figs. 1 and 2). The various ways by which quantification of polypeptides thus selected can be achieved has been addressed in Section 5.3. Our laboratory has used the GIST

Based on lectin affinity selection at the protein level and relative amounts
of non-glycosylated peptides.

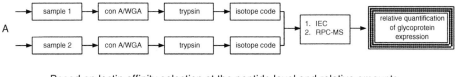

Based on lectin affinity selection at the peptide level and relative amounts
of glycosylated peptides.

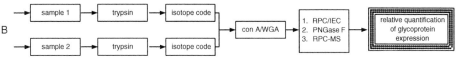

Figure 4. Glycoprotein expression analysis.

stable isotope coding method because it was the first lectin-based quantification strategy available for glycoproteomics, and measurement errors do not exceed 8% irrespective of the number of steps in the analytical process.

A strategy for quantifying expression at both the protein and peptide levels using stable isotope coding and lectin selection is shown in Fig. 4. A major feature of stable isotope based quantification is that differentially coded control and experimental samples are analyzed simultaneously, precluding MSI from impacting quantification. The upper half of the figure shows a protocol for quantification of expression based on protein selection. Following independent selection of *N*-glycosylated proteins with broad selectivity lectin columns, the selected glycoprotein samples are individually tryptic digested, differentially isotope coded, mixed, and then analyzed by LC/MS. The MUDPIT separation approach is suggested to achieve high resolution of the tryptic peptides. Relative differences in expression are determined from the isotope ratio of any set of peptide isotopomers derived from a protein.

The bottom half of Fig. 4 illustrates a protocol for expression analysis based on peptide selection. In this case, samples are first individually trypsin digested and differentially isotope coded before they are mixed and lectin selected. Because of the necessity for deglycosylation in the peptide-based approach, this method can only be used to examine *N*-glycosylation. Following lectin selection peptides are deglycosylated with PNGase F and then fractionated by the MUDPIT method or an equivalent high resolution reversed phase chromatography method. Relative concentration of peptide isotopomers and peptide identification is achieved by ESI-MS. The relative advantages and disadvantages of the glycoprotein versus the glycopeptide selection methods in Fig. 4 are the same as in Section 5.1 (Figs. 1 and 2).

A global protocol for *O*-glycoprotein expression analysis has not been developed, perhaps for several reasons. One is the lack of a set of broad selectivity lectins that will globally capture all *O*-glycoproteins without simultaneously selecting some *N*-glycoproteins as well. The second is that some *O*-glycosylated proteins also occur as non-glycosylated forms. Expression analysis requires that all forms of a protein be quantified. When the focus of a study is on a single glycoprotein or a small set of glycoproteins, Jacalin might be of utility. Removal of *N*-glycoproteins with Con A and WGA would allow Jacalin to be used in cataloging and expression analysis of a small set of *O*-glycosylated proteins as suggested in Fig. 1.

5.6. Quantifying differences in glycosylation

A distinguishing feature of glycosylation relative to other forms of PTM is the large amount of heterogeneity at a single site in a protein. Moreover, the fact that glycan heterogeneity seems to be regulated and can become aberrant in the disease state has led to the realization that site specific glycan signatures on particular proteins may be associated with particular pathologies or even lead to them. This means that the glycan or glycoform signatures on a protein may be a better disease marker than changes in polypeptide expression.

One way to quantify differences in glycoforms is through a SLAC strategy in which broad selectivity and narrow selectivity lectins are used together [51]. With *N*-glycosylated proteins, for example, it is possible to globally quantify all expressed glycoproteins as described in Fig. 4. A global analysis of the fraction of these glycoproteins that carry some smaller, unique feature can then be determined by using a narrow selectivity lectin. One approach by which this was achieved using glycopeptide selection and stable isotope coding is illustrated in Fig. 5. This could also be achieved using protein level lectin selection (Fig. 1) as well. Having established the normal ratio of features within the glycoproteome, it is quite easy to recognize changes among hundreds of glycoproteins. A major advantage of this approach is that it does not require elucidation of the glycan structure on large numbers of glycopeptides to notice changes in glycosylation. The disadvantage of the method is that it cannot quantitatively differentiate between differing numbers of modifications. For example, it is difficult to recognize changes in the number of sialic acid groups on a glycan based on selection with the SNA/MAL lectin set.

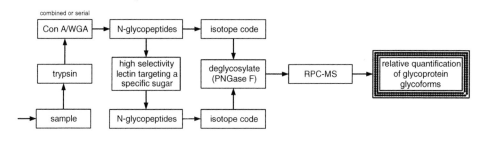

Figure 5. Relative quantification of glycoforms.

5.7. Analysis of glycosylation with narrow selectivity lectins

Lectins of narrow selectivity have been used for decades in the histochemical detection of changes in glycosylation, often to the level of single sugar residues [52, 53]. In fact, lectin-based histochemistry gave the first clues that there were large changes in fucosylation and sialylation in cancer. Unfortunately, histochemical methods did not allow identification of the proteins involved. With the advent of lectin-based glycoproteomics, it is now possible to go back and identify these proteins. Recent studies with lectin affinity chromatography columns, that target either fucose or sialic acid, have allowed identification of large numbers of glycoproteins with these unique glycan features that change substantially in concentration in association with specific types of cancer. An example of the stable isotope coding method used to quantify changes in fucosylation in canine lymphosarcoma associated with chemotherapy is shown in Fig. 6. Serum samples were trypsin

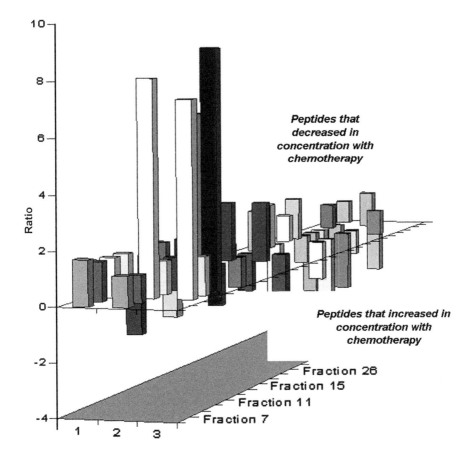

Figure 6. Fucosylated peptides that were elevated in canines with lymphosarcoma and decreased or increased in concentration during chemotherapy.

digested and following differential isotope coding, mixed and affinity selected with LTA [31]. Glycopeptides thus selected were deglycosylated and identified by ESI-MS. Most of the proteins identified were associated with cell adhesion and metastasis. Selection and quantification can also be achieved at the protein level.

Having noted the power of combining old histochemical observations with modern proteomics methods, it is important to examine both the strength and weakness of narrow selectivity lectin targeting. The obvious strength is that clinical correlations with disease progression are quickly and readily made. They will be of great value in translational and clinical proteomics studies. The disadvantage of using single, narrow selectivity lectins is that they do not differentiate between changes in expression and PTM. Narrow selectivity lectins determine isoforms of a glycoprotein. Changes in the amount of an isoform could arise as a result of (1) a change in the total amount of glycoprotein expressed without a change in isoform ratio, (2) a change the ratio of isoforms without increased expression, or (3) changes in both expression and the ratio of isoforms.

6. The Future

Cells are often covered with a diverse array of glycoproteins, of which the glycan components play a major role in mediating their interaction and colonization of the matrices in which they live. The diversity of the glycans involved reflect differences between the types of cells involved, their developmental state, and even their progression into a disease state as seen with cancer above. Unfortunately, we have no idea how diverse these carbohydrate chains actually are or why they are so complex. Why are there so many glycan isoforms at a single glycosylation site on a glycoprotein? What information is coded in these structures and how is it being used in the regulation of biological systems.

Unfortunately, these questions are beyond the reach of current technology. A new generation of tools must be developed that enhance the ionization efficiency of glycopeptides while at the same time allowing simultaneous structure analysis of both the polypeptide and glycan portions of glycopeptides and glycoproteins [54]. Although much of this review dealt with separation technology, continuing improvement is needed to deal with glycan complexity [55–57] and to push limits of detection down to the point that small amounts of biological fluids and pieces of tissue can be examined.

References

[1] N.B. Cech and C.G. Enke, Mass Spectrom. Rev., 20 (2001) 362–387.
[2] E. Krause, H. Wenschuh and P.R. Jungblut, Anal. Chem., 71 (1999) 4160–4165.
[3] R.D. Smith, Int. J. Mass Spectrom., 200 (2000) 509–544.

[4] S.P. Gygi, B. Rist, S.A. Gerber, F. Turecek, M.H. Gelb and R. Aebersold, Nat. Biotechnol., 17 (1999) 994–999.

[5] B.-S. Yoo and F.E. Regnier, Electrophoresis, 25 (2004) 1334–1341.

[6] D. Ren, N.A. Penner, B.E. Slentz, H. Mirzaei and F.E. Regnier, J. Proteome Res., 2 (2003) 321–329.

[7] M.V. Turkina and A.V. Vener, Methods Mol. Biol., 355 (2007) 305–316.

[8] H. Mirzaei and F.Regnier, J. Chromatogr. B: Analyt. Technol. Biomed. Life Sci., 817 (2005) 23–34.

[9] E.R. Stadtman and B.S. Berlett, Chem. Res. Toxi., 10 (1997) 485–494.

[10] T.H. Patwa, J. Zhao, M.A. Anderson, D.M. Simeone and D.M. Lubman, Anal. Chem., 78 (2006) 6411–6421.

[11] Z. Yang and W.S. Hancock, J. Chromatogr. A, 1053 (2004) 79–88.

[12] M.E. Taylor and K. Drickamer. In: M.E. Taylor and K. Drickamer (Eds.), Introduction to glycobiology, Oxford University Press, Oxford, 2003.

[13] J.C. Tercero and T. Diaz-Maurino, Anal. Biochem., 174 (1988) 128–136.

[14] U. Schumacher and B. Willershausen, Lectins: Biol. Biochem. Clin. Biochem., 7 (1990) 391–393.

[15] R.D. Cummings and S. Kornfeld, J. Biol. Chem., 259 (1984) 6253–6260.

[16] D. Hoja-Lukowicz, A. Litynska and B.Wojczyk, J. Chromatogr. B: Biomed. Sci. Appl., 755 (2001) 173–183.

[17] Y. Miura, V.S. Perkel, K.A. Papenberg, M.J. Johnson and J.A. Magner, J. Clin. Endocrinol. Metab., 69 (1989) 985–995.

[18] A.J. Morris, J.T. Gallagher and T.M. Dexter, BMC, 1 (1986) 41–47.

[19] H. Sakai, F. Yamagishi, M. Miura, K. Hata, I. Koyama, Y. Sakagishi and T. Komoda, Tumour Biol., 15 (1994) 230–235.

[20] O. Saitoh, W.C. Wang, R. Lotan and M. Fukuda, J. Biol. Chem., 267 (1992) 5700–5711.

[21] S. Sumi, K. Arai, S.Kitahara and K.-I. Yoshida, J. Chromatogr. B: Biomed. Sci. Appl., 727 (1999) 9–14.

[22] J. Zhao, D.M. Simeone, D. Heidt, M.A. Anderson and D.M. Lubman, J. Proteome Res., 5 (2006) 1792–1802.

[23] T. Takahashi, Y. Ikeda, E. Miyoshi, Y. Yaginuma, M. Ishikawa and N. Taniguchi, Int. J. Cancer, 88 (2000) 914–919.

[24] M. Ohno, A. Nishikawa, M. Koketsu, H. Taga, Y. Endo, T. Hada and K.T. Higashino, Int. J. Cancer, 51 (1992) 315–317.

[25] L.A. Smets and W.P. Van Beek, Biochem. Biophys. Acta, 738 (1984) 237–249.

[26] S.I. Hakomori, Adv. Cancer Res., 52 (1989) 257–261.

[27] G.A. Turner, Clin. Chim. Acta, 298 (1992) 149–171.

[28] C.J. Dimitroff, J.Y. Lee, S. Rafii, R.C Fuhlbrigge and R. Sackstein, J. Cell Biol., 153 (2001) 1277–1286.

[29] M.T. Goodarzi and G.A. Turner, Clin. Chim. Acta, 236 (1995) 161–167.

[30] E.V. Chandrasekaran, J. Xue, J. Xia, R. Chawda, C. Piskorz, R.D. Locke, S. Neelamegham and K.L. Matta, Biochemistry, 44 (2005) 15619–15635.

[31] L. Xiong, D. Andrews and F.E. Regnier, J. Proteome Res., 2 (2003) 618–625.

[32] J. Hirabayashi, Y. Arata and K.-I. Kasai, Proteomics, 1 (2001) 295–303.

[33] Z. Yang and W.S. Hancock, J. Chromatogr. A, 1070 (2005) 57–64.

[34] S.-I. Do and K.-Y. Lee, FEBS Lett., 421 (1998) 169–173.

[35] J.D. Pearson, W.C. Mahoney, M.A. Hermodson and F.E. Regnier, J. Chromatogr., 207 (1981) 325–332.

[36] L. Xiong and F.E. Regnier, J. Chromatogr. B, 782 (2002) 405–418.

[37] J.F. de la Mora, G.J. Van Berkel, C.G. Enke, R.B. Cole, M. Martinez-Sanchez and J.B. Fenn, J. Mass Spectrom., 35 (2000) 939–952.

[38] N.B. Cech and C.G. Enke, Anal. Chem., 73 (2001) 4632–4639.

[39] S. Gao, Z.-P. Zhang and H. T. Karnes, J. Chromatogr. B, 825 (2005) 98–110.

[40] H. Mirzaei and F.E. Regnier, Anal. Chem., 78 (2006) 4175–4183.

[41] R.-Q. Qiu and F. E. Regnier, Anal. Chem., 77 (2005) 2802–2809.

[42] S.-E. Ong and M. Mann, Nat. Chem. Biol., 1 (2005) 252–262.

[43] D.R. Goodlett and E.C. Yi, Trends Anal. Chem., 22 (2003) 282–290.

[44] D.S. Kirkpatrick, S.A. Gerber and S.P. Gygi, Methods, 35 (2005) 265–273.

[45] M.W. Linscheid, Anal. Bioanal. Chem., 381 (2005) 64–66.

[46] F.E. Regnier, L. Riggs, R. Zhang, L. Xiong, P. Liu, A. Chakraborty, E. Seeley, C. Sioma and R.A. Thompson, J. Mass Spectrom., 37 (2002) 133–145.

[47] K. Aggarwal, L.H. Choe and K.H. Lee, Brief. Funct. Genomic. Proteomic., 5 (2006) 112–120.

[48] X. Yao, C. Afonso and C. Fenselau, J. Proteome Res., 2 (2003) 147–152.

[49] P. Liu and F.E. Regnier, Anal. Chem., 75 (2003) 4956–4963.

[50] S. Julka and F.E. Regnier, Brief. Funct. Genomic. Proteomic., 4 (2005) 158–177.

[51] J. Hirabayashi, Glycoconj. J., 21 (2004) 35–40.

[52] J. Schrevel, D. Gros and M. Monsigny, Prog. Histochem. Cytochem., 14 (1981) 1–269.

[53] H. Lis and N. Sharon, Annu. Rev. Biochem., 55 (1986) 35–67.

[54] M.V. Novotny and Y. Mechref, J. Sep. Sci., 28 (2005) 1956–1968.

[55] J. Hirabayashi, Y. Arata, K. Hayama and K.-i. Kasai, Trends Glycosci. Glycotechnol., 13 (2001) 533–549.

[56] J. Hirabayashi, K. Hayama, H. Kaji, T. Isobe and K.-I. Kasai, J. Biochem. (Tokyo), 132 (2002) 103–114.

[57] J. Hirabayashi and K.-I. Kasai, J. Chromatogr. B, 771 (2002) 67–87.

Lectins: Analytical Technologies
C.L. Nilsson (Editor)
© 2007 Elsevier B.V. All rights reserved.

Chapter 9

Miniaturized Lectin Affinity Chromatography

Xiuli Mao, Jianhua Qin and Bingcheng Lin

Dalian Institute of Chemical Physics, CAS, Dalian 116023, China

1. Introduction

One of the distinguishing features of the proteome in eukaryotic cells is that the vast majority of proteins are subject to one or a variety of post-translational modifications (PTM). Modifications such as phosphorylation, glycosylation, sulfation, and ubiquitination have generated much attention in both the functional and structural proteomics research areas. Glycosylation is one of the more common and complex PTMs, which make the protein structure more complex. To analyze minute changes, which occur in glycoproteins, lectin affinity chromatography (LAC) is becoming a technique of choice for the fractionation and purification of glycoproteins and examination of their microheterogeneity to the level of a single sugar residue [1–3].

LAC is a technique of chromatography, where lectins are used to separate components within the sample. The process itself can be thought of as an entrapment, with the target molecule becoming trapped on a solid or stationary phase or medium. Lectins are a class of carbohydrate-specific proteins of nonimmune origin that have a selective affinity for a carbohydrate or a group of carbohydrates [4]. The specific binding of lectins to carbohydrate structures is enable to identify and characterize these structures. Using lectin affinity, it is easy to separate structural isomers and to isolate oligosaccharides of glycoprotein based on specific features. When used in conjunction with other separation techniques, LAC is able to assist in purifying oligosaccharides rapidly and provide substantial information about their structural features. Thus, LAC can be used for the selective isolation, purification, and concentration of glycoconjugates [5].

With the development of proteome, methodologies for the analysis of glycosylation have been focused on high sensitivity and specificity. This is probably due to the fact that the glycoproteins of interest might not be available in large quantities. Therefore, the developments of miniaturized LAC should contribute greatly to the quest for higher sensitivity for glycan and glycopeptide analysis.

Miniaturized high-performance liquid chromatography (HPLC) was developed in early 1980s [6–13]. The capillary HPLC with 75 μm diameter columns and 200 nL/min flow rates are the norm for many applications and result in three to four orders of magnitude reduction in the volume of such columns over the standard analytical 4.6 mm diameter format. In recent years, there has been considerable research devoted to the development of microfluidic platforms. Some progress has been made toward performing HPLC separations in chip-based platform [14–19]. The miniaturization of these chromatographic separation systems in both the pressure and the voltage driven mode promises to open new possibilities in analytical chemistry. The most important advantages of such micro- and nano-chromatography systems are the option for parallel processing (e.g., in high throughput screening) and the possibility to inject extremely small sample volumes. In addition and especially in combination with MS detection, higher mass sensitivities and lower detection limits can often be attained. Furthermore, the considerably smaller elution volumes of the miniaturized systems lead to lower solvent consumption [20]. Finally, the integrated system of HPLC on microfluidic chip results in simplification of solvents delivery system and interconnections. Presently, LAC is mainly miniaturized in a micrometer scale in the format of capillary and microfluidic channel [21–24]. In this chapter, we first describe the principles and preparations of the miniaturized LAC, then, we introduce its applications in protein analysis, and last, we discuss the perspectives of this technique for future development.

2. Principle of Miniaturized LAC

Miniaturized LAC operates on the principle of biological recognition, which is similar to conventional LAC. A lectin is immobilized on a micro-scale stationary phase. The sample of glycoconjugate is loaded into the lectin column and the fractions in sample, which have no specific affinity with the immobilized lectin, are washed out with binding buffer. The fractions which contain glycans specifically affinity bound with immobilized lectin are retained by the immobilized lectin and then released with a concentration or gradient hapten sugar solution.

The micro-scale stationary phase for lectin immobilization can be prepared in both micrometer scale capillary and the microfluidic channel. As previous section described, the most important difference between traditional and capillary LAC is the size of the lectin column, which results in the reduction of sample requirement and the flow rate of buffers. Usually, the buffer delivery systems for the capillary LAC are much the same as systems performing the conventional LAC. Interconnections, such as fittings and tubing, between different components should be carefully chosen to avoid or eliminate the dead volumes and leaks.

Comparatively, the miniaturized LAC in microfluidic channels largely extends the concept of LAC. LAC is not limited as a serial of pump and HPLC column.

The microfluidic chip can be designed to obtain and process all the LAC measurements from small volumes with efficiency and speed. Therefore, the LAC on a microfluidic chip is much more appealing. Besides the affinity binding and debinding glycoconjugates from immobilized lectin, the principle of the integrated LAC microfluidic chip also involves the design of microfluidic chip platform to facilitate the whole process of LAC automatically. Since the LAC on microfluidic chip is still under development, there is still no general mode for the design of an integrated LAC microfluidic chip. We will give an example to show how to design and prepare an integrated LAC microfluidic chip in our laboratory in Section 3.2.

3. Preparation and Applications of Miniaturized LAC

3.1. Miniaturized LAC in capillary

3.1.1. Preparation of LAC column in capillary
Stationary phases in capillary LAC are typically formed from either a porous polymeric monolith or a packed bed of functionalized silica beads.

3.1.1.1. Monolith based LAC column
Monoliths, also called continuous stationary phases, are important tools for bioseparation, biotransformation, and synthesis. Although the first experiments were reported in the late 1960s and early 1970s, the use of the monolith as a stationary phase in capillary was greatly expanded in the last few years [25]. The reason for that is the easy preparation, possibility of attachment to the glass surface, good mass transfer, absence of the packing procedure, and excellent performance [26]. However, monolith supports have been frequently used for immobilization of ligands in affinity chromatography [27–29]. This is probably due to the fact that the slow adsorption kinetics is the rate-limiting factor in affinity chromatography rather than the mass transfer. Thus, the monolithic support should not be superior to the conventional supports. But Schuster et al. compared gigaporous supports with monolithic materials. They immobilized a monoclonal antibody against the commonly used FLAG tag and found that the adsorption kinetics of this ligand on monolith material was extremely fast but the binding capacity did not change within the applied range of flow rates [30, 31].

Typically, the monolith supports for affinity chromatography are polymerized in situ and the ligand is then immobilized. For protein immobilization, the most widely used support is polymethacrylate monolith. The monolith support can be synthesized by both thermo and UV light initialized polymerization. Generally, for the preparation of a capillary column, thermo initialization is preferred since the capillary is usually covered with a layer of polyimide material. In the following section, we will introduce the procedure to prepare a lectin monolith column in quartz or glass capillary using the thermo-initialized polymerization of GMA-EDMA (Poly (glycidyl methacrylate-*co*-ethylene dimethacrylate)) as an example [23].

(a) Materials and chemicals: Capillary (40 cm), 0.1 M NaOH, 0.1 M HCl, deionized H_2O, methanol, 30% (v/v) 3-(trimethoxysilyl) propyl methacrylate solution in acetone, GMA, EDMA, cyclohexanol, dodecanol, 2,2'-azobisisobutyronitrile (ABIN), 0.2 M H_2SO_4, 0.1 M $NaIO_4$, 4 mg/mL lectin + 50 mM $NaBH_3CN$ + 0.1 M hapten monosaccharide in 0.1 M acetate buffer (pH 6.4), 0.1 M $NaHCO_3$ (pH 9.5), 1 M ethanolamine (in 0.1 M $NaHCO_3$, pH 9.5).

(b) Procedure:
(1) Take a new capillary. Wash the capillary with 0.1 M NaOH 30 min, deionized H_2O 10 min, 0.1 M HCl 30 min, and deionized H_2O 10 min in sequence. The capillary is flushed with methanol and dried with a stream of nitrogen.
(2) Fill the washed capillary with 30% v/v solution of 3-(trimethoxysilyl) propyl methacrylate to vinylize the inner wall. The reaction is continued overnight and unreacted reagent is washed out with methanol.
(3) Polymerization solution is prepared with EDMA 12 wt%, GMA 18 wt%, cyclohexanol 58.8 wt% and dodecanol 11.2 wt%. ABIN (1% with respect to monomers) is added to the polymerization solution as initiator. Thereafter, the solution is degassed by purging with nitrogen for 5 min.
(4) Then the modified capillary is filled with the polymerization solution up to 30 cm by immersing the inlet of the capillary in the solution vial and applying vacuum to the outlet. The capillary ends are then plugged with GC septum, and the capillary is submerged in a 50°C water bath for 24 h.
(5) The resulting monolith is washed with ACN and then water using an HPLC pump. Them monolith is again filled with porogen (cyclohexanol and dodecanol) and put in a GC oven for 16.5 h, with a temperature gradient from 30°C to 70°C at a rate of 0.5°C/min. Then the resulting monolith is rinsed with ACN and water using an HPLC pump.

Lectin can be immobilized on the monolith material in two ways. One is to covert the epoxy groups of the monolith support into aldehyde and then reacted with the amino group, and the other is to couple the epoxy group of the monolith with the amino group of lectin directly in basic condition. Procedures for these two methods for lectin immobilization are shown in the following:

i. The GMA monolith is first rinsed thoroughly with water and then filled with either 0.1 M HCl solution and heated at 50°C for 12 h or 0.2 M H_2SO_4 and heated at 80°C for 3 h to hydrolyze the epoxy groups. (Some reports show that the qualitative conversion of epoxy groups to diol functions with 0.2 M H_2SO_4 is higher than that with 0.1 M HCl.) The column is then rinsed with water followed by a freshly prepared solution of 0.1 M sodium periodate for 1 h to oxidize the ethylene glycol on the surface of the column to an aldehyde. Lectin is then immobilized to the monolith column by pumping a solution of lectin at 4 mg/mL in 0.1 M sodium acetate buffer (pH 6.4) with 0.1 M hapten monosaccharide and 50 mM sodium cyanoborohydride through the column overnight at room temperature [23].

ii. The GMA monolith is first washed with a solution of methanol/water (v/v 50%) and then washed with water thoroughly. The monolith is then washed with 0.1 M NaHCO$_3$ (pH 9.5) for 1 h. 2–10 mg/mL lectin in 0.1 M NaHCO$_3$ (pH 9.5) is pumped through the lectin column. The immobilization reaction is kept for 16 h at 30°C in a sealed setup [21].

The prepared lectin column is washed thoroughly with the tris-HCl buffer. Here GMA is used as monomer in the polymerization mixture since it contains a reactive epoxy group allowing the material to react with the ε-amino groups of lectin. EDMA acts as a cross-linker. Cyclohexanol and dodecanol are used as porogen. The reaction mechanism for the polymerization of GMA-EDMA is shown in Scheme 1 (1), and cross-linking is shown in Scheme 1 (2). The immobilization of lectin on the GMA-EDMA monolith is shown in Scheme 1 (3) and (4).

Although the composition of polymerization solution is given in the above protocol, the pore size of monolith material could be controlled through the optimization of the polymerization conditions, such as reaction temperature and time, quality of the free radial initiator, the percentage of porogenic solvent as well as cross-linking monomer, and so on.

3.1.1.2. Silica packing-based LAC column

Polysaccharides materials such as agarose matrix are commonly used for lectin column packing. Presently, there are some commercial agarose–lectin products available for use in LAC, and for application in high-sensitivity glycomic/gly-coproteomic structure analysis [5, 32, 33]. Although some small LAC columns with 1–5 mL bead volume are packed with agarose-based material [1, 34], the procedures for these columns are not compatible to minimize sample handling, contamination, and sample loss [35]. And it has also been proven that the microcolumn of lectin packed with silica-based material is usually more effective than agarose-based materials [22].

The procedure for preparation of a silica-based lectin microcolumn involving lectin immobilization and column packing is described in the following.

(a) Materials and chemicals: PEEK capillary (500 μm I.D.), stainless steel frit, macroporous silica LiChrospher Si 1000 (10 μm, 1,000 Å), 6 M HCl, deionized H$_2$O, dimethylsulfoxide (DMSO), 3-glycidoxpropyltrimethoxysilane, triethylamine, toluene, acetone, ether, 10 mM H$_2$SO$_4$, ethanol, acetic acid/water (90/10), sodium periodate, NaBH$_3$CN, NaBH$_4$, 0.1 M NaHCO$_3$ + 0.5 M NaCl, 10 mM tris-HCl + 0.5 M NaCl + 1 mM CaCl$_2$ + 1 mM MgCl$_2$ + 1 mM MnCl$_2$.

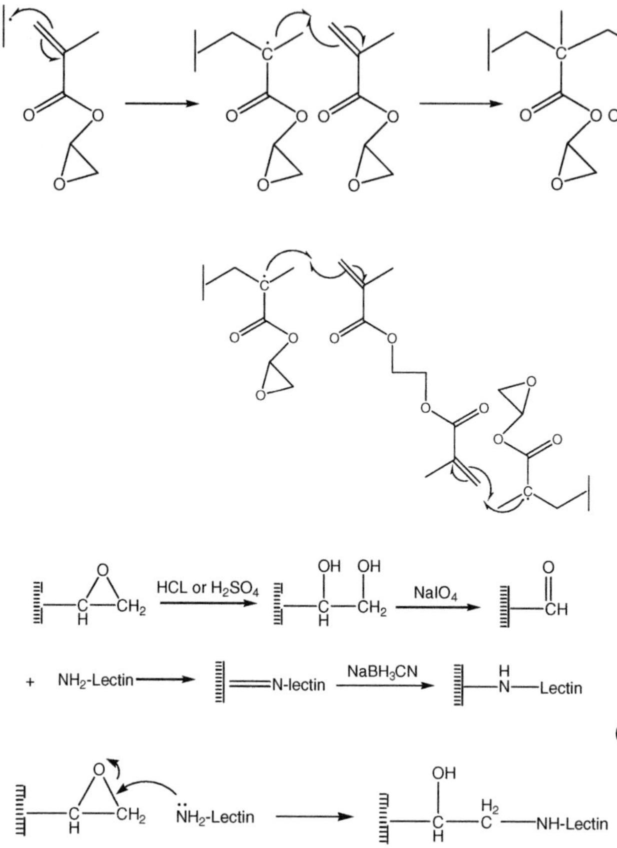

(1)

(2)

(3)

(4)

Scheme 1. Polymerization of GMA-EDMA monolith material and immobilization of lectin on the GMA-EDMA monolith support through the epoxy group of GMA.

(b) Procedure:

(1) 1 g of macroporous silica is washed sequentially with ~60 mL of water; 6 M HCl; and finally, with H_2O. The washed material is dried overnight at 150°C.

(2) Dried silica is suspended in 15 mL of dry toluene containing 200 µL of 3-glycidoxpropyltrimethoxysilane and 5 µL of triethylamine. The mixture is stirred at 105°C under reflux for 16 h.

(3) After the reaction, the silica is filtered and washed with toluene, acetone, and ether; the washed silica is then dried under vacuum.

(4) The epoxy silica is then suspended in 100 mL of 10 mM sulfuric acid and incubated at 90°C for 1 h. Then the silica is filtered; washed extensively with water, ethanol, and water; and dried in a vacuum.

(5) The diol-silica is then suspended in 20 mL of acetic acid/water (90/10) and adding 1 g sodium periodate. The mixture is stirred at room temperature for 2 h.

(6) The aldehyde silica is filtered; washed with water, ethanol, and ether.

(7) 1 mg of lectin is solubilized in 1 mL of 0.1 M sodium bicarbonate (pH 8.0) containing 0.5 M sodium chloride.

(8) 125 mg of freshly prepared aldehyde-modified silica is added to the lectin solution. The suspension is deaerated and sonicated for 5 min.

(9) 3 mg of sodium cyanoborohydride is added into the suspensions. The mixture is stirred at room temperature for 3 h.

(10) 5 mg of sodium borohydride is added, stepwise, over 30 min to reduce the unreacted aldehyde groups to diols, and the suspension is stirred for 1 h.

(11) The lectin-bound silica is filtered; washed twice with coupling buffer and once with 10 mM tris buffer (pH 7.4) containing 0.5 M sodium chloride, 1 mM calcium chloride, 1 mM manganese chloride, 1 mM magnesium chloride, 0.02% sodium azide; and stored in the same buffer at 4°C.

(12) Take a PEEK tubing, 500 μm I.D. The tubing is cut to 5 cm length, and 2 μm stainless steel frit is placed into the PEEK union that is connected to one end of the tubing. The other end is connected to a Capillary Perfusion Toolkit (Perseptive Biosystem Frammingham, MA).

(13) After addition of 1.5 mL of properly sonicated lectin silica slurry, a HPLC pump is connected to the apparatus.

(14) The slurry is packed to the tubing at 2,500 psi pressure using 10 mM tris buffer (pH 7.4) containing 0.5 M sodium chloride, 1 mM calcium chloride, 1 mM manganese chloride, 1 mM magnesium chloride, 0.02% sodium azide.

(15) Upon the packing completion, the system is allowed to depressurize while the packed tubing is disconnected from the apparatus and capped with another 2 μm frit placed inside a union.

(16) The lectin–silica microcolumn is washed with packing buffer at 10 μL/min for 20 min prior to their storage at 4°C.

The mechanism of immobilization of lectin on silica-based materials is similar to that on the GMA-EDMA porous monolith. But prior to coupling lectin, the surface of silica must be additionally derivatized since it only contains hydroxyl groups. The approach to modifying silica based on the formation of the aldehyde–silica surfaces which are eventually coupled to the ligand through reductive amination is proved fast, is not expected to adversely influence the lectin's activity, and is expected to offer relatively high coupling yield [22].

The reaction processes of silica modification and lectin immobilization are shown in Scheme 2.

There is another simpler way for lectin immobilization on the silica-based material through polyacylic acid given as follows. This method has been also widely used in the protein immobilization on silica material [36].

(a) Materials and chemicals: Lichrospher Si 1000, 6 M HCl, 5% 3-amino-propyl triethoxysilane, toluene, polyacylic acid (M 450 000), *N*-hydroxyl succinimide,

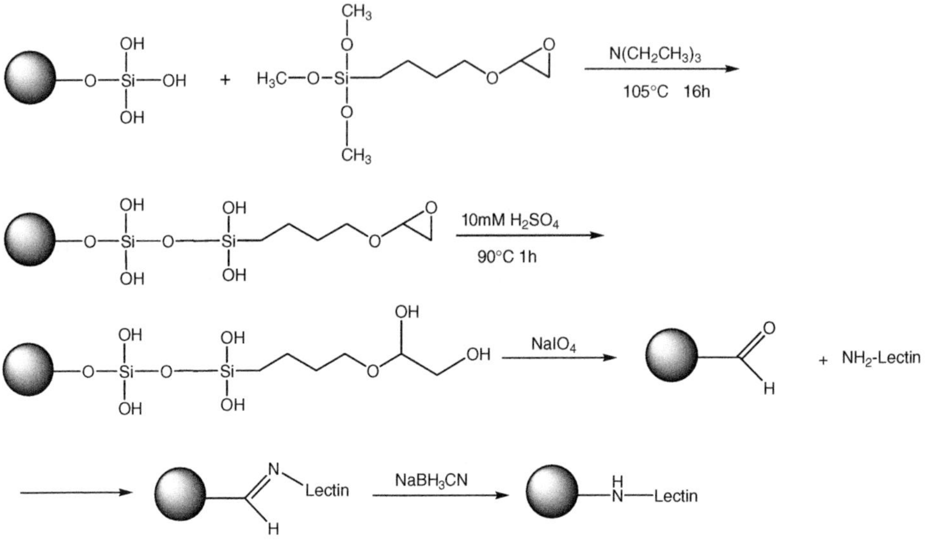

Scheme 2. The reaction processes of silica bead modification and lectin immobilization.

DMSO, 0.1 M NaHCO$_3$, 0.1 M tris-HCl + 0.1 M NaCl + 1 mM CaCl$_2$ + 1 mM MnCl$_2$ + 0.2% NaN$_3$ (pH 7.5).

(b) Procedure:
(1) 1 g Lichrospher Si 1000 is activated with 40 mL 6 M HCl under N$_2$ for 6 h at room temperature.
(2) After washing with water to neutral pH, the silica is filtered and dried in the oven overnight at 120°C.
(3) The resulting particles are treated with 5% 3-amino-propyl triethoxysilane in 10 mL of sodium-dried toluene for 5 h in 90°C to produce 3-aminopropyl silane (APS)-derivatized silica.
(4) Polyacrylic acid (0.503 g; MW 450,000) and N-hydroxyl succinimide (1.672 g) are dissolved in 50 mL DMSO, then 6.0 g dicyclohexylurea dissolved in 10 mL DMSO is added and stirred at room temperature. The resulting mixture is filtered to get rid of dicyclohexylurea, a white precipitate.
(5) The filtrate, which contains succinimidyl polyacrylate, is added to the APS-modified silica. After shaking at room temperature for 12 h, the silica is filtered and washed with 50 mL of DMSO, dioxane, and deionized H$_2$O, respectively. This step produces polyacrylate-coated silica with residual N-acyloxysuccinimide groups, specified as polyacrylic acid –N-acyloxysuccinimmide (PAA–NAS) silica.
(6) Lectin can be immobilized on the silica surface in the following way. Lectin (100 mg) is dissolved in 20 mL of 0.1 M NaHCO$_3$ buffer containing 0.2 M hapten monosaccharide and mixed with roughly 1 g of PAA–NAS silica.

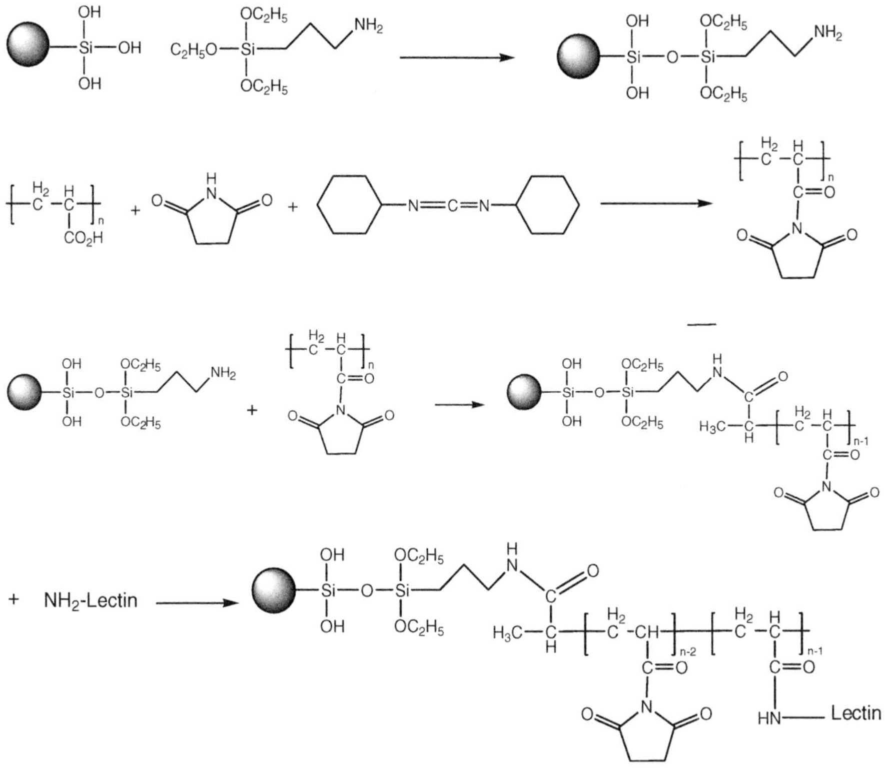

Scheme 3. The reaction processes of silica bead modification and lectin immobilization through polyacrylic acid.

The reaction is allowed to proceed with vigorous shaking for 12 h, after which the immobilized lectin sorbent is recovered by centrifugation.

(7) The remaining active succinimidyl ester sites are blocked with 0.1 M ethanolamine (pH 8.0) agitating for 30 min. The silica particles are then washed with 1 M NaCl to remove the uncoupled ligands.

(8) Finally the sorbent is washed with buffer (0.2 M Tris, 0.1 M NaCl, 1 mM $CaCl_2$, and 1 mM $MnCl_2$, 0.2% NaN_3, pH 7.5) and packed into a capillary as described in the previous section.

The reaction processes of silica modification and lectin immobilization are shown in Scheme 3. This procedure has not been used in the capillary LAC column. But it has been widely used in the LAC packing column.

3.1.2. Operation procedures of capillary LAC

There are several ways to operate capillary LAC, including singly, multidimensionally or in series by both nano-HPLC and capillary electrochromatography (CEC).

3.1.2.1. Single capillary LAC
The operation procedure of capillary LAC in nano-HPLC mode is similar with conventional LAC. Briefly the LAC column is equilibrated with binding buffer. Sample is injected into the LAC column, and then the column is washed with binding buffer. The unbound fractions in the sample are washed out. This process continues until no fraction is washed out. Then the bound fractions are eluted with a solution with hapten sugar. The operation process is more complicated for the capillary LAC in CEC mode.

Usually in the process of LAC, high concentration salt is added into the binding or elution buffer to prevent the nonspecific absorption between the sample and stationary phase. The high current caused by the high concentration salt under high electric field will result in high Joule heat which will destroy the activity of the immobilized lectin after a few running times. So Okanda and Rassi [24] developed a three-step process for capillary LAC in CEC mode. The whole process involved a nano-LC elution step to elute unretained species followed by a second nano-LC step to change from a high-ionic strength to a low-ionic strength mobile phase and a final CEC step for debinding the retained glycoproteins and achieving their separation. This process represents a superiority of affinity CEC over affinity nano-HPLC in the sense that in the latter all captured glycoproteins will co-elute by applying the hapten sugar whereas in the former once the glycoproteins are released from the immobilized lectin, they move differentially down the column according to their charge-to-mass ratio and separate under electric field. So the capillary LAC in CEC provides the simultaneous capturing and separation of different glycoproteins, thus emerging as a very suitable microcolumn separation technique for nanoglycomics/nanoproteomics.

3.1.2.2. Multidimensional system of capillary LAC
Capillary LAC can perform the enrichment of classes of glycoproteins and glycans by the immobilized lectin. Moreover, different lectins are able to specifically affinity bind different saccharidic sequences belonging to numerous glycoproteins. This is the reason that the immobilized lectins are often used to fractionate complex mixtures of glycoproteins. Therefore, capillary LAC can also be used as an effective first dimension in the purification and separation of glycoproteins followed by nano-RPLC. The manual operation of the two-dimensional capillary LAC/nano-RPLC can be described as follows [23]. First the capillary LAC column and nano-RPLC column are connected together, and sample is injected into the two-dimensional system. The lectin column retains the glycoproteins in the sample, which have specific affinity with the immobilized lectin, and other glycoproteins will pass through the LAC column and accumulate on the top of the nano-RPLC column. After this sample injection step, the two columns are disconnected, and the LAC column with adsorbed glycoproteins is taken out and let aside for further use. The accumulated glycoproteins in the nano-RPLC column are eluted and separated by step gradients of organic solvent in the mobile phase. After this step, the LAC column is reconnected with the nano-RPLC column and the glycoproteins retained by LAC column are subsequently eluted with the hapten sugar solution, a condition that moves the

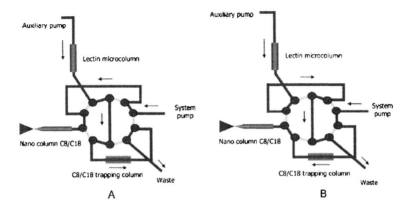

Figure 1. High-performance affinity chromatography setup coupled on line to electrospray mass spectrometry, including a system for loading of proteins/glycoproteins (A) and analysis of proteins/glycoproteins (B). Adapted from Ref. [22] with permission.

glycoproteins to the nano-RPLC column. Thereafter, the two columns are disconnected again and the glycoproteins, which have been absorbed on the LAC column, are eluted by step gradient and separated in nano-RPLC column. The advantage of this operation protocol is that no extra valve is required. But the manual operation of the switch between connection and disconnection of the two column is troublesome and causes bad reproducibility.

Fig. 1 shows a setup of the two-dimensional system of capillary LAC and nano-RPLC coupled through a 10-port valve [22]. The whole system includes an auxiliary isocratic pump connected with injection system, a lectin affinity column, a C8/C18 preconcentration cartridge needed for desalting a 10-port valve and a nano-RPLC C18 column. With this setup, the process of LAC and nano-RPLC for the analysis of protein/glycoprotein can be operated automatically. And the most important point, the capillary LAC can be coupled with mass spectrometry on line to obtain the detailed analysis of glycosylation. The whole operation procedure for glycosylation analysis with this system includes three steps. First the sample is injected into the lectin column and the C8/C18 trapping cartridges. This setup ensures that the sample matrix is removed after trapping glycoproteins/glycopeptides on the lectin column and proteins/peptides on the C8/C18 trapping cartridge. The C8/C18 trapped sample is sufficiently eluted prior to the MS analysis. Next, glycoproteins/glycopeptides are eluted from the lectin column by using the auto-sampler and the auxiliary pump to inject the elution buffer containing hapten sugar. Eluted glycoproteins/glycopeptides are then trapped again on the C8/C18 trapping cartridge. The C8/C18 trapped sample (now assigned as lectin-bound) is sufficiently washed to eliminate all sugar and salts prior to the MS analysis. During the MS analysis, the auxiliary pump is connected only to the lectin microcolumn and waste again, while the flow rate is increased to condition the lectin column for the next analysis.

These two-dimension capillary LAC/nano-RPLC separation schemes are very advantageous, especially for small amount of glycoproteins/glycoconjugates, since the separated analytes stay in the liquid phase and are transferred from column-to-column without sample loss and finally are eluted with a volatile hydro-organic mobile phase of typical RPLC.

3.1.2.3. Series capillary LAC

A single LAC column usually cannot completely realize the trapping of all glyco-conjugates since one lectin binds a class of glycosylation with specifically glycan sequence. The series capillary LAC columns in tandem are able to largely improve the separation efficiency. The series capillary LAC in nano-HPLC mode is similar with the conventional series LAC involving sample loading, and the elution steps for the lectin columns. Whereas, the series capillary LAC in CEC mode for the separation of glycoprotein requires more than three steps which are required for single capillary LAC. Two steps are nano-LC, involving loading and washing mobile phases, and the other step is CEC requiring eluting mobile phases for the lectin columns.

3.1.3. Applications of capillary LAC

To date, capillary LAC is mainly applied to the analysis of glycoprotein. One aim of capillary LAC is to minimize sample handling, contamination and sample losses. Besides, the miniaturized LAC also provides high sensitivity and high separation efficiency for protein glycosylation analysis. Although the capillary LAC is still under development, it has been applied to different areas of protein research ranging from glycoprotein separation, glycoprotein enrichment to glycoform separation, and so on.

3.1.3.1. Capillary LAC for glycoprotein separation

The separation of glycoprotein from non-glycoprotein with LAC is very useful for the selective analysis of glycoprotein for the proteome research. Bedair and Rassi [23] proved that a series of glycoproteins such as horseradish peroxidase, ribonuclease B, and ovalbumin could be separated from non-glycosylated protein using monolith-based capillary LAC. Two kinds of polymethacrylate monoliths were prepared, namely poly(GMA-*co*-EDMA) and poly(glycidyl methacrylated-*co*-ethylene dimethacrylate-co-[2-(methacryloyloxy) ethyl] trimethyl ammonium chloride) to yield neutral and cationic macroporous polymer, respectively. With standard glycoproteins and proteins as sample, they found that the neutral mono-liths with immobilized lectins exhibited lower permeability under pressure driven flow than the cationic monoliths indicating that the latter had wider flow-through pores than the former. Fig. 2 shows the chromatograms of horseradish peroxidase and ovalbumin in the presence of myoglobin and β-lactoglobulin performed on a Con A capillary monolith column by nano-liquid chromatography. The non-glycoproteins were eluted in the dead volume of the column with the binding

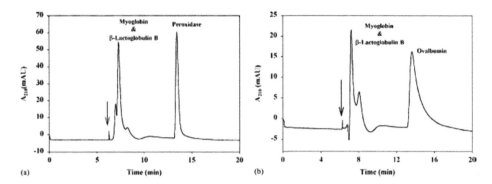

Figure 2. Chromatograms of horseradish peroxidase in (a) and ovalbumin in (b) in the presence of non-glycosylated myoglobin and β-lactoglobulin B, using Con A immobilized on a GMA-EDMA monolithic column, with 25 cm effective length, 33.5 cm total length × 100 μm I.D. Binding mobile phase: 20 mM BisTris, pH 6.0, containing 100 mM NaCl, 1 mM $CaCl_2$, 1 mM $MnCl_2$, and 1 mM $MgCl_2$; eluting mobile phase: 0.2 M methyl-α-D-mannose in the binding mobile phase, introduced at 6.0 min as indicated by the arrow. Pressure drop: 1.0 MPa for both running mobile phase and sample injection. Sample injection: 12 s. Adapted from Ref. [23] with permission.

mobile phase, while the glycoproteins were strongly retained by the Con A column and its elution necessitated the use of a mobile containing the hapten sugar.

3.1.3.2. Capillary LAC for glycoprotein enrichment

Due to the strong affinity binding ability, the capillary lectin columns allow the injection of relatively large volume of diluted samples of glycoproteins or show high effective preconcentration abilities and hundreds of level enrichment factors for glycoproteins and glycans [22]. Fig. 3 is the chromatogram of 1.58×10^{-8} M human α_1-acid glycoprotein concentrated on a wheat germ agglutinin (WGA) capillary monolith column with UV detection [23]. The α_1-acid glycoprotein solution was fed through the WGA-monolith column for 30 min, which results in ∼1.08 pmol of the glycoprotein accumulated at the inlet end of the column. The column was eluted with the eluting buffer containing the 0.2 M N-acetyl-D-glucosamine as the hapten sugar. The concentration factor is about 153.

The nano scale two-dimension systems of capillary LAC coupled with nano-RPLC or CEC have been proved useful for high sensitivity and high resolution of glycomics and proteomic research. Fig. 4 is the chromatogram of a mixture of five proteins, conalbumin, trypsinogen, α-lactalbumin, glucose oxidase, and transferrin, separated with a two-dimension system by coupling Con A column with a nano C17 column [23]. Among these proteins, conalbumin, trypsinogen, and α-lactalbumin have no affinity to Con A and firstly separated by C17 column. The obtained chromatogram is shown in Fig. 4a. Glucose oxidase and transferrin were captured by Con A column and latterly eluted by 0.2 M methyl-α-D-mannose into C17 column. Then they were further separated by C17 column. Their chromatogram is shown in Fig. 4b.

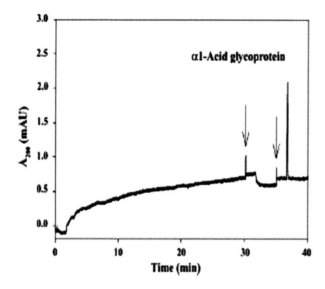

Figure 3. Chromatogram showing the isolation and concentration of 5×10^{-8} M human αl-acid glycoprotein using WGA immobilized on a poly (glycidyl methacrylate-*co*-ethylene dimethacrylate-*co*-[2-(methacryloyloxy) ethyl] trimethyl ammonium chloride) monolithic column, with 25 cm effective length, 33.5 cm total length × 100 μm I.D. Binding mobile phase: 20 mM BisTris, pH 6.0, containing 100 mM NaCl; eluting mobile phase: 0.2 M GlcNAc in the binding mobile phase. Pressure drop: 1.0 MPa for both running mobile phase and sample injection. Adapted from Ref. [23] with permission.

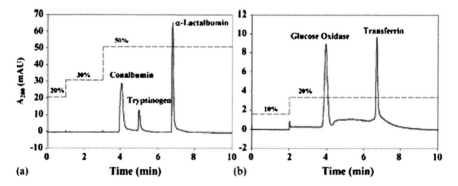

Figure 4. Two-dimensional separations of a mixture of some proteins using a Con A column (12 cm × 100 μm I.D.) in the first dimension followed by RPLC on a neutral C17 capillary column (25 cm effective length, 33.5 cm total length × 100 μm I.D.) in the second dimension. (a) Chromatogram obtained on the C17 column for the unretained proteins on the Con A column (passed through fraction); mobile phase used was 20 mM BisTris, pH 6.0, at 20% (v/v) acetonitrile (ACN) for 1 min and then the ACN content was increased stepwise to 30% at 1 min, and to 50% at 3.0 min. (b) Chromatogram obtained on the C17 column for the two glycoproteins transfered from the Con A column (i.e., previously captured by Con A column); mobile phase used was 20 mM BisTris, pH 6.0, at 10% (v/v) for 2 min and then the ACN content was increased to 20% at 2.0 min. Adapted from Ref. [23] with permission.

Although the sites of glycosylation on a glycoprotein are commonly determined by a scanning mass spectrometry, the weaker ionization of glycopeptides relative to other peptides makes this approach impractical in many cases and requires the use of large amount of sample, which in many cases may not be available. Therefore, it is desirable to isolate glycopeptides from the other peptides that are present in a tryptic digest, thus eliminating peptide interference. This can be achieved by utilizing LAC. In addition, the coupling of capillary LAC with nano-RPLC provides the possibility to connect the LAC with mass spectrometry on line directly.

Lectin trapping column, a Con A trapping column, which is selective for mannose-rich glycans, was used to characterize the site of glycosylation of fetuin. The glycopeptides of fetuin were analyzed by both conventional LC/MS and trapped on line through a Con A lectin trapping column prior to LC/MS [22]. As shown in Fig. 5, the peak intensity of the glycopeptide LCPDCPLLAPLNDSR possessing N_{159} glycosylation site analyzed using on line lectin trapping (Fig. 5a) is ~20-fold higher than that observed without the use of on line lectin trapping (Fig. 5b). The peak intensities of the other glycopeptides with the glycosylation sites at N_{176} and N_{99} were also higher when the on line lectin trapping was utilized (compare Fig. 5c and d and Fig. 5e and f respectively).

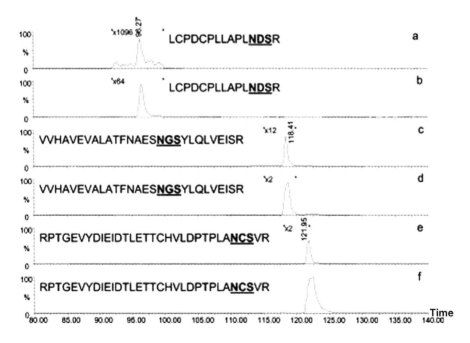

Figure 5. Extracted ion chromatograms of fetuin glycopeptides analyzed by LC/MS without (a, c, and e) and with on line Con A lectin trapping (b, d, and f). Adapted from Ref. [22] with permission.

3.1.3.3. Series capillary LAC for glycoprotein separation

Fig. 6 shows the separation of six different glycoproteins, α-acid glycoprotein (AGP), fetuin, κ-casein, avidin, human transferrin (HT) and collagen, in about 35 min by CEC using two lectin columns coupled in series in the order WGA → LCA (lens culinaris agglutinin) [24]. WGA exhibits affinity toward fetuin and κ-casein and these two glycoproteins are eluted first from WGA column as two separate peaks in third step (CEC-WGA). Avidin, collagen, and HT, which have affinity to both WGA and LCA, are eluted in the fourth step (CEC-LCA) from LCA column as three separate peaks. AGP yield three peaks corresponding to different glycoforms. One peak has affinity to WGA while the other two peaks correspond to glycoforms with affinity to LCA. This also shows the potential of series capillary LAC to separate the glycoforms of a given glycoprotein.

3.2. Miniaturized LAC on microfluidic chip

The ability to perform laboratory operations on a small scale using miniaturized devices is very appealing. Small volumes reduce the time taken to synthesize and analyze a product; the unique behavior of liquids at the microscale allows greater

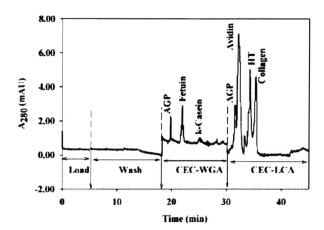

Figure 6. Electrochromatogram of AGP, HT, collagen, κ-casein, avidin, and fetuin, obtained on coupled lectin columns in the order WGA and LCA. Column dimension: 25 cm effective length, 33.5 cm total length × 100 μm I.D. composed of two segments connected butt to butt with a zero dead volume teflon tubing, where the WGA monolith occupies the first segment (12.5 cm) and the LCA monolith the second segment and has an effective length of 12.5 cm and an open portion of 8.5 cm; binding mobile phase: 10 mM ethylenediamine tetraacetic acid (EDTA), pH 6.0, containing 100 mM NaCl, 1 mM $CaCl_2$, 1 mM $MnCl_2$, and 1 mM $MgCl_2$; washing mobile phase: 2.5 mM EDTA, pH 6.0 introduced at 5 min; eluting mobile phase: 0.1 M α-methyl-D-mannose and 0.1 M N-acetyl-β-glucosamine in 2.5 mM phosphate, pH 3.0 introduced at 18.0 and 30.0 min, respectively, as indicated by the arrows; ΔP, 1.0 MPa for both running mobile phase and sample injection; running voltage, 220 kV for eluting mobile phases; sample injection, 6 s. Adapted from Ref. [24] with permission.

control of molecular concentrations and interactions; and reagent costs and the amount of chemical waste can be much reduced. Compact devices also allow samples to be analyzed at the point of need rather than a centralized laboratory [37].

Microfluidic chips or lab on a chip are small platforms comprising channel systems connected to liquid reservoirs by, for example, tubing systems in turn linked to syringes. The size of the channels is in the range of a few micrometers, which greatly facilitates handling of volumes much smaller than a microlitre. Appropriate channel design and integrated tools such as electrodes or a specific surface pattern are now facilitating the incorporation of many operational steps by allowing molecular compounds to pass each unit successively. These individual steps include sampling, sample enrichment (preconcentration and preconditioning steps, such as filtering), mixing, and reaction modules (e.g., different heating zones), product separation, isolation, and analysis [38]. As mentioned in the beginning, LAC has been miniaturized in the form of microfluidic chip and shown to have potential abilities for glycoform separation rapidly and automatically. In the following, we will introduce the fabrication of microfluidic chip and the design, preparation, and operation procedure for the miniaturized LAC on microfluidic chip.

3.2.1. Fabrication of microfluidic chip
Diverse fabrication procedures, originally developed for micro-electro–mechanical systems (MEMS) and for the microelectronic industry, are applied to the fabrication of component for the microfluidic chip.

The fabrication of microfluidic chips usually requires clean room facilities and specific equipment. A great variety of microchip fabrication techniques and materials are available for producing highly sophisticated two- and three-dimensional microstructures with integrated modules. Pumps and valves, mixers, motors, and other functional units that assist chemists in the macroscopic lab have been miniaturized [39–44]. Likewise, sensors and detectors – even optical components – can be integrated on chip [45, 46]. The vast majority of microfluidic devices are, however, simple planar microchips fabricated by photolithography on substrates such as glass, silicon or polymers [47–49].

Glasses are transparent in a wide wavelength range and they are insulators with a high breakdown voltage. Metallization is also relatively easy with them. Another benefit is that the electroosmotic flow on glass surface is well defined and has good reproducibility. The absorbance of glycoprotein on the glass surface can be eliminated by surface modification with well-developed methods [40]. For these reasons, glasses have been the most common materials for the microfluidic chip of protein/glycoprotein analysis.

The standard photolithography and wet etching procedure for glass microfluidic chip fabrication includes three steps, thin film deposition, lithography, and etching. The whole procedure is shown in Fig. 7. First a bare glass wafer is covered with a thin film of metal, typically Cr, with chemical vapor deposition or sputtering. Then the glass wafer is coated with a thin film of photoresist over the metal

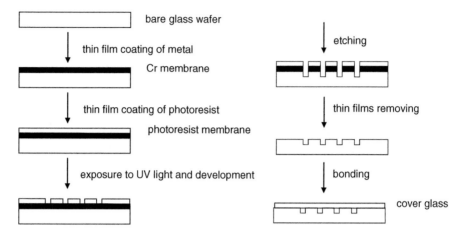

Figure 7. The procedure of fabrication of glass microfluidic chip.

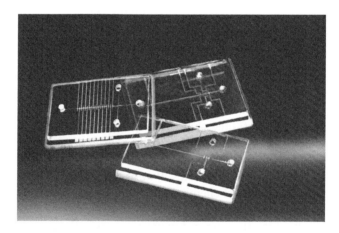

Figure 8. Some typical glass microfluidic chips. Adapted from Ref. [50] with permission.

layer. The photoresist coated glass wafer is exposed to UV light with a mask covering on it and developed. Usually, the mask can be designed with computer-aided drawing (CAD) software and printed with high-resolution printer. Then the glass wafer is immersed in HF–HNO$_3$ solution to be etched. After the etching process, the thin films of photoresist and Cr are removed. The preparation of sample and reagent reservoir holes into a microfluidic chip is generally done by etching or mechanical/electrical drilling. Then the glass wafer is cleaned thoroughly. The glass wafer with etched components is then bonded with a cover glass wafer to form a microfluidic chip. Fig. 8 shows some glass microfluidic chips made in our laboratory [50].

3.2.2. Design and preparation of an LAC microfluidic chip

The whole procedure of LAC includes the sample injection, affinity binding, and elute of bound fractions. A perfect LAC microfluidic chip should perform all these processes automatically. Since the microfluidic chip realizes these analytical processes through microchannels, the design of the chip for this goal should include a channel for the lectin column, a channel to deliver sample, a channel to deliver the binding buffer, and a channel to deliver the elute buffer. The main challenge is how to connect these channels. We design a lectin affinity microfluidic chip (Fig. 9) to fractionate the glycoforms of a given glycoprotein. The chip includes binding buffer channel, eluting buffer channel, sample channel, and lectin column channel.

Protein immobilization techniques have been recently extended into microfluidic format for various bioanalytical procedures, such as immunoabsorption [51, 52], enzymatic reactions [53–55], and so on. Various strategies of protein immobilization have also been demonstrated, including chemical modification of the microchannel surface [51, 52, 56], packing of biochemical-coated beads [57–59], and packing with a biochemical-bearing porous monolithic slab [55]. These works provide basement for the development of chip-based LAC. Among the methods of protein immobilization on microfluidic chip, we find that the monolith matrix is favorable because of its easy preparation, high immobilization efficiency, and most important in situ polymerization.

Here we selected GMA-EDMA monolith as the support of lectin column. The monolith support should be synthesized by UV light initialized polymerization since the monolith matrix is required to specifically locate in the chip channel.

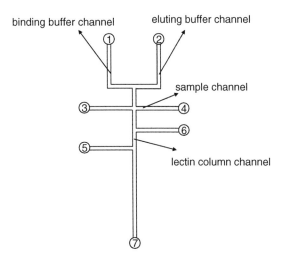

Figure 9. Diagram of lectin affinity microfluidic chip. 1, running buffer reservoir; 2, eluent buffer reservoir; 3, sample reservoir; 4, sample waste reservoir; 5,6, washing reservoir; 7, waste reservoir.

The procedure of monolith preparation with UV light initialization is shown in Fig. 10. The mixture of monolith solution for this column was optimized as 24% GMA + 15% GMA + 60% dodecanol + 1% 2,2-dimethoxy-2-phenylacetophenone (UV light initiator). The monolith mixture is filled in the channel and then covered with a mask. The chip is exposed to a UV light of 365 nm and kept for 6 min. Then a micrometer scale (500 μm) monolith column is formed in the chip channel (Fig. 11a).

The inner structure of the monolith column in a chip channel is shown in Fig. 11b. Lectin was immobilized on the monolith in basic buffer (0.1 M NaHCO$_3$) through the reaction of epoxy group and ε-amide group using the procedure described in Section 3.1.1.

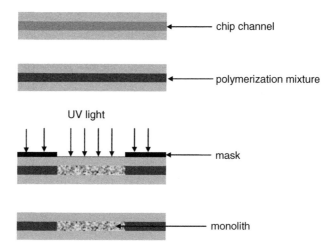

Figure 10. Processes of monolith column preparation in the channel of a microfluidic chip.

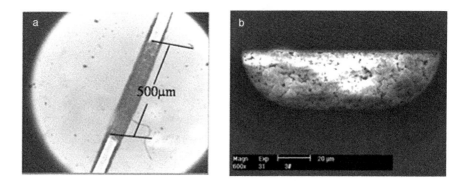

Figure 11. Lectin column in a microfluidic chip channel. (a) Photograph of the lectin column; (b) scanning electron microscope image of the lectin column inner structure.

3.2.3. Operation procedure of LAC microfluidic chip

One of the advantages of glass microfluidic chip is that electroosmotic flow can be generated. Electroosmotically driven flow can deliver reagents and a route of transport in the channel network can be controlled by the potentials on reservoir wells without any valve. With electroosmotically driven flow, the whole process of LAC on a microfluidic chip involves five steps, sample injection 1, sample injection 2, binding, washing, elution 1, and elution 2, as shown in Fig. 12.

Sample is put into reservoir 3 (as in Fig. 9). Other reservoirs are filled with binding buffer except reservoir 2, which is used for elution buffer. In sample injection 1 step, the sample is filled in the sample channel by applying voltages on the reservoir 3 and 4. Switching the voltages to reservoir 1 and 7, the sample plug on the cross section will move in the direction of the lectin column and enter it. The voltages are kept on reservoir 1 and 7 to wash the unbound fractions. This process is continued until no fraction is detected. Then, the voltages are applied on reservoir 2 and 7 to elute the bound fractions from the lectin column with elution buffer. The bound fractions can be further separated with different concentrations of hapten sugar.

As previously mentioned, nonspecific interactions can occur in the process of LAC. Usually, the nonspecific interaction can be eliminated by high concentration of salt, typically 0.1–1.0 M sodium hydrochloride in buffers. But for the electroosmotically driving LAC on a microfluidic chip, high salt concentration will greatly increase the current in the running process. The high Joule heating caused by high current will denature the immobilized lectin and thus destroy lectin affinity after a few runs. Some nonionic surfactants, such as Triton X-100, Tween 20, and NP-40 have no effects on the activity of lectins and can eliminate the

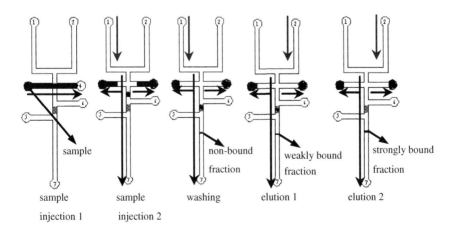

Figure 12. Operation procedure of electroosmotically driven LAC on a microfluidic chip. The arrows indicate the movement of solutions. Adapted from Ref. [21] with permission.

nonspecific interactions between sample and lectin column. So in the electroosmotically driving LAC on chip, all the buffers can contain the nonionic surfactants
instead of high concentration salts.

3.2.4. Applications of LAC microfluidic chip

In theory, the LAC microfluidic chip can realize all the analytical functions that
capillary LAC does such as glycoprotein separation, enrichment, and so on. But
due to the underdevelopment of microfluidic chip, there is few work about its
application. We used *Pisum sativum* agglutinin (PSA) affinity microfluidic chip
(PSA-AMC) to fractionate glycoforms [21].

With the previously described chip and procedure, glycoforms of three glycoproteins were separated by the PSA-AMC. Fig. 13 shows the electrochromatograms of turkey ovalbumin separated with PSA affinity microfluidic chip
running four times. All glycoforms were absorbed onto the PSA monolith
column. When the absorbed glycoforms eluted with the 10 mM and 500 mM
α-methyl-D-mannose solutions, two fractions, the weakly bound and the strongly
bound fractions, were obtained. The relative amounts of the two fractions were
$47.8 \pm 1.8\%$ and $52.2 \pm 1.8\%$, respectively, according to the peak areas ($n = 4$).
Two other glycoproteins from egg, chicken ovalbumin, and ovomucoid were
studied by LAC as shown in Fig. 14. Three fractions of chicken ovalbumin,
corresponding to nonbound, weakly bound and strongly bound glycoforms,
were obtained. The relative amounts of the three fractions can also be calculated
as $10.8 \pm 3.9\%$, $42.8 \pm 7.9\%$, and $46.4 \pm 4.8\%$ ($n = 4$), respectively. Only two
fractions, the nonbound and the strongly bound glycoforms, were obtained for
ovomucoid. The relative amounts of the two fractions were 91.6 ± 2.4 and

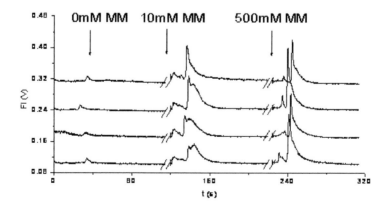

Figure 13. Electrochromatograms of turkey ovalbumin separated on PSA affinity microfluidic chip.
MMα-methyl-D-mannose. Modified from Ref. [21] with permission.

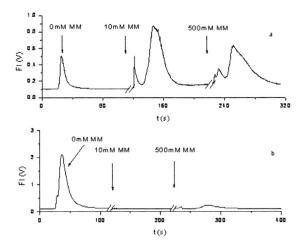

Figure 14. Electrochromatograms of chicken ovalbumin (a) and ovomucoid (b) separated on PSA affinity microfluidic chip. Modified from Ref. [21] with permission.

8.4 ± 2.4% ($n = 4$), respectively. These results show that the reproducibility of this LAC chip is good enough for the analysis of glycoforms distribution.

PSA has affinity specificity toward terminal α-D-mannosyl residues or an *N*-acetylchitobiose-linked fucose residue contained in its receptors. No glycoform of turkey ovalbumin passed through the PSA monolith column after being washed by 0 mM α-methyl-D-mannose, which possibly means that all the glycans contain terminal mannosyl residue or an *N*-acetylchitobiose-linked fucose residue. There is no report about the detailed glycan structures of turkey ovalbumin. But according to our experiments of glycans from turkey ovalbumin on matrix-assisted laser desorption/ionization mass spectrometry, it does not contain fucose and its structure is very different from that of chicken ovalbumin. So we can conclude that the major glycans from turkey ovalbumin contain terminal mannosyl residue(s).

Chicken ovalbumin is known to have various kinds of *N*-glycans at *N*-293 of a long polypeptide core. The 13 kinds of *N*-glycans can be grouped into the high mannose and hybrid types, and the latter can be further divided into two categories on the basis of the presence or absence of the galactose residue at the reducing termini [60]. These two types of glycoforms were separated into three fractions, nonbound, weakly bound, and strongly bound fractions toward PSA. Some glycoforms of chicken ovalbumin could not bind to the PSA monolith column, which means that glycans linked to these glycoforms have no terminal mannosyl residues. These glycoforms might correspond to the hybrid-type glycoforms with no terminal mannosyl residues. Ovomucoid, another glycoprotein from hen egg white, contains a trimannosyl pentasaccharide chain, and other complex sugar chains [61, 62]. It has at least 12 kinds of glycans, and only one of them is a high-mannose-type glycan with a trimannosyl pentasaccharide chain. The glycoforms with complex-type glycans should pass through the PSA monolith without affinity binding, whereas the glycoform with a trimannosyl pentasaccharide chain is bound. So when ovomucoid was separated on PSA-AMC, the amount of the glycoforms that passed through PSA monolith column was much higher than that of glycoform bound on the PSA monolith column. According to our experimental results, only two glycoform fractions were obtained by PSA, and apparently the amount of the nonbound fraction is much more than that of the strongly bound fraction.

All of the three glycoproteins were successfully separated into several fractions, which demonstrates that the LAC microfluidic chip is capable of further analysis of glycoforms. The existence of the weakly bound and the strongly bound fractions demonstrates that glycoforms of a glycoprotein can be separated on a single LAC microfluidic chip by the elution with different concentration of haten sugar.

As previously mentioned, the LAC on a microfluidic chip largely reduces the analysis time. The whole analytical process can be completed within 400 s, whereas it usually needs a few of hours for conventional LAC and tens of minutes for capillary LAC. Glycoproteins can be separated into fractions according to their affinity ability toward immobilized lectin in the microchannel of a microfluidic chip.

4. Perspective

Miniaturized LAC can be realized in the form of micro-scale capillary and microfluidic channel. Due to the minimization of sample requirement, the analytical time has been largely reduced. And in virtue of microfluidic chip, the processes for LAC including sample loading, washing, and elution can be integrated on a small chip and operated automatically by electroosmotically driven without any valve. The potential it exhibits in less sample consumption, rapid analytical process, and integration demonstrates that it will make important contribution to the proteome and glycome research.

Despite these achievements of miniaturized LAC, it is still underway. The application in the real sample analysis has not been explored. Efforts should be directed toward the use of different lectin microcolumns in the fractionation of complex mixtures of modified proteins or peptides as well as a selective preconcentration of glycoconjugates and the automatic system for the multidimensional miniaturized LAC.

Acknowledgment

This project was supported by NSFC (No. 20575067) of China.

References

[1] P.R. Satish and A. Surolia, J. Biochem. Biophys. Methods, 49 (2001) 625.

[2] D. Ghosh, O. Krokhin, M. Antonovici, W. Ens, K.G. Standing, R.C. Beavis and J.A. Wilkins, J. Proteome Res., 3 (2004) 841.

[3] R.Q. Qiu and F.E. Regnier, Anal. Chem., 77 (2005) 2802.

[4] H.B.F. Dixon, Nature, 292 (1981) 192.

[5] J. Hirabayashi and K. Kasai, J. Chromatogr. B, 771 (2002) 67.

[6] T. Takeuchi and D. Ishii, J. Chromatogr. A, 190 (1980) 150.

[7] T.T.A.D. Ishii, J. Chromatogr. A, 213 (1981) 25.

[8] T. Takeuchi and D. Ishii, J. Chromatogr. A, 218 (1981) 199.

[9] T. Takeuchi and D. Ishii, J. Chromatogr. A, 238 (1982) 409.

[10] F.J. Yang, J. Chromatogr. A, 236 (1982) 265.

[11] D.C. Shelly, J.C. Gluckman and M.V. Novotny, Anal. Chem., 56 (1984) 2990.

[12] K. E. Karlsson and M. Novotny, Anal. Chem., 60 (1988) 1662.

[13] R.T. Kennedy and J.W. Jorgenson, Anal. Chem., 61 (1989) 1128.

[14] S. Ekstrom, J. Malmstrom, L. Wallman, M. Lofgren, J. Nilsson, T. Laurell and G. Marko-Varga, Proteomics, 2 (2002) 413.

[15] M.H. Fortier, E. Bonneil, P. Goodley and P. Thibault, Anal. Chem., 77 (2005) 1631.

[16] B. He and F. Regnier, J. Pharm. Biomed. Anal., 17 (1998) 925.

[17] J.J. Li, T. LeRiche, T.L. Tremblay, C. Wang, E. Bonneil, D.J. Harrison and P. Thibault, Mol. Cell Proteomics, 1 (2002) 157.

[18] J. Xie, Y.N. Miao, J. Shih, Y.C. Tai and T.D. Lee, Anal. Chem., 77 (2005) 6947.

[19] N.F. Yin, K. Killeen, R. Brennen, D. Sobek, M. Werlich and T.V. van de Goor, Anal. Chem., 77 (2005) 527.

[20] R. Freitag, J. Chromatogr. A, 1033 (2004) 267.

[21] X.L. Mao, Y. Luo, Z.P. Dai, K.Y. Wang, Y.G. Du and B.C. Lin, Anal. Chem., 76 (2004) 6941.

[22] M. Madera, Y. Mechref and M.V. Novotny, Anal. Chem., 77 (2005) 4081.

[23] M. Bedair and Z. El Rassi, J. Chromatogr. A, 1079 (2005) 236.

[24] F.M. Okanda and Z. El Rassi, Electrophoresis, 27 (2006) 1020.

[25] F. Svec, E.C. Peters, D. S'ykora and J.M.J. Fr'echet, J. Chromatogr. A, 887 (2000) 3.

[26] H.F. Zou, X.D. Huang, M.L. Ye and Q.Z. Luo, J. Chromatogr. A, 954 (2002) 5.

[27] D. Josic and A. Buchacher, J. Biochem. Biophys. Methods, 49 (2001) 153.

[28] D. Josic, A. Buchacher and A. Jungbauer, J. Chromatogr. B, 752 (2001) 191.

[29] D. Josic and A. Strancar, Ind. Eng. Chem. Res., 38 (1999) 333.

[30] M. Schuster, E. Wasserbauer, A. Neubauer and A. Jungbauer, Bioseparation, 9 (2000) 259.

[31] A. Einhauer and A. Jungbauer, J. Biochem. Biophys. Methods, 49 (2001) 455.

[32] R.Q. Qiu and F.E. Regnier, Anal. Chem., 77 (2005) 7225.

[33] Z.P. Yang and W.S. Hancock, J. Chromatogr. A, 1070 (2005) 57.

[34] K. Taketa, Electrophoresis, 19 (1998) 2595.

[35] M.V. Novotny and Y. Mechref, J. Sep. Sci., 28 (2005) 1956.

[36] X. Li, D. Andrews and F. Regnier, J. Proteome. Res., 2 (2003) 618.

[37] R. Daw and J. Finkelstein, Nature, 442 (2006) 367.

[38] P.S. Dittrich and A. Manz, Nat. Rev. Drug Discovery, 5 (2006) 210.

[39] J.H. Qin, N.N. Ye, X. Liu and B.C. Lin, Electrophoresis, 26 (2005) 3780.

[40] H.Q. Huang, F. Xu, Z.P. Dai and B.C. Lin, Electrophoresis, 26 (2005) 2254.

[41] X.M. Zhou, D.Y. Liu, R.T. Zhong, Z.P. Dai, D.P. Wu, H. Wang, Y.G. Du, Z.N. Xia, L.P. Zhang, X.D. Mei and B.C. Lin, Electrophoresis, 25 (2004) 3032.

[42] H.W. Gai, Y. Li, Z. Silber-Li, Y.F. Ma and B.C. Lin, Lab Chip, 5 (2005) 443.

[43] D.J. Laser and J.G. Santiago, J. Micromech. Microeng., 14 (2004) R35.

[44] N.T. Nguyen and Z.G. Wu, J. Micromech. Microeng., 15 (2005) R1.

[45] E. Verpoorte, Lab. Chip, 3 (2003) 42N.

[46] L. Jiang, Y. Lu, Z.P. Dai, M.H. Xie and B.C. Lin, Lab. Chip, 5 (2005) 930.

[47] H. Becker and C. Gartner, Electrophoresis, 21 (2000) 12.

[48] J.C. McDonald, D.C. Duffy, J.R. Anderson, D.T. Chiu, H.K. Wu, O.J.A. Schueller, and G.M. Whitesides, Electrophoresis, 21 (2000) 27.

[49] Y.N. Xia and G.M. Whitesides, Annu. Rev. Mater. Sci., 28 (1998) 153.

[50] B.C. Lin and J.H. Qin, Microfluidics based laboratory on a chip, Science Press, Beijing, 2006, p. 2.

[51] A. Dodge, K. Fluri, E. Verpoorte and N.F. de Rooij, Anal. Chem., 73 (2001) 3400.

[52] V. Linder, E. Verpoorte, W. Thormann, N.F. de Rooij and M. Sigrist, Anal. Chem., 73 (2001) 4181.

[53] I.M. Lazar, R.S. Ramsey and J.M. Ramsey, Anal. Chem., 73 (2001) 1733.

[54] H.B. Mao, T.L. Yang and P.S. Cremer, Anal. Chem., 74 (2002) 379.

[55] D.S. Peterson, T. Rohr, F. Svec and J.M.J. Frechet, Anal. Chem., 74 (2002) 4081.

[56] J. Lahann, I.S. Choi, J. Lee, K.F. Jenson and R. Langer, Angew Chem. Int. Ed., 40 (2001) 3166.

[57] T. Buranda, J.M. Huang, V.H. Perez-Luna, B. Schreyer, L.A. Sklar and G.P. Lopez, Anal. Chem., 74 (2002) 1149.

[58] N. Malmstadt, P. Yager, A.S. Hoffman and P.S. Stayton, Anal. Chem., 75 (2003) 2943.

[59] H. Andersson, W. van der Wijngaart, P. Enoksson and G. Stemme, Sens. Actuators B, 67 (2000) 203.

[60] D.J. Harvey, D.R. Wing, B. Kuster and I.B.H. Wilson, J. Am. Soc. Mass Spectrom, 11 (2000) 564.

[61] K. Yamashita, J.P. Kamerling and A. Kobata, J. Biol. Chem., 257 (1982) 12809.

[62] K. Yamashita, J.P. Kamerling and A. Kobata, J. Biol. Chem., 258 (1983) 3099.

Lectins: Analytical Technologies
C.L. Nilsson (Editor)
© 2007 Elsevier B.V. All rights reserved.

Chapter 10

Frontal Affinity Chromatography: Systematization for Quantitative Interaction Analysis Between Lectins and Glycans

Sachiko Nakamura-Tsuruta[a], Noboru Uchiyama[a], Junko Kominami[a,b] and
Jun Hirabayashi[a]

[a]*Research Center for Glycoscience, National Institute of Advanced Industrial Science and Technology (AIST), Central 2,1-1-1, Umezono, Tsukuba, Ibaraki 305-8568, Japan*
[b]*Fine Chemical & Foods Laboratories, J-Oil Mills, Inc., 11, Kagetoricho, Totsuka-ku, Yokohama 245-0064, Japan*

1. Introduction

Comprehensive interaction analysis of biomolecules is a critical issue in proteomics. Among a number of post-translational modifications occurring in eukaryotic proteins, glycosylation is the most frequent event of biological significance. This is exemplified by the facts that more than 50% of proteins are estimated to be glycosylated [1] and that glycans have extensive effects on protein solubility, stability, destination, and affinity for ligand. As a result, glycosylated proteins are involved in a variety of biological phenomena. To investigate the diversity and quantity of glycans expressed on the cell surface, sugar-binding proteins, i.e., lectins, have been utilized for more than 100 years. In order to understand the fundamental features of lectin affinity, systematic analysis of lectin–glycan interactions is an absolute requirement.

It is widely known that binding affinities of lectins to oligosaccharides are relatively low, particularly in comparison with antibodies: dissociation constants (K_ds) of the former are typically in the range of over $10^{-3} - 10^{-6}$ M, while those of the latter are $<10^{-7}$ M. It is therefore often difficult to obtain precise K_ds, or affinity constants (K_as, where $K_a = 1/K_d$) in a systematic manner. For this purpose, frontal affinity chromatography (FAC) has emerged as perhaps the best choice among various interaction analysis methods.

FAC was originally developed by Kasai et al. in the 1970s as a quantitative method to analyze enzyme–substrate analog interactions (for a review, see Ref. [2]).

Since its introduction, the method has acquired a solid reputation for its clarity of principle, simplicity in operation, and accuracy in determining biomolecular interactions. FAC is usually performed using an open column having a relatively large bed volume (typically, 1 ml), resulting in long analysis times (at least 1 h) and difficulties with reproducibility. In 2000, the FAC system was dramatically modified by incorporating a high-performance liquid chromatography (HPLC) system and pyridylaminated (PA)-oligosaccharides [3]. A similar degree of improvement was achieved by Schriemer and Hindsgaul, who modified FAC for high-throughput screening of small compounds by means of LC-MS [4]. In comparison, our system was developed for enhanced, high-throughput interaction analysis between immobilized lectins and fluorescently derivatized oligosaccharides, with the emphasis on precise quantitative determination of the binding affinity [5, 6]. In this context, the "hect-by-hect" project, involving the determination of interactions between 100 lectins and 100 glycans, has now been completed. As a result, FAC has become a widely accepted technique for the comprehensive analysis of lectin–oligosaccharide interactions [7].

Though use of the original manual injection system allows the determination of several tens of K_d values per day [5], a dramatic improvement in throughput was achieved by the introduction of an automated instrument. (e.g., FAC-1 and FAC-T; described below) [8]. Use of the automated injection system allows the determination of more than 100 lectin–oligosaccharide interaction analyses per day [7, 8].

In the first part of this article, we briefly describe the overall features and principle of FAC. Recent publications should be consulted for a more detailed description [6, 8]. The major improvements introduced in the automated machines, i.e., FAC-1 and FAC-T, are described here in detail, together with examples of their application. These include development of data analysis software, results of specificity analyses of both conventional and recently discovered lectins, and an example of analysis using fluorescently labeled glycopeptides.

2. General Features and Principle

Since the principle and properties of FAC have been described in detail elsewhere [6, 8], we emphasize here some practical advantages of FAC from a theoretical point of view.

2.1. Major advantages

Clarity of the principle: The separation principle of FAC is based on "Langmuir's adsorption isotherm". The principle is simple for the treatment of interaction analysis of various biomolecules. In practice, FAC analysis involves an experimentally obtained parameter, $V-V_0$. According to the basic equation of FAC, this $V-V_0$ value is proportional to the affinity constant (K_a) under certain conditions (described below). K_a (or K_d) values are simply obtained from $V-V_0$ and B_t, another parameter representing the effective ligand content of a given column. The results thus obtained are easily understood in terms of affinity strength.

Application to weak interactions: As has been widely acknowledged, affinities between lectins and oligosaccharides are relatively weak ($K_d < 10^{-6}$M). Usual methods (e.g., inhibition assay of hemagglutination) therefore require large amounts of oligosaccharide, whereas FAC requires minimal amounts of labeled glycan. In order to detect the very weak affinities in FAC, preparation of a "high-content" lectin column is necessary. Otherwise, the column size would need to be enlarged substantially to detect a significant difference or "leakage" of analyte in terms of elution volume ($V-V_0$).

Requirement for a low analyte concentration: This has practical and economic advantages when one performs FAC using oligosaccharides. From a theoretical viewpoint, the use of a low enough concentration of analyte, $[A]_0$, relative to K_d ($[A]_0 << K_d$), is important, because under such conditions the basic equation of FAC is simplified, and interpretation of results becomes easier. To realize such conditions, the use of fluorescently labeled glycans (enabling a detection limit <1 nM; most K_d values between lectins and glycans are in the range $10^{-3}-10^{-6}$M) is necessary.

Suitability to high-throughput analysis: As described above, once the B_t value for a given column is known using an appropriate analyte, K_d values for all other analytes (e.g., PA-glycans) can be determined directly from $V-V_0$ values according to the basic equation of FAC.

Accuracy and reproducibility by a simple operation: Since FAC is performable essentially by a simple isocratic elution from an appropriate lectin column, highly reproducible and precise determination of V values is possible for a series of analytes (e.g., PA-glycans). Under the condition where $[A]_0 << K_d$, $V-V_0$ values become substantially independent of $[A]_0$ (described below). This is of great practical advantage. The elution volume can now be calculated automatically by processing the integrated multiple data points [6], where the influence of noise is minimized.

Simplicity of the equipment: Ordinary laboratory equipment, i.e., a conventional HPLC system consisting of a pump, an injector and a detector, is directly used for FAC. For high-throughput analysis, automated (FAC-1) or more advanced systems (FAC-T) are now commercially available (Shimadzu Co., Kyoto, Japan; described later). These automated machines allow the analysis of more than 100 interactions within 10 h.

Commercial availability of glycans: Utilization of labeled glycans is a very important factor in systematic FAC analysis. Although it is in general difficult to collect a panel of (e.g., >100) purified (>95% purity), structurally defined glycans, it is possible for PA-oligosaccharides, as they are commercially available from several companies at reasonable prices.

2.2. Basic equation of FAC

The separation principle involved in FAC was extensively described by Kasai et al. in the 1970s. Originally, it was investigated as a technique for analyzing interactions between enzymes and their substrate analogues [2]. It was then applied to carbohydrate-binding proteins, lectins, and their counterpart molecules, sugars, which include simple saccharides [9] and radio-labeled oligosaccharides [10]. However, more systematic study was first accomplished for galectins, a group of animal lectins, using a series of PA-oligosaccharides [5]. Since the principle of FAC has been described in detail [6], we briefly mention here the fundamental information essential for FAC operation from a practical viewpoint.

Initially, two items, a lectin–immobilized column and set of glycan solutions, need to be prepared. In FAC, an excess volume of glycan solution is continuously applied to a lectin–immobilized column (Fig. 1A, *left*), where the volume of glycan solution depends on the column size and the affinity strength between the lectin and analyzed glycan. The larger the column size, and the higher the affinity, the more glycan solution should be prepared.

If a glycan has no affinity to an immobilized lectin, it will pass through the column with no retardation of the elution front. This results in "curve I" in Fig. 1B. On the other hand, if a glycan has some affinity for the lectin, it should repeatedly interact with the immobilized lectin when passing through the column (Fig. 1A, *right*). If the amount of glycan exceeds the capacity of the column, some "leakage" should occur, and thus, retardation of the elution front will be observed. Finally, the glycan concentration will reach a "plateau", which corresponds to the initial concentration of the analyte glycan, i.e., $[A]_0$ (curve II in Fig. 1B). The volume of the elution front (V) of each glycan is precisely calculated according to the method described by Arata et al. [11]. Retardation of the elution front ($V-V_0$) is calculated by subtracting V_0 from V, where V_0 is the elution volume of an appropriate standard oligosaccharide (or monosaccharide) having no affinity to the lectin (Fig. 1B). With the $V-V_0$ values thus obtained, a series of K_d values is produced, using the following equation:

$$K_d = \frac{B_t}{V - V_0} - [A]_0 \qquad (1)$$

(A) (B)

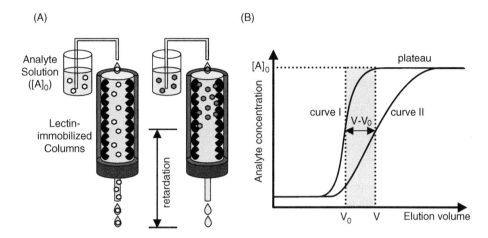

Figure 1. Scheme of frontal affinity chromatography (FAC). (A) Appropriately labeled glycan solution, of which initial concentration is represented by $[A]_0$, is continuously applied to a lectin-immobilized column. When the glycan has no affinity to the immobilized lectin, its elution front is observed immediately (*left*). In contrast, when the glycan has some affinity, its elution front should be retarded depending on the strength of interaction (*right*). (B) Curve I indicates the elution profile of a glycan without affinity, while curve II indicates that with some affinity. The elution volumes of the glycans without and with affinity are indicated as V_0 and V, respectively.

where B_t is the effective ligand content (expressed in mol) and $[A]_0$ the initial concentration of PA-oligosaccharide. Eq. (1) can be simplified to Eq. (2), when $[A]_0$ is negligibly small compared with K_d.

$$K_d = \frac{B_t}{V - V_0} \quad \text{if } K_d \gg [A]_0 \tag{2}$$

In most combinations of PA-oligosaccharides ($[A]_0 < 10^{-8}$ M) and lectins (e.g., $K_d > 10^{-6}$ M), this requirement is fulfilled.

It is often useful to discuss lectin–oligosaccharide interactions in terms of affinity constant (K_a) rather than K_d. The two equilibrium constants are related thus:

$$K_d = \frac{1}{K_a} \tag{3}$$

$$K_a = \frac{V - V_0}{B_t} \tag{4}$$

Here, $V - V_0$ is proportional to K_a.

Once the B_t value of a given column is determined using a particular glycan, K_d values for all other glycans can be automatically obtained by direct substitution of $V-V_0$ values in Eq. (2). For optimal performance of FAC, accurate determination of B_t by concentration-dependence analysis is important.

In practice, concentration-dependence analysis is carried out by the following procedure: Eq. (1) is transposed to another form, i.e., a Woolf–Hofstee type equation.

$$[A]_0 (V - V_0) = - K_d(V - V_0) + B_t \qquad (5)$$

If various concentrations ($[A]_0$) of appropriately labeled oligosaccharides (e.g., p-nitrophenyl (pNP)-glycans) are applied to a lectin column, the resulting $V-V_0$ values should vary. It is apparent from Eq. (1) that $V-V_0$ values become smaller with increasing concentrations of $[A]_0$. Application of a Woolf–Hofstee type plot results in $-K_d$ and B_t values from the slope and the intercept on the ordinate, respectively. An example is described in a later paragraph.

2.3. Data analysis software "FAC-Analyzer"

Since recently developed FAC machines (described later) enable high-throughput analysis (up to 200 interaction analyses per day), the concurrent development of software is important. In this context, the "FAC-Analyzer" was developed according to the data processing method originally described by Arata et al. [11]. This software enables one to process FAC results in an automated and high-throughput manner. The FAC-Analyzer comprises two stages: calculation of $V-V_0$ values (based on visual basic) and representation of elution profiles (based on Microsoft Excel). The scheme for the FAC-Analyzer is shown in Fig. 2.

The former stage deals with raw data collected by HPLC (manual FAC or automated FAC systems (e.g., FAC-1; see below)).

(1) Collected raw data are converted to ASCII format, and imported into the FAC-Analyzer. After subtraction of the background from the raw intensity (Fig. 3A *left* and *center*), the plateau point is searched automatically (Fig. 3A *right*).
(2) The V value is calculated according to the method described by Arata et al. [11].
(3) The net intensity is normalized and expressed as a percentage relative to the plateau point (Fig. 3B). This is repeated until calculation of all the samples is completed. The latest version of the FAC-Analyzer (ver. 5.1.2) can process up to 120 samples.

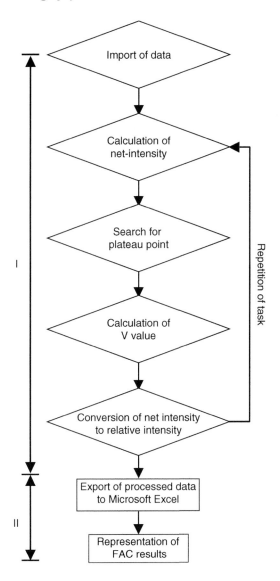

Figure 2. A scheme to summarize FAC-Analyzer. For high-throughput data processing, the "FAC-Analyzer" was developed. The software was originally produced according to the method described by Arata et al. [11]. The FAC-Analyzer comprises (I) calculation of $V - V_0$ values (based on visual basic) and (II) representation of graphs (Microsoft Excel).

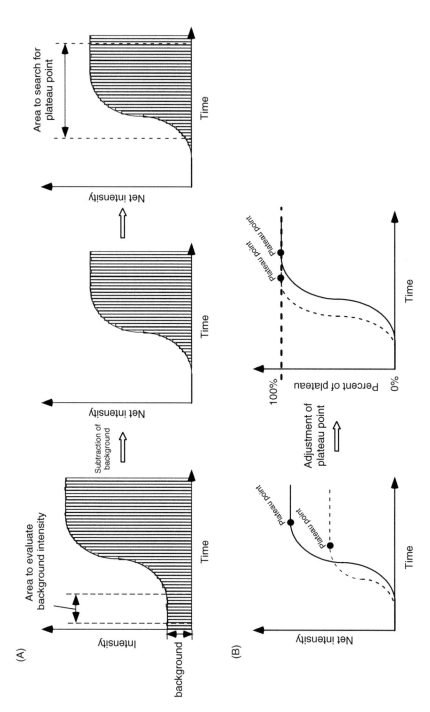

Figure 3. Schematic drawing of data processing. (A) Subtraction of background and search for the "plateau" point. (B) Normalization of the signal intensities in chromatograms for pair-wise comparison. Usually, a chromatogram obtained for an appropriate reference glycan (or monosaccharide) giving proper V_0 (a negative control) is overlaid on that for a test sample.

The latter stage of the FAC-Analyzer was devised for effective representation of FAC results. The analyzer can show a panel of elution profiles of all glycans analyzed, each overlaid with that of a reference glycan. Actual procedures are as follows:

(1) The above-processed data are exported to a template constructed on Microsoft Excel, and $V - V_0$ values are calculated by subtracting V_0 (V of a glycan having no affinity) from V obtained for each glycan.
(2) Once the B_t value of the column has been entered, K_a values are calculated automatically. A typical report format is shown in Fig. 4A, where the affinity strength of each glycan is differentiated by color. Bar graph representation on a separate sheet is also available (Fig. 4B).

3. Operation

3.1. Overall strategy

A typical scheme of FAC for elucidation of sugar-binding specificity of lectins is shown in Fig. 5. A lectin is first immobilized on an appropriate resin (e.g., agarose) and packed into a miniature column. After equilibrium of the lectin column with an appropriate buffer for interaction analysis, a series of diluted PA-glycans is injected sequentially in an isocratic manner. Regeneration of the column is achieved by washing with the same buffer used for sample dilution. Elution of PA-glycans from the column is monitored by fluorescence, and $V - V_0$ values are calculated by the FAC-Analyzer for each glycan. According to B_t values determined by "concentration-dependence analysis", $V - V_0$ values are converted into K_d values. Detailed methods are described below. Other fluorescence groups, such as 2-aminobenzamide (2-AB) and 2-aminobezoic acid (2-AA) may be used with no apparent non-specific binding to the line system, whereas 2-aminoacridone (AMAC) and pyrene derivatives are inappropriate due to significant adsorption to the line system consisting of a stainless column, resin (agarose) and line tubes (PEEK).

3.2. Preparations

3.2.1. Lectin columns
For FAC analysis, either lectin or glycan can be immobilized. From an economic viewpoint, however, immobilization of complex glycans or glycopeptides (i.e., reverse-FAC) is not feasible, so it is more practical to immobilize lectins, to which a panel of fluorescently labeled glycans is applied. NHS- and CNBr-activated agarose are commercially available for this purpose. These packings immobilize proteins via lysyl and N-terminal amino groups. It should be noted that lysine

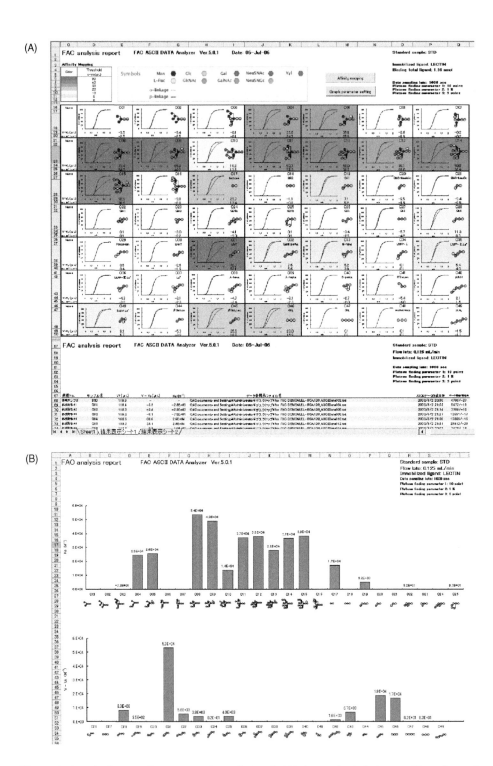

Figure 4. A typical report format on Microsoft Excel. (A) A panel of elution profiles of sample glycans (up to 120) overlaid with that of a reference glycan. Affinity strength of each glycan is differentiated by colors. (B) A bar graph representation is made on a different sheet.

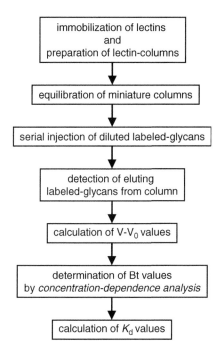

Figure 5. A representative scheme of FAC analysis to elucidate sugar-binding specificities of lectins.

residues can locate around sugar-binding sites of some lectins, and such immobilization results in substantial reduction of lectin affinity or availability. Such deactivation can be avoided by either adding a relatively high concentration of inhibitory saccharide or simply shortening the coupling time. Alternative immobilization methods, e.g., via SH groups, glycans, or Tag (e.g., His-Tag and FLAG-Tag), are available (Fig. 6). As a matter of routine, the most effective strategy for the immobilization of each lectin should be investigated.

In general, affinity chromatography aiming at purification of glycoconjugates requires a high density of immobilized lectin (e.g., 5 mg/ml agarose) in order to capture target glycans completely. On the other hand, FAC requires a moderate concentration of lectin–agarose. For FAC analysis, it is most important to prepare "appropriate" lectin columns. Since the desired lectin concentration depends on the affinity strength of immobilized lectins to a set of glycans, it is difficult to fix an optimal protein concentration. Initially, a trial experiment with 3–5 mg lectin per ml resin, where K_d values are assumed in the range of 10^{-4}–10^{-5} M, should be conducted. Generally, this corresponds to 10–500 µg of protein, which is required for immobilization using 100 µl of agarose resin (*Note*: the bed volume of a miniature column described below is 31.4 µl).

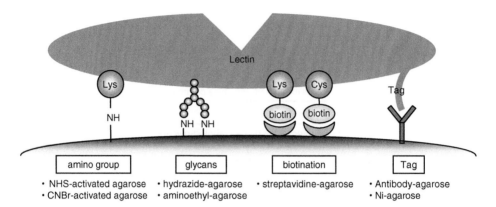

Figure 6. Various strategies for lectin immobilization. An appropriate method should be selected for each lectin. Resins with various kinds of active groups are commercially available.

A typical procedure for lectin immobilization is as follows: purified lectin (300–500 μg) dissolved in 500 μl to 1 ml of coupling buffer (10 mM NaHCO$_3$ buffer, pH 8.3, containing 0.5 M NaCl), is added to 100 μl of NHS-activated Sepharose 4FF (Amersham) according to the manufacturer's instructions. After incubating for 0.5–2 h at room temperature or overnight at 4°C with continuous rotation, excess NHS groups are deactivated with 1 M monoethanolamine. After washing with the coupling buffer and, if necessary, further with 10 mM sodium acetate buffer, pH 4.0, containing 0.5 M NaCl (*Note*: this procedure may inactivate certain acid-labile lectins), the lectin–agarose slurry is suspended in an equal volume of appropriate analysis buffer (e.g., 10 mM Tris-HCl, pH 7.4, containing 1 mM CaCl$_2$ and 0.1 mM MnCl$_2$).

The lectin–agarose resin thus prepared is packed into an empty column. For this purpose, a capsule-type miniature column has been designed for FAC (Shimadzu Co.; Fig. 7A). Alternatively, a small column generally used as guard column in HPLC systems (e.g., 4.0 × 10 mm; bed volume, 126 μl, GL Science) is suitable. We now describe a method optimized for the capsule-type miniature column (Fig. 7B). First, the column is filled with appropriate buffer. Any bubbles appearing are removed by aspiration with a syringe connected to the outlet of the column. A 50% slurry of lectin–agarose is poured into the column, until the full volume of the resin is packed (*Note*: If necessary, repeat the process of pouring the resin suspension and gentle aspiration with the syringe). The inlet of the column is closed with a cap-type filter (manufactured by Shimadzu Co.) with care taken not to leave an air space between the resin and the filter. The resulting lectin column is slotted into a stainless steel holder and connected to the FAC instrument. In most cases, lectin columns thus packed retain activity in an appropriate preservation buffer at 4°C for years, with minimal reduction of lectin activity.

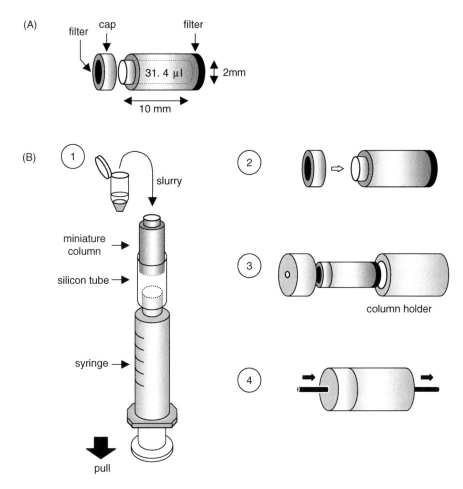

Figure 7. (A) A schematic diagram of the miniature column used for automated FAC. (B) A typical procedure for the preparation of a lectin column. (1) Lectin-immobilized resin (suspension) is added to a column filled with buffer, by gentle aspiration from the outlet with a syringe. After addition of a sufficient amount of resin, residual buffer is completely removed. (2) The inlet is closed with a cap-type filter. (3) The resulting lectin column (capsule) is slotted into a stainless steel holder, and (4) is connected to the line system of the FAC instrument.

3.2.2. Labeled glycans

In FAC, to obtain a solid "elution front" and following "plateau" region, an excess volume (i.e., >10 times relative to the column bed volume) of glycan solution is injected. This volume corresponds to >300 μl and >1.3 ml, when miniature (31.4 μl) or guard columns (126 μl) are used, respectively. It is strongly recommended that a low concentration of glycan solution compared with K_d values be used, as under such conditions the elution front (V) does not substantially depend on the initial concentration of glycan (i.e., $[A]_0$) according to the basic equation of FAC (Eq. (2)). Since most lectin–carbohydrate interactions are likely to be in the range 10^{-4}–10^{-6} M in terms of K_d, the use of fluorescently labeled glycans (detectable at nanomolar order) is of practical importance.

Labeling is essential to detect a relatively low concentration of glycan. In our system, either fluorescence-labeled (e.g., 2-aminopyridine, 2-AA, and 4-methyl-umbelliferone) or UV-labeled (e.g., pNP) glycans can be used. Considering reasonable sensitivity, total absence of non-specific interaction with resins, and ease of purification by HPLC as well as the benefit of the established 2D/3D-mapping system, pyridilaminated (PA)-glycans are the best for this purpose. The stock of PA-glycans generally available in our laboratory is shown in Fig. 8. Most of the listed PA-glycans are commercial products from Takara Bio Inc. and Seikagaku Co. Non-labeled glycans, which are available from several companies, e.g., Funakoshi Co., Dextra Laboratories Ltd. and Calbiochem, and those purified from biological materials, can also be used after pyridylamination [12]. Considering the detection limit, the use of 2.5 nM (N-linked glycans) and 5 nM (other glycans) solutions meets practical purposes for both economy and accuracy in routine FAC analysis. For the preparation of diluted PA-glycans, 1 ml of running buffer is pipetted into appropriate glass vials, followed by 2.5 or 5 µl of the stock solution (1 pmol/µl). Considering loss incurred by the use of an auto-sampling system, the injection volume along with 200 µl of PA-glycan solution should be prepared. In the case of a manual injection system, the volume of sample loop along with 50 µl is usually required to fill the loop completely.

3.3. Systems

3.3.1. Manual injection system for FAC
An FAC system can be constructed with ordinary laboratory equipment (e.g., a conventional HPLC system). A diagram of the manual FAC system is shown in Fig. 9. The system comprises an isocratic pump, a manual injector connected to a sample loop of relatively large volume (e.g., 0.5–4 ml), a miniature column, a column oven (or water bath), fluorescence- or UV-detector, and a system controller linked to a PC. Ligand solution (0.5–4 ml + 50 µl) is injected into the sample loop using a syringe. Throughout the operation, newly opened plastic tools (e.g., tubes and syringes) should be washed extensively with distilled water, as a trace amount of fluorescent material on the plastic wall may interfere with the fluorescence detection.

A lectin column connected to the HPLC system is equilibrated with an appropriate buffer, which should be chosen considering the nature of each lectin. Inclusion of any detergent should be avoided, as a standard draw-and-push

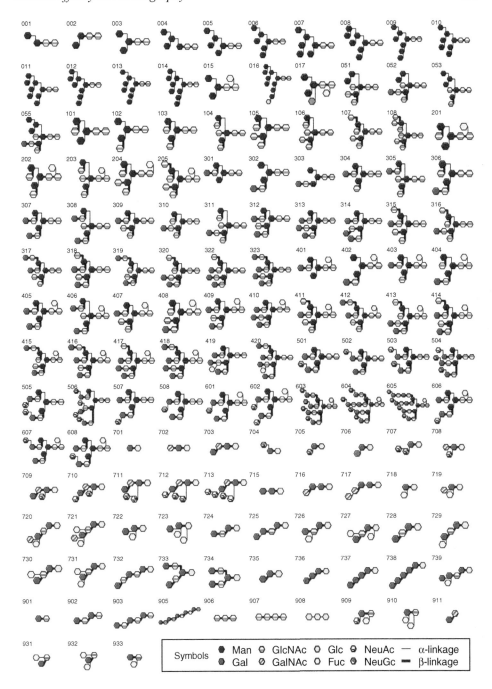

Figure 8. Schematic representation of oligosaccharide structures used for comprehensive FAC analysis. Note that reducing terminal monosaccharides are pyridylaminated. Symbols used to represent pyranose rings of monosaccharides are shown in the box at the bottom of the figure.

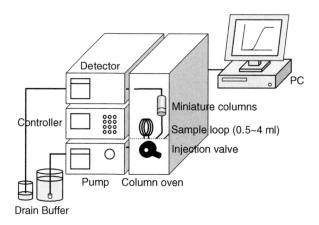

Figure 9. Diagram of a manual FAC system. Essentially, a conventional HPLC system is used, comprising an isocratic pump, a manual injector connected to a sample loop of relatively large volume (e.g., 0.5 – 4 ml), a miniature column, a column oven (or water bath), either a fluorescent- or a UV-detector, and a system controller linked to PC.

pumping system cannot cope with bubbles. Examples of running buffers are as follows:

- 10 mM Tris-HCl buffer, pH 7.4, containing 0.8% NaCl with or without 1 mM CaCl$_2$ (TBS)
- 10 mM phosphate buffer, pH 7.0, containing 0.8% NaCl (PBS)
- 10 mM HEPES buffer, pH 7.4, containing 0.8% NaCl with or without 0.1 mM CaCl$_2$ and 0.01 mM MnCl$_2$

The flow rate and column temperature are kept at 0.125 ml/min and 25°C, respectively. Before analysis, the sample loop with an appropriate capacity (more than 10 times the column volume) is filled with a ligand solution. The injector is turned to the "load" position, and the sample loop is completely evacuated by injection of air using a syringe. This process makes a sharp boundary between the elution buffer and the glycan (analyte) solution, which results in minimal dilution of the analyte. The injection volume is set to a slightly larger amount (e.g., 50 µl) of the analyte solution than the capacity of the sample loop in order to fill the loop completely. When the injector is turned to the "inject" position, the glycan solution in the sample loop is introduced to the column. The glycan eluted from the column is monitored with a fluorescence detector (PA-oligosaccharides; excitation and emission wavelengths of 310 nm and 380 nm, respectively), or UV-detector (*p*NP-oligosaccharides; 280 nm). Signals are collected every second, sent to a computer, and processed by "FAC-Analyzer" software. After each analysis, no regeneration step is necessary.

The overall procedure to operate the manual FAC system is summarized:

(1) Connect a column to an HPLC system.
(2) Equilibrate the column with an appropriate running buffer. (e.g., TBS and PBS)
(3) Turn the injector to the "load" position and evacuate the sample loop.
(4) Fill the sample loop with glycan (analyte) solution. (volume of the loop capacity + 50 μl)
(5) Turn the injector to the "inject" position.
(6) Detect the analyte being eluted from the column.
(7) Process the obtained data on a PC (FAC-Analyzer)

3.3.2. Automated injection system for FAC

For high-throughput analysis, an automated FAC machine, designated "FAC-1", equipped with auto-sampling and parallel column switching facilities was developed in collaboration with Shimadzu Co. (Fig. 10A *left*). The FAC-1 is linked to either fluorescence (Shimadzu, RF10AXL) or UV detectors (Shimadzu, SPD-10A VP). The obtained signals are processed by a PC workstation loaded with "LCsolution" software [7, 13]. A diagram of the system is shown in Fig. 10B. The FAC-1 consists of two isocratic pumps (pump A, for analysis; pump B, for washing), an auto-sampling system (up to 210 samples), a column oven, and two miniature columns connected in parallel to either a fluorescence or a UV detector. Owing to efficiency of the "parallel" column system, the time required for analysis is minimized: i.e., cycle analysis time is reduced to almost 50% compared with the standard one-column system, which requires "tandem" analysis-washing procedures. Use of a capsule-type, miniature column (Fig. 7A) also contributes to the realization of high-throughput analysis by reducing both injection and analysis time. More recently, a further advanced, automated system, designated "FAC-T", was developed in collaboration with Shimadzu (Fig. 10A *right*). FAC-1 and FAC-T are similar in concept and overall features, while the latter system is equipped with four columns in a "twin parallel" (2×2) manner.

FAC-1 deals with two columns in parallel in an efficient arrangement, but the basic operation is essentially the same as in the manual FAC described above. After equilibration of the columns with an appropriate running buffer for more than 10 min, an excess volume of glycan dissolved in the running buffer is successively injected via the auto-sampling system according to the programmed order on a batch file. Injection volume is also set in the same file (*Note*: the injection volume will vary depending on the purpose of analysis and the strength of interactions; i.e., when large retardation is observed, a large volume, e.g., 0.8 ml, must be applied). Elution of PA-oligosaccharides is monitored by fluorescence

(A)

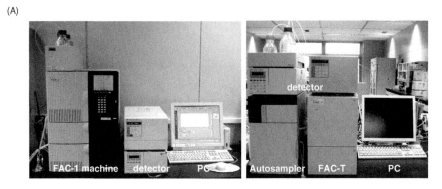

(B) **FAC-1 machine**

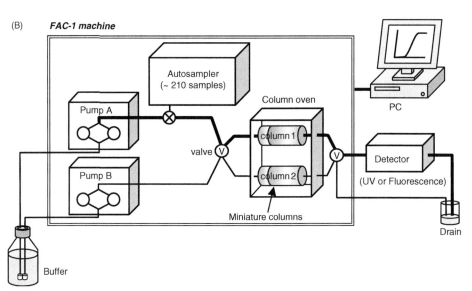

Figure 10. (A) Outside of FAC-1 (*left*) and FAC-T (*right*). (B) A diagram of FAC-1. FAC-1 consists of two isocratic pumps, an auto-sampling system (up to 210 samples), a column oven, and two miniature columns connected in parallel to either a fluorescence- or a UV-detector. The parallel column system reduces analysis time by operating two columns in a sequential manner.

(excitation and emission wavelengths of 310 nm and 380 nm, respectively), whereas that of *p*NP-glycosides is detected by UV (280 nm). $V-V_0$ values are automatically calculated using "FAC-Analyzer" as mentioned above. Since all of the injected glycans are finally eluted from the columns by continuous washing with running buffer, no particular step for regeneration is necessary between analyses.

The overall procedure to operate the FAC-1 system is summarized below:

(1) Connect a pair of columns to the FAC-1 system.
(2) Equilibrate the columns with an appropriate running buffer (e.g., TBS and PBS).
(3) Set a series of sample vials on the auto-sampling system. (total injection volume + 200 μl).
(4) Prepare the time schedule (batch file) to direct the order of analysis and the volume of injection.
(5) Run a batch file (start a sequence of analyses).
(6) Detect glycans being eluted from the column.
(7) Process the obtained data on a PC (FAC-Analyzer).

4. Applications

4.1. Profiling the sugar-binding specificity of a particular lectin

An example of detailed analysis is described for *Erythrina cristagalli* agglutinin (ECA). For systematic analysis, 103 PA-glycans were used and their binding to ECA was investigated by means of FAC-1 (Fig. 11) [8], in which two ECA columns with different lectin–protein concentrations (0.6 and 3.0 mg/ml) were prepared to cover a wide range of affinity. $V-V_0$ values obtained using the latter column (0.6 mg/ml) were combined with those obtained using the former column (3 mg/ml). The *right* vertical axis of Fig. 11 indicates the combined $V-V_0$ values thus obtained for the two columns. As a result of so-called "concentration-dependence analysis" using various concentrations (8–50 μM) of *p*NP-labeled lactose, which shows substantial affinity to ECA (Fig. 12A), and following a Woolf–Hofstee type plot (Fig. 12B), effective ligand content of the column (B_t) was determined. Using the basic equation of FAC Eq. (2) and the obtained B_t value, $V-V_0$ values were converted into K_a values (Fig. 11, the *left* vertical axis).

Detailed analysis using *N*-linked glycans indicated that ECA showed significant affinity for complex-type glycans, whereas it had no apparent affinity for high-mannose type and completely agalactosylated glycans (Fig. 11A). Since ECA is reported to recognize the Galβ1-4GlcNAc unit specifically [14], this result is quite reasonable. Notably, its affinity substantially increased with the increase in the branching number, i.e., **323** and **418** (tetra-antennary) > **313** and **410** (tri-antennary) > **307** and **405** (di-antennary) > **301** and **401** (mono-antennary). These results are consistent with previous observations [15]. Obviously, the affinity strength depended on the number of Galβ1-4GlcNAc units, e.g., **402** (mono-antennary, Galβ1-4GlcNAc x 1) = **404** (bi-antennary, Galβ1-4GlcNAc x 1) > **405** (bi-antennary,

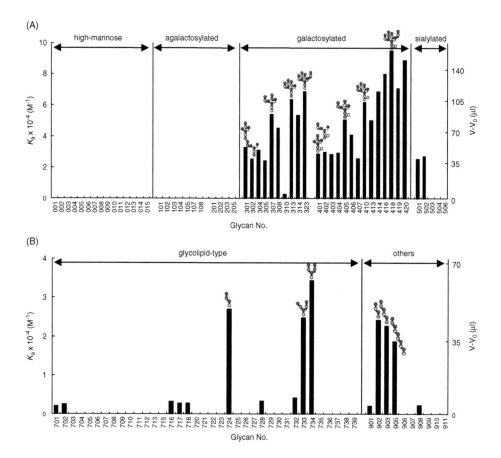

Figure 11. Bar graph representation of $V-V_0$ values (*right* vertical axis) and affinity constants (K_a, *left* vertical axis) of ECA toward (A) 55 *N*-linked glycans and (B) 48 glycolipid-type glycans and others. Glycan numbers correspond to glycan structures shown in Fig. 8.

Galβ1-4GlcNAc x 2). Judging from **401** vs. **402** and **403** vs. **404**, branched positions are not important for ECA recognition. Since ECA preferred **313** (Galβ1-4GlcNAc x 3) to **314** (Galβ1-4GlcNAc x 2 and Galβ1-3GlcNAc x 1), and the latter showed almost the same affinity for **307** (bi-antennary glycan, Galβ1-4GlcNAc x 2), it recognizes only Galβ1-4GlcNAc (type 2 chain), but not Galβ1-3GlcNAc (type 1). As observed for **307** vs. **308** and **405** vs. **406**, the presence of bisecting GlcNAc (β1-4GlcNAc) is somewhat dimini-shed, but did not totally eliminate the affinity. α2-6 Sialylation also cancelled the efficiency of LacNAc recognition (**307** vs. **501** and **502**).

ECA also showed significant affinity for glycolipid-type glycans and other-type glycans (Fig. 11B). Similarly for *N*-linked glycans, all of the glycans having significant affinity for ECA (**724**, **733**, **734**, **902**, **903**, and **905**) contained a

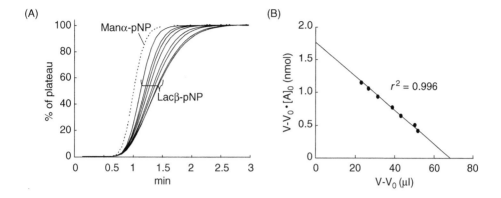

Figure 12. An example of "concentration-dependence analysis". For determination of B_t values of ECA-immobilized column, elution of various concentrations (8 – 50 μM) of *p*NP-labeled lactose is examined (A) and Woolf–Hofstee type plots (B) are carried out. (A) The solid and dotted lines indicate elution profiles of *p*NP-lactose and control sugar (*p*NP-mannose), respectively. (B) Woolf–Hofstee type plots are made by using $V-V_0$ values.

non-modified Galβ1-4GlcNAc unit. Repetition of the Galβ1-4GlcNAc unit did not enhance but rather reduced the affinity (**902 > 903 > 905**). The observation that affinity for **734** (Galβ1-4GlcNAc x 2) was significantly higher than that for **733** (Galβ1-4GlcNAc x 1, Galβ1-3GlcNAc x 1) confirmed the above observation that ECA prefers a type 2 chain.

4.2. Pair-wise comparison of lectin sugar-binding specificities

FAC also has high potential to discriminate between detailed sugar-binding specificities among related lectins. Here, we describe an example of comparative analysis between two lectins with similar molecular properties and sugar-binding specificities.

Two jacalin-related lectins (JRLs), *Castanea crenata* agglutinin (CCA) and *Cycas revoluta* leaf lectin (CRLL) belong to the mannose-binding subgroup of JRLs (mJRLs), and are known to show high affinity for manno-oligosaccharides [16, 17]. Although their fine specificities have not been clarified, recent FAC analysis revealed their shared and distinct aspects of sugar-binding specificity [13]. Purified proteins were immobilized on NHS-activated Sepharose 4FF, and the resulting resins were packed into miniature columns (31.4 μl). The amounts of immobilized CCA and CRLL were 2.0 and 1.1 mg/ml gel, respectively. B_t values of the columns were determined to be 1.49 nmol and 0.81 nmol, respectively, by concentration-dependence analysis using methotrexate (MTX)-derivatized Man$_3$GlcNAc$_2$ (M3GN2-MTX) [13].

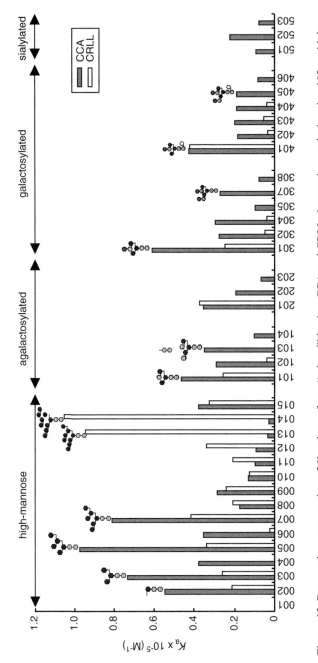

Figure 13. Bar graph representation of K_a values of two "related" lectins, CCA and CRLL. Interaction analysis using 102 pyridylaminated (PA)-glycans is carried out, and glycans having affinity for at least one of the two lectins are extracted. Glycan numbers correspond to glycan structures shown in Fig. 8.

In order to profile sugar-binding specificities of CCA and CRLL, interaction analysis using 102 PA-glycans, which include 55 *N*-linked glycans and 38 glycolipid-type glycans, was carried out [13]. When overall specificities of the lectins were compared, several common features were clarified: i.e., both CCA and CRLL show affinity for a wide range of *N*-linked glycans, but not at all for glycolipid-type glycans. In addition, α1-3Man of core mannotriose is found to be essential for binding of both lectins (**001** vs. **002**). Data from glycans having affinity for at least one of the two lectins were extracted to construct a bar graph. (Fig. 13). Among high-mannose type glycans, CCA show higher affinity for "smaller" (Man2–Man6; **002**–**007**) high-mannose type glycans, whereas CRLL better recognizes "larger" (Man7–Man9; **008**–**014**) ones. The best ligands for CCA and CRLL are **005** (Man3) and **014** (Man9), respectively. By careful examination of these results, it becomes evident that the binding of CCA is greatly diminished by the addition of α1-2Man residue(s) to the α1-3Man recognition unit, whereas that of CRLL is enhanced by α1-2Man extension in the Manα1-2Manα1-2Manα1-3Manβ branch. Distinct features are also found in their specificity for complex-type glycans. CCA can bind to mono- (e.g., **101** and **301**) and bi-antennary *N*-linked glycans (e.g., **103** and **307**) with varied affinities, but not to tri- and tetra-antennary ones. On the other hand, CRLL can bind only to α1-6 branched mono-antennary, complex-type *N*-linked glycans (**101**, **201**, **301**, and **401**), but never to α1-3 branched mono- (**102**, **302**, and **402**), bi-, tri- and tetra-antennary glycans.

Thus, it is clearly demonstrated that CCA and CRLL bound to restricted members of *N*-linked glycans with high specificities. The information obtained here may also give clues for understanding their physiological functions.

4.3. Profiling the branching features of N-linked glycans

In the course of the "hect-by-hect" project, which was intended to analyze comprehensive interactions between 100 lectins and 100 glycans by means of FAC [7], it became evident that several lectins are highly useful for glycan profiling. Among them, *Agaricus bisporus* agglutinin (ABA) [18], *Boletopsis leucomelas* lectin (BLL) and *Griffonia simplicifolia* lectin-II (GSL-II) [19] were found to have the best potential for profiling the branching features of *N*-linked glycans. FAC analysis clearly demonstrates that all three lectins recognize agalactosylated (GlcNAc exposed), complex-type *N*-linked glycans, but their preferences for branching features are markedly different.

For more systematic discussion, we propose a new representation format for *N*-linked glycan structures designated "GRYP code", which defines non-reducing terminal features by different colors; i.e., green (GlcNAc), red (galactose), yellow (α1-6 fucose), and purple (NeuAc) [19]. In this book, the GRYP code is converted to black and white format (Fig. 14A); i.e., gray (galactose), diagonal lines

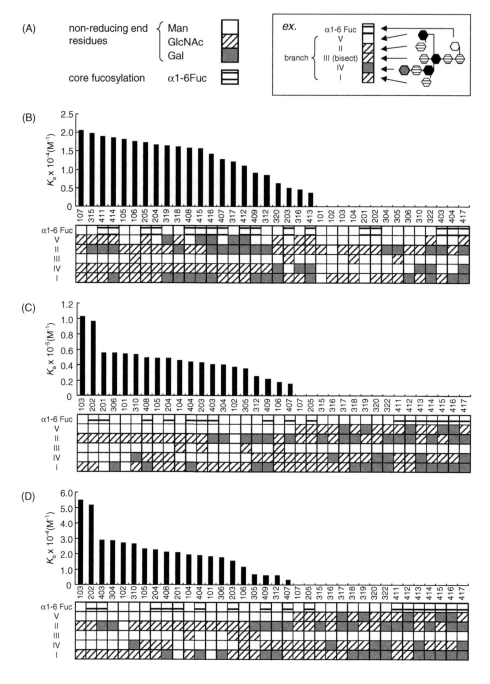

Figure 14. (A) Definition of "GRYP code" to represent branch positions and non-reducing end residues. Each branch is numbered from I to V corresponding to GlcNAc transferases [19]. Examples of glycan codes expressed by this system. Non-reducing end sugar is shown in different patterns: gray, galactose; diagonal lines, GlcNAc; horizontal stripes, α1-6 fucose; white, mannose (no elongation). (B–D) Bar graph representation of affinity constants (K_a) of three "related" lectins, GSL-II (B), BLL (C) and ABA (D), toward completely/partially agalactosylated N-linked glycans. Numbers at the bottom of the bar graphs correspond to sugar numbers indicated in Fig. 8. Glycans are arranged in the order of affinity strength from the left. Corresponding GRYP codes are shown under the graphs.

(GlcNAc), horizontal stripes (α1-6 fucose), and white (mannose, non-modified). Branch positions are numbered I–V according to the conventional nomenclature system applied for mammalian GlcNAc transferases (Fig. 14A). In order to clarify a structural element required for each lectin, K_a values determined by FAC are arranged in the order of strength in a vertical bar graph together with the GRYP codes (Fig. 14B–D).

As a result, it becomes evident that GSL-II strongly binds to tri- and tetra-antennary glycans (**107–320**, **316**, and **413** in Fig. 14B), but has no or only trace affinity for mono- and bi-antennary ones (**203** and **101–417** in Fig. 14B). More importantly, GSL-II shows the strongest affinity for a set of glycans (**107–312** in Fig. 14B) having non-reducing terminal GlcNAc at the *branch-IV* (diagonal lines in *box-IV*). In fact, addition of Gal to this GlcNAc (gray-colored in *box-IV*) abolishes the affinity. Therefore, an essential requirement for GSL-II binding is the presence of an intact GlcNAc residue in *branch-IV*.

Similarly to GSL-II, BLL shows substantial affinity for a series of partially and completely agalactosylated, complex-type glycans (Fig. 14C). However, structural elements required for BLL-binding are obviously different from those of GSL-II. As is evident from Fig. 14C, BLL shows the strongest affinity for bi-antennary, agalactosylated *N*-linked glycans (**103** and **202**). Mono-antennary, agalactosylated glycans are also good ligands for BLL, but with significant branch-specificity: i.e., *branch-II* is more preferred than *branch-I* (**304** vs. **306**). Although tri-antennary, agalactosylated glycans (**310–204** and **312–407** in Fig. 14C) are also good ligands, none of the tetra-antennary glycans (**107–417** in Fig. 14C) showed significant affinity for BLL. A GlcNAc residue at *branch-V* has a detrimental effect for BLL recognition.

ABA can also distinguish branching patterns. Clearly, its binding features are similar to those of BLL. As shown in Fig. 14D, bi-antennary glycans (**103** and **202**) are the best ligands for ABA. Mono- and tri-antennary glycans are also well recognized (**102–407** in Fig. 14D), whereas tetra-antennary glycans (**107–417** in Fig. 14D) are not recognized at all. A significant difference between ABA and BLL is in the preference for *branch-I* and *branch-II*. As mentioned, BLL prefers *branch-II* to *branch-I*, while the opposite is the case for ABA (Fig. 14C, D).

Importantly, branching features are often associated with biological functions of glycoproteins [20–22]. In order to profile *N*-linked glycans, it is very important to know their frame structures consisting of five GlcNAc residues discussed above. It was demonstrated by FAC for the first time that two types of GlcNAc-binding lectins (i.e., GSL-II and BLL/ABA) are useful in discriminating branching features of agalactosylated, *N*-linked glycans. It is important to note that the specificities of these lectins are almost complementary. These lectins should therefore contribute to the development of highly useful glycan profilers in the emerging field of glycomics [23].

4.4. Interaction analysis of fluorescently labeled glycopeptides

Since FAC basically accepts any biomolecular interactions, glycopeptides are also good analytes. For sensitive detection, glycopeptides are usually labeled with fluorescent reagents via amino groups of either N-terminal or inner Lys residues. This kind of analysis was made for a Gal/GalNAc-binding lectin, Jacalin and AlexaFluoro® 488 (AF)-labeled hinge-region peptides (HPs). An example is shown in Fig. 15 [24]. AF-labeled HPs with/without glycans, i.e., Tn (GalNAcα) and Core1 (Galβ1-3GalNAcα), are dissolved in TBS, and are injected into the Jacalin–agarose column (0.5 mg/ml). Elution of glycopeptides is detected by measuring fluorescence (excitation and emission wavelengths of 495 nm and 519 nm, respectively) (Fig. 15), and their interactions are analyzed as described for labeled glycans. However, one should bear in mind that there is possibility of non-specific binding between resin and (glyco)peptides, which is attributable to hydrophobic interaction. In the case of Jacalin, even non-glycosylated HP shows a slight but significant retardation, an indication of non-specific binding. In this case, one cannot determine precise K_a values simply from the apparent $V-V_0$. Nevertheless, it is clear from Fig. 15 that Jacalin has significant affinity toward AF-Tn-HP and AF-Core1-HP. Thus, FAC has now been proved to be widely applicable to lectin–carbohydrate interactions in various fields of glycomics.

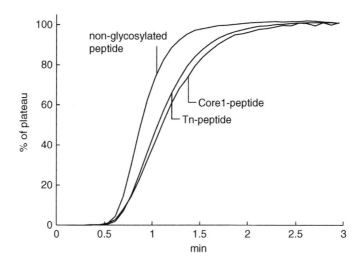

Figure 15. Elution profiles of AlexaFluoro® (AF)-labeled hinge-region peptides with and without sugars. Chromatograms of (glyco)peptides obtained with Jacalin–agarose columns are overlaid.

5. Concluding Remarks

The importance of glycans in physiological phenomena has now been widely accepted. However, the structural complexity and poor availability of functionally important glycans continue to make their systematic profiling difficult. To overcome these difficulties, the use of lectins, which have diverse specificities with a wide range of binding affinity, is promising. To make the best of their ability, elucidation of detailed lectin specificities is absolutely necessary. In this context, we carried out the "hect-by-hect" project to determine at least 10,000 interactions between >100 lectins and >100 glycans. FAC is suitable for this purpose. During the course of the project, FAC proved to be an extremely precise and powerful technique to define the sugar-binding specificities of lectins. At the same time, we were able to re-investigate useful properties of some conventional lectins (as described for GSL-II in this chapter). When the FAC method is combined with other established methods, more sophisticated lectin/glycan profiling will be realized. In this context, we recently reported that the combination of FAC and MS^2 profiling clearly differentiated between type1 and type2 structural isomers with high accuracy [25].

References

[1] R. Apweiler, H. Hermjakob and N. Sharon, Biochim. Biophys. Acta, 1473 (1999) 4–8.
[2] K. Kasai, Y. Oda, M. Nishikawa and S. Ishii, J. Chromatogr. A., 379 (1986) 33–47.
[3] J. Hirabayashi, Y. Arata and K. Kasai, J Chromatogr. A., 890 (2000) 261–271.
[4] D.C. Schriemer, Anal. Chem., 76 (2004) 440–448.
[5] J. Hirabayashi, T. Hashidate, Y. Arata, N. Nishi, T. Nakamura, M. Hirashima, T. Urashima, T. Oka, M. Futai, W.E. Muller, F. Yagi and K. Kasai, Biochem. Biophys. Acta, 1572 (2002) 232–254.
[6] J. Hirabayashi, Y. Arata and K. Kasai, Methods Enzymol., 362 (2003) 353–368.
[7] J. Hirabayashi, Glycoconj. J., 21 (2004) 35–40.
[8] S. Nakamura-Tsuruta, N. Uchiyama and J. Hirabayashi, Methods Enzymol., 415 (2006) 311–325.
[9] Y. Oda, K. Kasai and S. Ishii, J. Biochem., 89 (1981) 285–296.
[10] Y. Ohyama, K. Kasai, H. Nomoto and Y. Inoue,. J. Biol. Chem., 260 (1985) 6882–6887.
[11] Y. Arata, J. Hirabayashi and K. Kasai, J. Chromatogr. A., 905 (2001) 337–343.
[12] S. Hase, Methods Enzymol., 230 (1994) 225–237.
[13] S. Nakamura, F. Yagi, F. Totani, Y. Ito and J. Hirabayashi, FEBS J., 272 (2005) 2784–2799.
[14] J.L. Iglesias, L. Halina and N. Sharon, Eur. J. Biochem., 123 (1982) 247–252.

[15] P.M. Kaladas, E.A. Kabat, J.L. Iglesias, H. Lis and N. Sharon, Arch. Biochem. Biophys., 217 (1982) 624–637.

[16] K. Nomura, S. Nakamura, M. Fujitake and T. Nakanishi, Biochem. Biophys. Res. Commun., 276 (1) (2000) 23–28.

[17] F. Yagi, T. Iwaya, T. Haraguchi and I.J. Goldstein, Eur. J. Biochem., 269 (2002) 4335–4341.

[18] S. Nakamura-Tsuruta, J. Kominami, A. Kuno and J. Hirabayashi, Biochem. Biophys. Res. Commun., 347 (2006) 215–220.

[19] S. Nakamura-Tsuruta, J. Kominami, M. Kamei, Y. Koyama, T. Suzuki, M. Isemura and J. Hirabayashi, J. Biochem., 140 (2006) 285–291.

[20] K. Ohtsubo, S. Takamatsu, M.T. Minowa, A. Yoshida, M. Takeuchi and J.D. Marth, Cell, 123 (2005) 1307–1321.

[21] M. Demetriou, M. Granovsky, S. Quaggin and J.W. Dennis, Nature, 409 (2001) 733–739.

[22] E.A. Partridge, C. Le Roy, G.M.Di. Guglielmo, J. Pawling, P. Cheung, M. Granovsky, I.R. Nabi, J.L. Wrana and J.W. Dennis, Science, 306 (2004) 120–124.

[23] A. Kuno, N. Uchiyama, S. Koseki-Kuno, Y. Ebe, S. Takashima, M. Yamada and J. Hirabayashi, Nat. Methods, 2 (2005) 851–856.

[24] K. Tachibana, S. Nakamura, H. Wang, H. Iwasaki, K. Tachibana, K. Maebara, L. Cheng, J. Hirabayashi and H. Narimatsu, Glycobiology, 16 (2006) 46–53.

[25] N. Kamekawa, K. Hayama, S. Nakamura, A. Kuno and J. Hirabayashi, J. Biochem., 140 (2006) 337–347.

Lectins: Analytical Technologies
C.L. Nilsson (Editor)

Chapter 11

Technical Aspects of Glycoprotein Enrichment

Yehia Mechref, Milan Madera and Milos V. Novotny

Department of Chemistry, Indiana University, Bloomington, IN 47405, USA

1. Introduction

Glycosylation occurs with proteins small and large. With an increase of molecular size, the occurrence of multiple glycosylation sites typically increases, adding to the complexity of glycosylation patterns and the usual difficulty of resolving fine structural differences in a large biopolymer. The successful attempts to resolve partially or completely different glycoforms of native glycoproteins are thus generally confined to relatively small proteins or glycopeptides. In biological materials such as cellular extracts and physiological fluids, glycoproteins are often encountered in minute quantities, placing high demands on both the measurement sensitivity and proper isolation procedures. A combination of orthogonal separation techniques and the use of affinity principles are the most commonly practiced isolation/enrichment strategies. Miniaturization of these separation methodologies represents a general trend in bioanalysis and is thus applicable to all stages of modern glycoanalysis.

Glycoproteins of interest may be encountered as either soluble or membrane-bound molecules. The isolation strategies will vary, depending upon whether such glycoproteins occur in cytosolic space, nucleus, extracellular space, cellular or subcellular membranes, etc. Also, any isolation protocol has to be adjusted to take into account the physicochemical properties of a given glycoprotein or a glycoprotein class. For example, detergents must be included in the extraction buffer to yield membrane-bound glycoproteins. For the extraction and fractionation of soluble glycoproteins, detergents are usually not needed. However, the addition of certain detergents can occasionally increase extraction yields and reduce contamination during purification.

The methods used in any glycoprotein isolation and enrichment protocols are very dependent on whether or not a biological activity must be retained for further studies. Many glycoproteins, including a number of membrane-bound molecules, can be easily denatured on contact with the surface of glassware or a chromatographic packing. If the sample is to be subjected to a protease cleavage or a release of glycans, isolation, or purification under harsher conditions is unlikely to affect adversely the final analysis. Generally, glycoproteins can be purified by most conventional protein separation methodologies, including gel electrophoresis and various chromatographic forms such as ion-exchange, size exclusion, reversed phase using C_{18}, C_8, or C_4 columns, hydrophobic interaction, and affinity [1, 2]. A most useful, specific isolation principle is currently the use of lectins that are immobilized on chromatographic resins.

Electrophoretic separations in gels are now widely applied to a variety of problems in the isolation and analysis of proteins. Simplicity of the apparatus and robustness are the chief selling points of both sodium dodecylsulfate polyacrylamide gel electrophoresis (SDS-PAGE) and two-dimensional electrophoresis (2-DE) in gels that combines the separation mechanisms based on molecular size and isoelectric points. Both techniques have found their wide application in glycoprotein isolation/analysis. Due to the recently advanced sensitivity of MS measurements, it has become feasible to analyze the contents of individual spots on a gel.

Mostly, SDS-PAGE has adequate resolving power to migrate proteins according to their size and to provide a rough estimate of their molecular weights. It is particularly informative to compare the "profiles" of proteins in a complex sample with those processed through a lectin chromatographic column, indicating which components of a mixture are glycosylated. Typical gel/buffer systems used in this procedure are essentially those described in the pioneering work by Laemmli [3]. However, SDS-PAGE of glycoproteins suffers from certain pitfalls. A positive bias in the estimation of molecular weights can be observed due to a lower SDS binding in presence of carbohydrate structures. Conversely, negative deviations are experienced due to the presence of charged sialic acids that contribute to the overall electrophoretic mobilities of glycoproteins. The glycoprotein bands observed on gels are often broad due to the tendency to resolve microheterogeneities.

In 2-DE separations, glycoproteins tend to be translocated into "trains" of spots as depicted in Fig. 1, reflecting their differences in both molecular mass and isoelectric points [4–9]. In various proteomic applications, the in-gel digestion of the isolated spots and a subsequent analysis by MALDI-MS or ESI-MS have become common in providing peptide fingerprints that can be matched with database entries. Although this appears relatively simple, one shortcoming with glycoproteins is an inadequate separation of glycoforms.

In the ultrasensitive measurements of glycobiology, an extreme caution must be exercised in manipulating the minute quantities of biomolecules. Glycoproteins at the microgram and lower levels, while becoming measurable with the modern instrumental techniques, can easily be adsorbed on the surface of glassware before such measurements. A sample loss during ultrafiltration, dialysis, lyophilization, etc., can easily become a bottleneck of the entire analysis. Another problem with working at such a reduced scale is contamination (dust, solvents, reagent impurities, etc.). It is

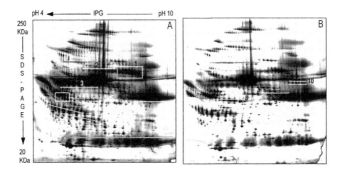

Figure 1. Silver-stained serum/plasma proteins after 2-D PAGE from an alcoholic patient (A) and a control subject (B). (1) Transferrin, (2) IgM μ chain, (3) IgA α chain, (4) α1-antitrypsin, (5) haptoglobin β chain and haptoglobin cleaved β chain, (6) Ig light chains, (7) IgG γ chain, (8) fibrinogen g chain, (9) fibrinogen b chain, (10) fibrinogen a chain. Reproduced from [4], with permission.

thus crucial to minimize the number of handling and transfer steps during the analysis. In terms of reduced column diameters, solvent flow rates and the overall surface area that a glycoprotein sample may encounter during analysis, miniaturized forms of separation are becoming significant in high-sensitivity work.

2. Practical Aspects of Lectin Affinity Chromatography

Lectins are specialized proteins that have been isolated from various plants and animal sources. For a number of years, they have been employed to study interactions with glycoconjugates [4, 10]. Lectins have also been widely used to isolate, purify, and characterize glycoproteins and glycolipids in various modes of affinity chromatography [10]. The lectin affinity separations in conventional columns have now been developed into a powerful means for purification of glycoproteins prior to structural studies [11]. These techniques are based on a reversible biospecific interaction of certain glycoproteins with the lectins immobilized to a solid support. In this mode of purification (Fig. 2), a sample containing both non-glycosylated and glycosylated proteins is usually exposed to the immobilized lectin, ideally causing just the glycoproteins to bind to the affinity medium. Non-glycosylated proteins are washed with an appropriate binding buffer, while the bound glycoproteins are subsequently displaced through washing with an appropriate elution buffer, which typically contains displacing hapten sugars. Over the years, various lectins with a high specificity toward oligosaccharides have been immobilized to agarose and other separation matrices. Some of these materials have been commercially available to isolate glycoproteins on the basis of their different glycan structures (see Table 1).

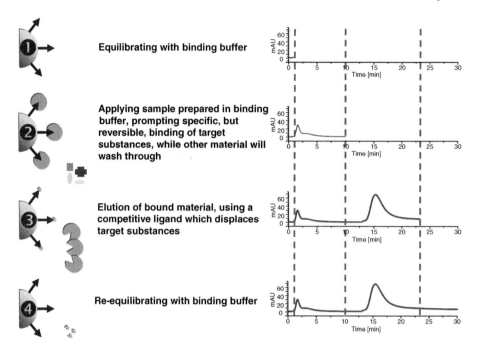

Figure 2. Typical lectin procedure.

The choice of affinity matrix is often critical and may depend on a particular application. Generally, a medium has to be insoluble in the commonly used aqueous buffers. It must either possess the reactive groups allowing a direct attachment of a lectin protein, or at least there must be a possibility of a facile derivatization of the surface to introduce such reactive groups into the matrix. This process is called the activation of a matrix. Additionally, the appropriate affinity support should be of a hydrophilic nature, causing minimum non-specific binding. Particle matrices are additionally required to be macroporous in order to allow the interaction of high-molecular-weight proteins with the immobilized lectins that are located predominantly inside the pores. However, in certain cases (when the capacity of affinity media is not an issue), lectin can be immobilized onto totally non-porous particles or solid surfaces to achieve fast, small-scale interactions. A list of the most commonly used affinity matrices that are commercially available is provided in Table 1. The list includes both the media that have been utilized for immobilization of lectins and those that constitute immobilized lectins as a commercial product.

Immobilization of a lectin to any type of a support involves several steps: (1) activation of a solid support; (2) immobilization of the lectin onto the activated matrix; and (3) blocking of excessive, unreacted active groups. The activation of

Table 1. Most commonly used materials for immobilization of lectin, their functional group, and their commercial sources.

Type of Sorbent	Product Name	Functionalization	Manufacturer
Agarose	Sepharose	NHS, CNBr, NH$_2$, Con A, WGA Variety of lectins NHS, SH, CNBr, epoxy, thiopropyl Con A, lentil lectin Variety of lectins Con A, WGA NH$_2$, SH, CHO, bromoacetyl Jackalin, Con A, WGA	Sigma-Aldrich EY Laboratories GE Healthcare Vector Laboratories EMD Biosciences MP Biomedicals
Cross-linked dextran	Sephadex		Sigma-Aldrich GE Healthcare
Polyacrylamide	Bio-Gel		Bio-Rad
Polyhydroxyethyl-methacrylate	HEMA, HEMA-BIO	epoxy, NH$_2$, DVS	Tessek
Bead cellulose Magnetic bead cellulose	IONTOSORB	NH$_2$, cyaunuric chloride, tosyl	Iontosorb
Polystyrene	SPHERO Polystyrene particles OptiLink	NH$_2$, COOH, epoxy COOH	Spherotech Seradyn
Magnetic polystyrene	SPHERO Magnetic Polystyrene particles SeraMag	NH$_2$, COOH, epoxy COOH	Spherotech Seradyn
Magnetic hydrophilic polymer	Dynabeads	tosyl, COOH, HN$_2$, epoxy NH$_2$, COOH	Invitrogen Kisker Biotech
Porous silica	LiChrospher NUCLEOSIL	NH$_2$, diol NH$_2$, diol	EMD Chemicals Macherey-Nagel
Non-porous silica Colloidal gold	SPHERO Silica	NH$_2$, COOH, epoxy NH$_2$, COOH Variety of lectins Maleimide, NH$_2$, sulfo-NHS	Spherotech Kisker Biotech EY Laboratories Nanoprobes
Rigid polymer	TSK-Gel AF Pak	tresyl, NH$_2$, epoxy, COOH, CHO Con A, WGA, RCA-I (packed columns only)	Tosoh-Biosep Shodex

solid supports is fairly specific, depending on the type of a used affinity matrix. Therefore, several types of derivatization procedures will briefly be discussed in the following sections, describing the use of specific matrices.

A covalent attachment of lectins to the activated surface or beads is generally achieved by mixing the lectin solution in an appropriate buffer and agitation with the solid for a certain period of time. Since lectins are usually sensitive to the extremes of pH, it is important to perform the coupling reaction in buffers with pH 6–8. It is also recommended to add sodium chloride to the coupling buffer, preventing non-specific adsorption of the ligand to the support surface. In order to protect a lectin-binding site and prevent its hindrance during the coupling process, the addition of a complementary sugar to the coupling solution is also recommended. Selection of the buffers, which are suitable for immobilization of a lectin, can be very broad; however, carbonate or phosphate buffers are most commonly used. Generally, a coupling buffer must be free of substances that would interfere with binding. For example, if a lectin is intended to be immobilized via its free amino groups, it is not possible to use TRIS buffers, as the primary amine of TRIS would be competing for the active immobilization sites.

The efficiency of lectin immobilization is a function of free lectin concentration, reaction time, and temperature. Increasing the concentration of lectin in the coupling solution usually leads to higher yields as long as the saturation of the matrix is not reached. Using higher lectin loads usually enhances the capacity of affinity media, but it can also introduce a significant source of non-specific binding, originating mainly from hydrophobic protein–protein interactions. In addition, binding constants expressing the affinity strength toward various carbohydrates are different for various types of lectins, so that using higher loads may in some cases aggravate elution of the bound glycoproteins.

Immobilization temperature and time are additional important factors that can also affect the resulting lectin load. In general, coupling reactions run faster at higher temperatures, but since the lectins (being proteins) can be sensitive to elevated temperatures, it is usually desirable to perform the coupling step at lower temperatures (e.g., 4°C) in order to protect biological activity. Naturally, this requires substantially longer reaction times, which sometimes can be as long as 48 h.

Upon completion of the coupling reaction, there are still unreacted active groups that need to be blocked with suitable low-molecular-weight compounds. Otherwise, irreversible, non-specific interaction with the matrix and a sample might take place. The choice of blocking reagents depends on the type of a coupling reaction, but the procedure is usually completed by agitating an immobilized lectin in TRIS or ethanolamine buffer at room temperature for 3 h. The matrix with an attached lectin is then washed with a high ionic strength buffer, resuspended in the buffer preventing lectin activity, and stored at 4°C until used.

3. Lectins Immobilized to Agarose Gels

The first attempts to use agarose as a chromatographic support for low-pressure lectin affinity chromatography in enrichment of glycoproteins date back to the late 1980s, with mainly *Concanavalin A* (Con A), wheat germ agglutinin (WGA), or *Len culinaris* (lentil lectin) employed for the affinity purification of membrane glycoproteins [12] or certain types of cell lines [13]. Agarose, as a linear polysaccharide consisting of alternating D-galactose and 3,6-anhydro-L-galactose units, connected by β(1,4) *O*-glycosidic bond, offers several properties meeting the critical requirements for being used as an affinity matrix: hydrophilicity, low tendency to non-specific binding, high permeability, and exclusion limit. Due to the polysaccharide backbone, the agarose surface features a substantial number of free hydroxyl groups which do not usually permit a direct attachment of lectins as affinity ligands, so they have to be modified prior to the immobilization of lectins.

Free polysaccharide hydroxyl groups can be derivatized through several types of chemical modifications. These modifications will be mentioned here briefly only, since a wide variety of agarose products with different surface modifications are commercially available already. Additionally, the commonly used activation procedures are summarized in Table 2 and discussed in detail elsewhere [11, 14].

Activation with cyanogen bromide (CNBr), devised by Axén and co-workers [15] and later improved by Kohn and Wilchek [16], has been very popular despite the high toxicity of CNBr. This reaction converts hydroxyl groups into reactive cyclic imidocarbonates, permitting a covalent attachment of proteins and peptides via free amino group. Activation with epoxiranes introduces a reactive epoxy ring to the agarose surface, permitting immobilization of the ligands containing free hydroxyl, amino, and

Table 2. Chemical procedures employed for the activation of agarose gels.

Reagent for Activation	Solvent	Activation pH	Functional Group in Lectin
CNBr (titration)	Aqueous, strong base	11–12	NH_2
CNBr (buffer)	Aqueous, buffer	9–10	NH_2
CDAP	Organic/aqueous	–	NH_2
DSC	Organic	–	NH_2
CDI	Organic	–	NH_2
Tosylchloride	Organic	4–5	NH_2, SH
Tresylchloride	Organic	–	NH_2, SH
Bisoxiranes	Aqueous	13–14; 8–10	OH, NH_2, SH
Epichlorhydrine	Aqueous	13–14	OH, NH_2, SH
DVS	Aqueous	13–14	OH, NH_2, SH

Abbreviations: CNBr = cyanogen bromide, CDAP = 1-cyano-4-(dimethylamino)-pyridiniumtetrafluoroborate, DSC = disuccinimidylcarbonate, CDI = carbonyldiimidazol.

thiol groups [14]. Although this reaction is much slower than the CNBr-based method, it provides high yields with the simultaneous incorporation of a hydrophilic spacer arm to the surface, allowing immobilization of low-molecular-weight ligands. Divinylsulphone (DVS) is yet another reagent used for the modification of the agarose surface, providing a highly reactive product that permits the attachment of ligands with free amino, hydroxyl, and thiol groups [14]. The enhanced reactivity of DVS-activated agarose allows immobilization procedures to be carried out at a lower pH in comparison with the bisoxirane-activated products. Modification with *N*-hydroxysuccinimide is based on the treatment of agarose with *N,N'*-disuccinimidyl carbonate (DSC). The resulting product is suitable for the immobilization of amino-containing ligands under very mild conditions [17].

As mentioned in the previous section, coupling of lectins is usually accomplished through mixing the activated agarose with a solution of lectin in an appropriate coupling buffer. It is generally recommended to use agarose, activated with CNBr or DSC, because in comparison with the epoxy-activated matrix, these derivatives provide substantially higher yields due to the low reactivity of epoxy ring under the neutral pH required for coupling of lectins. The amount of immobilized lectin can be modified by the initial concentration of lectin in the coupling solution, and is limited only by the saturation of matrix, or the solubility of lectin. However, due to the high price of some lectins, the difference in their binding constants and a higher tendency of adsorbents with excessive ligand loads toward non-specific binding, only moderate amounts of immobilized lectin are usually deemed sufficient for an efficient enrichment of glycoproteins. The list of commonly used lectins, immobilized on agarose gels together with their recommended loads, is shown in Table 3.

3.1. Activation of agarose gels

3.1.1. Activation of agarose with cyanogen bromide
According to Porath [14], spherical beads of Sepharose 4B are washed with 2 M potassium phosphate buffer at pH 12.1. A 10 g aliquot of a gel is commonly suspended in 10 ml of the same buffer. The suspension is then diluted with distilled water to a volume of 20 ml. A 4 ml of solution containing 100 mg of cyanogen bromide per milliliter is added in steps over 2 min. The temperature is maintained between 5 and 10°C under gentle stirring for 10 min. The product is washed on a glass filter with cold distilled water to neutrality.

3.1.2. Activation of agarose with N,N'-disuccinimidyl carbonate (DSC)
This procedure was introduced by Wilchek and Miron [17]. It involves a sequential washing of a 10-g aliquot of agarose gel with 100 ml portions of graded acetone–water mixtures (1:3, 1:1, and 3:1), culminating with dry acetone. Filtered gel

Table 3. Commonly used agarose-based lectins.

Name or Abbreviation	Source	Amount of Immobilized Lectin (mg/ml)
Con A (Concanavalin A)	*Canavalia ensiformis*	8–10
SNA-I	*Sambucus nigra*	2–3
MAA	*Maackia amurensis*	2–3
UEA-I	*Ulex europaeus*	4–5
Jacalin	*Artocarpus integrifolia*	2–3
PHA-L	*Phaseolus vulgaris*	4–5
Lotus	*Lotus tetragonolobus*	4–5
HPA	*Helix pomatia*	1–2
WGA (Wheat germ agglutinin)	*Triticum vulgaris*	4–5
RCA-I (Ricin)	*Ricinus communis*	4–5
LcH (Lentil)	*Lens culinaris*	4–5

EY Laboratories (St. Mateo, CA, USA).

is then resuspended in 10 ml dry acetone containing 800 mg DSC (at 4°C), followed by slow addition of either 0.75 ml triethylamine in 10 ml pyridine or 650 mg dimethylaminopyridine in 10 ml acetone under gentle stirring. After 1 h, the suspension is filtered, washed consecutively with cold acetone, 5% acetic acid in dioxane, methanol, isopropanol, and stored at 4°C in isopropanol, which assures the stability for at least several months.

3.1.3. Activation of agarose with bisepoxirane
This procedure was described in detail by Sundberg and Porath [18]. A 1-g aliquot of vacuum-dried agarose was washed with water and mixed with 1 ml of bisepoxirane and 1 ml of 0.6 M sodium hydroxide solution containing 2 mg/ml of sodium borohydride. The suspension was mixed by rotation for 8 h at 25°C, and the reaction was then stopped by washing the gel with large volumes of water (500 ml).

3.1.4. Immobilization of lectin to CNBr-activated agarose
As described by March et al. [19], 1-ml volume of activated agarose is washed with 0.2 M sodium bicarbonate (pH 8.1) and mixed with 1 ml solution of a lectin in the same buffer containing 0.5 M sodium chloride and 0.2 M methylmannoside. After gentle agitation at 4°C for 20 h, 1 M glycine is added to block the excessive unreacted groups, while the reaction is allowed to continue for additional 4 h. Subsequently, the lectin–agarose product is washed with the coupling buffer, water, an appropriate lectin-binding buffer, and stored at 4°C.

3.2. Selected applications of lectin immobilized to agarose gels

3.2.1. Isolation of glycoproteins from VNO

As a typical example for lectin affinity chromatography using agarose gels, vomeromodulin, a putative pheromone transporter of the rat vomeronasal organ (VNO), was isolated by lectin chromatography, purified, and subjected to an MS analysis [20]. The vomeronasal organs of adult female and male rats were dissected and broken into pieces on dry ice. The equivalents of approximately 1 mg of tissues were dissected in dry ice and homogenized in a polytron tissue disruptor with nine volumes of homogenizing buffer (50 mM TRIS-HCl, pH 7.5) containing 50 ml of protease inhibitor cocktail. The homogenizing buffer with the extract was centrifuged at 10,000g for 1 h. The supernatant was lyophilized and reconstituted in 2 ml of deionized water prior to lectin chromatography. Lectin chromatography was performed utilizing a column packed with 1 ml Con A–Sepharose. A binding buffer consisting of 10-mM TRIS buffer, pH 7.8, and 0.5 M sodium chloride, was pumped through the column at a flow rate of 200 ml/min. The same buffer containing 0.5 M α-D-methylglucopyranoside was used for elution. Separation was performed at 4°C. The lectin-enriched sample was then subjected to SDS-PAGE, as shown in Fig. 3. The SDS-PAGE run of the Con A-enriched fraction was shown to correspond to three bands, which included a 70-kDa protein corresponding to vomeromodulin.

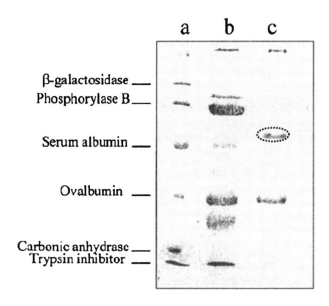

Figure 3. SDS–polyacrylamide gel electrophoresis of the vomeronasal tissue extract after lectin chromatography: (a) protein Mr markers; (b) proteins unretained by Concanavalin A; (c) retained fraction. Circled protein is vomeromodulin. Reproduced from Ref. [20], with permission.

3.2.2. Isolation of MUP glycoproteins from mouse urine

In another application, lectin chromatography was used to enrich a minor component of the major urinary protein (MUP) complex of the house mouse, which was then ascertained to be a previously suspected glycoprotein [21]. The minor glycosylated protein from the MUP complex was isolated and enriched using two chromatographic steps: (1) fractionation on a Superdex 75 column (prompting fractionation on the basis of molecular size) and (2) lectin chromatography on Con A column (enriching glycoproteins). The glycosylated protein was at trace levels compared to the other MUPs. Therefore, its isolation and enrichment was only possible using eleven 50-μl male urine aliquots. This protein is more acidic than the other MUPs, as indicated from its isoelectric focusing ($pI = 4.0$). This agrees with the fact that sialic acid residues were associated with this protein. Both the ion-exchange (upper trace) and lectin affinity chromatograms (lower trace) are depicted in Fig. 4. A MALDI mass spectrum of the enriched glycoprotein is shown in the figure inset in which the observed mass-to-charge

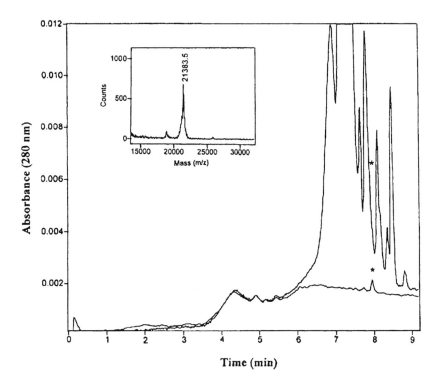

Figure 4. Anion-exchange chromatogram of the isolated MUP components (upper trace) and the Concanavalin A bound fraction (lower trace); inset is the mass spectrum of isolated glycoprotein, as indicated with asterisk. Reproduced from Ref. [21], with permission.

value is higher than the mass predicted from cDNA sequence, suggesting the presence of posttranslational modification (glycosylation), which was eventually established as tri- and tetra-antennary complex type with a wide heterogeneity, partially contributed to both the degree of sialylation and the linkages of galactose residues.

3.2.3. Enrichment of glycoproteins from snake venom using Con A–Sepharose

Recently, lectin affinity chromatography was employed to enrich glycoproteins from a snake venom [22]. A lyophilized sample was solubilized in a lectin-binding buffer (50 mM TRIS-HCl, 150 mM NaCl pH 7.5) to the concentration of 10 mg/ml, and 1 ml of the solution was added to 0.1 ml of Con A–Sepharose gel. The mixture was incubated overnight with continuous shaking. The gel was then washed five times with 2 ml of lectin-binding buffer to remove the unbound proteins, while the bound glycoproteins were displaced from the gel with 1 ml of binding buffer containing 0.2 M α-D-methylmannoside (Fig. 5).

3.2.4. Fractionation of glycoproteins from human blood serum using multi-lectin affinity chromatography

The multi-lectin affinity column was prepared by filling 0.5 ml Con A–agarose, 0.5 ml WGA–agarose, and 0.5 ml Jacalin–agarose into a disposable PD10 column [23]. Hundred microliters of human blood serum sample was diluted with the binding buffer (20 mM TRIS, 0.15 M NaCl, 1 mM Mn^{2+}, and 1 mM Ca^{2+}, pH 7.4) to 1 ml and applied onto the multi-lectin affinity column. After 15 min, the unbound proteins were eluted with 10 ml of binding buffer. Proteins bound to the Jacalin lectin were first released with 4 ml of 0.8 M galactose in 20 mM TRIS buffer (pH 7.4) containing 0.5 M NaCl. Then, the Con A-selected proteins were released with 4 ml of 0.5 M methyl-α-D-mannopyranoside in a 20 mM TRIS buffer (pH 7.4) containing 0.5 M NaCl. Finally, the WGA-bound proteins were released with 4 ml of 0.5 M N-acetyl-glucosamine in 20 mM TRIS buffer (pH 7.4) containing 0.5 M NaCl (Fig. 6).

4. Lectins Immobilized to Porous Silica

Due to its mechanical and chemical stability, hydrophilicity, and porosity combined with a well-defined spherical shape of particles, porous silica has become a popular chromatographic matrix utilized in many high-resolution separation techniques. These distinctive properties also permit porous silica to be employed as a stationary phase for high-performance affinity chromatography, as was demonstrated already in the late 1980s as an interesting alternative to low-pressure fractionation techniques. In comparison to its low-pressure counterpart, the

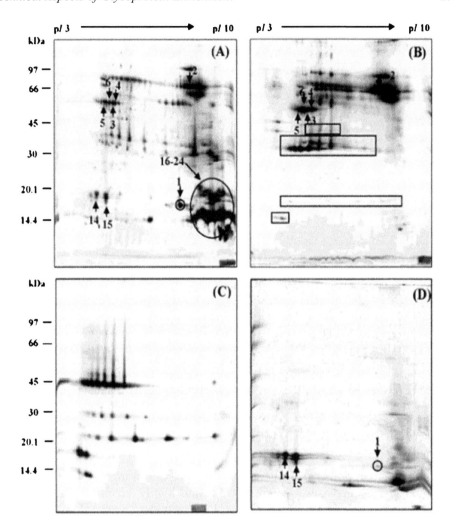

Figure 5. 2-DE gel images of *N. naja kaouthia* venom: (A) whole venom; (B) venom from Con A–agarose binding fraction; (C) control, Con A–agarose extracted with lysis buffer; and (D) venom from Con A–agarose non-binding fraction. All samples were dissolved in the same lysis buffer and use the same conditions for 2-DE. Squared spots are Con A–agarose proteins. Numbers are in order of identified proteins. Reproduced from Ref. [22], with permission.

silica-based high-performance lectin affinity chromatography (HPLAC) offers many advantages, such as sharper elution profiles, faster analyses, and the possibility of interfacing to various high-pressure, valve-based systems that may provide high throughput and automation. However, despite these evident merits, silica-based affinity matrices suffer from several disadvantages, which may limit their utility in certain applications. First, although the stability of porous silica

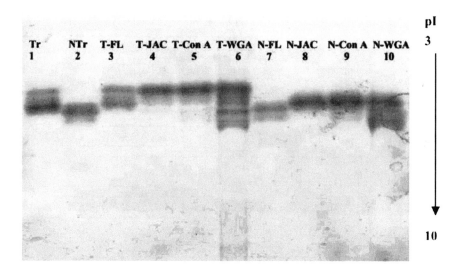

Figure 6. IEF gel separation profile of transferrin, asialotransferrin, and their fractions collected from M-LAC. Novex IEF gel, pH 3–10; loaded 5 µg protein in each well (native conditions); voltage: 100 V, 1 h; 200 V, 1 h; 500V, 30 min. Transferrin (Tr) and asialotransferrin (NTr) were loaded on two M-LAC columns. The unbound proteins were collected in the flow through (lanes 3 and 7, respectively). By sequentially using three displacers, the proteins captured by Jacalin, Con A, and WGA were separated and collected as three displacer fractions. Lanes 4–6 were the fractions from transferrin, and lanes 8–10 were the three fractions from asialotransferrin. In this profile, the displacement fractions from asialotransferrin showed a higher p*I*-value than those from the original transferrin, which indicated the success of removing sialic acids from transferrin by the use of neuraminidase. Reproduced from Ref. [23], with permission.

under acidic conditions is very good, shifting the pH towards basic values (above 8.0) significantly increases its solubility. It is thus highly recommended to perform coupling and fractionation procedures in buffers with the pH-values maintained below 8.0 to prevent a ligand leakage and protect the stability of siliceous matrices. Second, the presence of surface silanol groups introduces a significant source of non-specific binding, as based mainly on their ionic interactions. For this reason, and also due to the low reactivity, silanol groups have to be modified prior to the immobilization of lectins, which usually requires harsher conditions accompanied by longer reaction times.

The activation of silica surface is done exclusively through the silanization process, meaning that the reactive groups capable of forming a covalent bond with the lectin are introduced by the treatment of silica with functionalized silanes in either organic [24] or aqueous [25] solvents. A derivatized silica can then be subsequently modified in order to increase the number of possibilities for lectin attachment, as each particular way offers different coupling yields. Such a derivatization scheme, comprising several subsequent or parallel reaction steps, has been

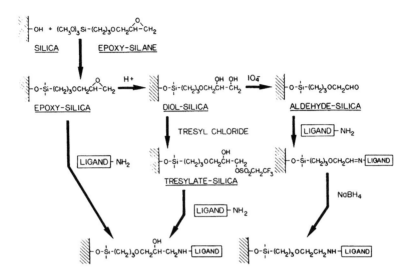

Figure 7. Different chemistries utilized for the immobilization of lectin to silica material. Reproduced from Ref. [26], with permission.

proposed by Larsson [26], as is depicted in Fig. 7. Bulk macroporous silica is first treated with 3-glycidoxypropyltrimethoxysilane to introduce active epoxy groups onto the surface. This reaction is usually done in dry toluene, in presence of a trace of triethylamine as a catalyst, and at elevated temperatures under reflux. A stream of nitrogen is continuously provided to maintain the anhydrous conditions, since epoxysilane is highly susceptible to hydrolysis in aqueous media. The resulting epoxy-silica can either react with the lectin directly, permitting a covalent attachment via free amino, hydroxyl and thiol groups, or it can be further converted to aldehyde- and tresyl-silica, which provide an alternative lectin attachment.

The bulk silica surface can also be modified through its coating with a functionalized polymer, which results in a uniform polymeric film covering the beads [27, 28]. This approach completely eliminates the tendency to non-specific binding, which is caused mainly by the presence of excessive silanols that did not completely react with the functionalized silane. The polymer-coated beads are then subjected to a further derivatization to gain the capability of a reaction with affinity ligands.

As shown in Fig. 7, multiple ways of derivatization of silica surface, starting with introduction of the active epoxy groups through treatment with 3-glycidoxypropyltrimethoxysilane, offer different coupling yields. A direct attachment of lectins to epoxy-activated silica is not very popular primarily due to its low reaction efficiency and also due to the fact that a lectin can be conjugated via more than one functional group, a possibility which may sterically hinder the

binding site. Conversely, an immobilization of lectins to aldehyde- and tresyl-
activated silicas has been demonstrated to result in coupling yields up to 97% and
used successfully in a for the large-scale fractionation of human blood serum glycoproteins [29] or structurally different oligosaccharides [30].

The efficiency of the above described coupling reaction is commonly illustrated
through the use of FITC-labeled Con A lectin and Texas red-labeled ovalbumin.
Fig. 8 depicts a microscopic picture of the derivatized silica (A) under white light,
(B) under the light of a wavelength suitable for the excitation of FITC, and
(C) under the light of a wavelength suitable for the excitation of Texas red. The
even distribution of the emission observed in Fig. 8B demonstrates an efficient
lectin coupling, whereas the light observed in Fig. 8C illustrates the efficiency of
glycoprotein trapping. The performance of the silica-based lectin materials was

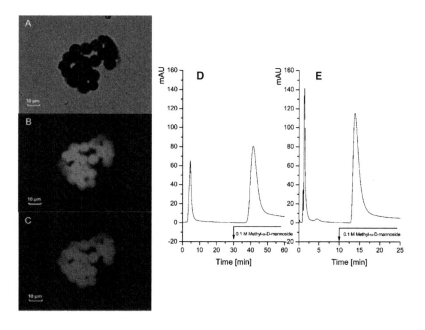

Figure 8. Microscopic photographs of the lectin beads coupled to FITC-Con A and treated with
Texas red-ovalbumin. Pictures were acquired under (A) white light, (B) light with excitation wave-
length of 480 nm suitable for FITC, and (C) light with excitation wavelength of 545 nm, suitable for
Texas red. Binding capacity of Con A Sepharose lectin (D) vs. Con A silica-based lectin (E).
Conditions: Columns, Con A Sepharose (1 mm × 50 mm, 10 μm, 1,000 Å) (D) and silica-based Con
A lectin with 60 mg Con A/g silica (1 mm × 50 mm, 10 μm, 1,000 Å) (E); buffers: loading buffer,
10 mM TRIS-HCl (pH 7.4), 0.5 M sodium chloride, 1 mM calcium chloride, 1 mM manganese chloride, 1 mM magnesium chloride, 0.02% sodium azide; elution buffer, the same as the loading buffer
containing 0.1 M methyl-α-D-mannoside; flow rate, 50 μl/min; 50 μg injected; UV detection at
280 nm. Reproduced from Ref. [53], with permission.

also evaluated against their Sepharose counterpart. The trapping capabilities of both the Sepharose-based Con A material (Fig. 8D) and silica-based Con A material (Fig. 8E) with ribonuclease B as the substrate were compared. The trapping efficiencies of both materials are very comparable, as was deduced from the similar peak areas resulting from these analyses. However, a shorter analysis time observed for the silica-based lectin can be attributed to the ability of these materials to withstand high back-pressures without adversely affecting their trapping efficiencies. This feature cannot be offered by the Sepharose-based material which endures gel shrinkage at high backpressures, thus lengthening the analysis time.

Despite some popularity of silica-based lectin columns since the late 1980s, the number of recent applications employing siliceous matrices is continuously being augmented by the emerging polymer monolithic column technologies, which will be discussed next. Examples of some applications utilizing various lectins covalently attached to derivatized porous silica beads are given in Table 4.

The advantages of silica-based lectin of small-scale affinity enrichment have recently been demonstrated by Madera et al. [31]. In this study, silica-based affinity microcolumns loaded with different lectins have been utilized in the sequential enrichment of human blood serum glycoproteins. Due to enormous complexity and diversity of glycan structures covalently attached to the glycoproteins in mammalian systems, four lectins having both broad and narrow specificities have been chosen to ensure the maximum enrichment. In addition, the difference in specificity characterizing the used lectins allowed profiling of human blood serum glycoproteome and creating maps that describe the changes in affinities of various glycoproteins to selected lectins. Con A lectin from *Canavalia ensiformis* offers rather broad specificity demonstrated by the preference to the glycans containing mannose, glucose, and galactose. For this reason, Con A would very likely interact with the majority of *N*-linked carbohydrates through the common chitobiose core. On the other hand, *Sambucus nigra* lectin (SNA), *Ulex europaeus* lectin (UEA-I) and *Phaseolus vulgaris* lectin (PHA-L) have very limited specificity, recognizing $\alpha(2,6)$ sialylated structures, $\alpha(1,2)$ fucose, and Galβ(1,4)GalNAcβ(1,2)Man trisaccharide, respectively. This limited specificity allows their use in monitoring certain glycosylation associated with glycoproteins. The study utilized these lectin materials to enrich glycoproteins prior to enzymatic digestion subsequent to peptide fractionation on C18 microcolumn prior to LC-MS/MS analysis. This study did not only demonstrated the efficiency of this approach, but also allowed the characterization of 108 glycoproteins from just 20 µl of human blood serum. In addition, this study illustrated the potential of small-scale lectin profiling as a valuable tool in proteomics investigations aimed at the discovery of glycosylated biomarkers from small sample amounts.

Table 4. Silica-based lectin affinity chromatography applications.

Lectin	Application	Reference
Con A (*Canavalia ensiformis*), SNA-I (*Sambucus nigra*)	Interfacing microcolumn lectin affinity chromatography on-line to high-resolution separation and detection techniques	[53]
	High-performance lectin affinity chromatography of membrane proteins	[54, 55]
	Affinity chromatography of glycoproteins on a lectin immobilized via metal interactions	[56]
	Separation of blood group A active oligosaccharides	[57]
PHA (*Phaseolus vulgaris*)	Combination of high-performance lectin affinity chromatography and size-exclusion chromatography of human serum glycoproteins	[29]
RCA-I, RCA-II (*Ricinus communis*)	Study of the interactions of *N*-glycanase-released oligosaccharides by means of high-performance lectin affinity chromatography	[58]
Con A, BSA-II (*Bandeiraea simplicifolia*)	High-performance lectin affinity enrichment of glycopeptides derived from human blood serum and cancer cell lines	[27]
LTA (*Lotus tetranogolobus*)	Comparative proteomics of dog serum glycoproteins exhibiting aberrant fucosylation	[57]
	Monitoring of unusual glycosylation in human serum glycoproteins	[59]
WGA (Wheat germ agglutinin)	Affinity separation of liver and bone alkaline phosphatase isoenzymes in human blood serum	[60, 61]
	High-performance lectin affinity chromatography of oligosaccharides	[62]
Polyporus squamosus	Rapid determination of binding constants using frontal affinity chromatography	[63]
	Rapid micropelicullar affinity purifications of glycoproteins	[54]
WGA, Con A	Lectin affinity enrichment of ovalbumin-derived glycopeptides	[64]
Con A, PSA (*Pisum sativum*)	Fractionation of membrane proteins	[65]

Enrichment and high-sensitivity profiling of human blood serum glycoproteins relying upon sequential microcolumn lectin affinity chromatography have also been used in a more automated fashion and implemented into the multimethodological approach involving immunoaffinity depletion, high-temperature Poroshell reversed-phase fractionation, and LC-MS/MS [32]. The different steps involved in this type of analysis are summarized in Fig. 9. Human blood serum depleted from the six most abundant proteins was introduced into the automated system comprising sample injector, silica-based lectin microcolumn connected on-line to C4 trapping cartridge and C8 Poroshell column through 10-port valve. Human serum sample was directed to the lectin microcolumn, where glycoproteins were bound. Next, bound glycoproteins were displaced from the lectin affinity media with the injection of appropriate elution buffer, on-line desalted on C4 trapping cartridge and fractionated on Poroshell C8 microcolumn maintained at 70°C. Fractions were collected into a 96-well plate and subjected to trypsin

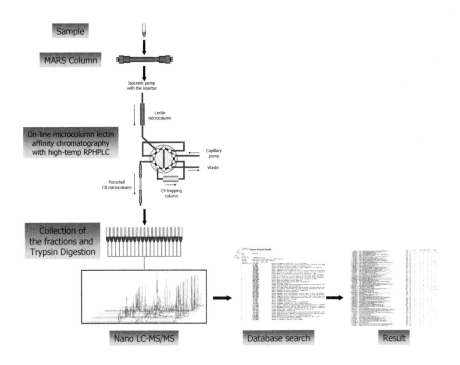

Figure 9. Experimental workflow chart. Human blood serum depleted from the major proteins is loaded on lectin microcolumn packed with silica-based lectin. The enriched glycoproteins are then separated on a pososhell C8 column and fractionated to 96-well plate. The collected fractions (20 fractions over 20 min) are then reduced, alkylated, and tryptically digested prior to LC/MSMS analysis performed using pulled-tip C18 capillary column (75 μm × 15 cm). Results are tabulated after database searching using MASCOT. Reproduced from Ref. [32], with permission.

digestion prior to nano-LC-MS/MS analysis. This multimethodological approach based on microcolumn lectin affinity chromatography employing the same four lectins described in the previous study [32] permitted identification of 271 unique glycoproteins form 16 μg of depleted blood serum. The fact that the concentration levels of some identified glycoproteins were determined to be nearly pg/ml range advocated this methodology to be an invaluable proteomic tool in biomarker discoveries at small-scale, and high-throughput.

5. Lectins Immobilized to Monoliths

The versatile features of monolithic stationary-phases have prompted their wide use as chromatographic media in many capillary electrochromatography (CEC) and nano-LC applications [33–42]. These desirable features include: (i) nearly limitless avenues to explore in polymer design and modification; (ii) the ease of fabrication, allowing a monolith confinement in large-diameter columns, capillaries, and microfluidic channels; and (iii) improved solute mass transfer due to the formation of macroporous structures, allowing the use of high flow rates without a penalty in resolution. The monolithic media are commonly designed in such a way that the mobile phase flows freely through macropores, while the surface area necessary for the separation of analytes is provided by the mesopores. Thus far, there have been only a limited number of applications demonstrating the utility of monolithic columns with immobilized affinity ligands for performing affinity-based analysis. Pan et al. [43] and Bedair and Rassi [44, 45] have recently demonstrated some potentials of affinity monoliths in nano-LC and nano-LC and CEC, respectively. Bediar and Rassi addressed the possibility of using monoliths for the preparation of lectin affinity-based media for nano-LC glycoprotein enrichment [45]. The application of lectin affinity monoliths was further expanded by Mao et al. [46] to include affinity-based separations in microfluidic chips.

Typically, monolithic stationary phases for affinity separations are prepared using ethylene dimethyl acrylate (EDMA) or trimethylopropane trimethylacrylate (TRIM) as the cross-linking monomer and glycidyl methacrylate (GMA) as the active monomer for a subsequent attachment. Zou and co-workers prepared monolithic capillary columns (50 mm × 4.6 mm i.d.) for affinity chromatography by an in situ polymerization procedure using GMA as a monomer and TRIM and EDMA as cross-linkers [43]. The porous properties of TRIM- and EDMA-based monoliths were different. More importantly, the rod prepared with GMA and TRIM was more mechanically stable than that prepared with GMA and EDMA. The GMA/TRIM column was subsequently employed to immobilize protein A and to determine the hIgG concentration in human serum. Additionally, yeast *Saccharomyces cerevisiae* mannan was immobilized to GMA/EDMA monolithic capillaries and utilized to isolate and separate mannose- or mannan-binding proteins or lectins through nano-LC and CEC [44].

At this date, monolithic materials incorporating lectins have been conservatively demonstrated. Con A and WGA were immobilized to the monolithic capillary columns prepared through polymerization of GMA with EDMA or [2-(methacryloyloxy)ethyl]trimethylammonium chloride (MAETA) to yield neutral and cationic macroporous polymers, respectively [45]. Wider macropores were observed for the neutral monoliths with an immobilized lectin. For the fabrication of neutral monoliths, the polymerization solutions weighing 2 g each were prepared from GMA 18% (w/w), EDMA 12% (w/w), cyclohexanol 58.8% (w/w), and dodecanol 11.2% (w/w). For the preparation of positively charged monoliths, MAETA 1% (w/w) was added to the polymerization solution (2 g) consisting of GMA 17% (w/w), EDMA 12% (w/w), cyclohexanol 59.5% (w/w), and dodecanol 10.5% (w/w). In all cases, 2,2′-azobisisobutyronitrile (AIBN, 1.0%, w/w, with respect to monomers) was added to the polymerization solution as the initiator. The solution was then degassed by purging with nitrogen for 5 min. A 40-cm long pretreated capillary was filled with the polymerization solution, up to 30 cm, by immersing the inlet of the capillary in the solution vial and applying vacuum to the outlet. The capillary ends were then sealed with a rubber septum, and the capillary submerged in a 50°C water bath for 24 h. The resulting monolithic column was washed with acetonitrile and then with water using an HPLC pump. The immobilization of lectin was achieved through the reaction between the epoxy groups found in the polymer structure and the ε-amino groups of the lectin. The glycidyl methacrylate monolithic column was first rinsed thoroughly with water and then filled with 0.1 M HCl solution and heated at 50°C for 12 h to hydrolyze the epoxy groups. The column was then rinsed with water, followed by a freshly prepared solution of 0.1 M sodium periodate for 1 h to oxidize the ethylene glycol on the surface of the column to an aldehyde. Lectin was then immobilized on the monolithic material by pumping a solution of lectin at 5 mg/ml in 0.1 M sodium acetate, pH 6.4, containing 1 mM of $CaCl_2$, 1 mM $MgCl_2$, 1 mM $MnCl_2$, 0.1 M, and 50 mM sodium cyanoborohydride through the column overnight at room temperature. The resulting column was then rinsed with water and cut to an effective length of 25 cm and a total length of 33.5 cm. The monolithic columns were then utilized in conjunction with nano-LC in a two-dimensional format to isolate (enrich) and separate glycoproteins [45]. This was demonstrated for a mixture of five model proteins, including glucose oxidase, human transferrin, conalbumin (an ovotransferrin), trypsinogen, and lactalbumin. The first three are glycoproteins. The mixture was introduced into the 2D system containing a short Con A column connected to a C17 monolith. While glucose oxidase and human transferrin have affinity to Con A, ovotransferrin does not. Therefore, glucose oxidase and human transferrin were captured by the Con A material, while the other proteins passed through and accumulated on the top of the C17 monolith. After this sample introduction step, the two columns were disconnected, and the C17 monolithic column was kept in the cartridge holder, while the lectin column with the adsorbed glycoproteins was taken out and let aside for further use. Elution from the C17

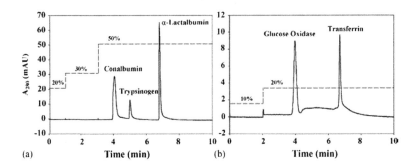

Figure 10. Two-dimensional separations of a mixture of some proteins using a Con A column (12 cm × 100 μm i.d.) in the first dimension followed by reversed-phase LC (RPLC) on a neutral C17 capillary column (25 cm effective length, 33.5 cm total length × 100 μm i.d.) in the second dimension. (a) Chromatogram obtained on the C17 column for the unretained proteins on the Con A column (passed through fraction); mobile phase used was 20 mM BisTRIS, pH 6.0, at 20% (v/v) acetonitrile for 1 min and then the acetonitrile content was increased stepwise to 30% at 1 min, and to 50% at 3.0 min. (b) Chromatogram obtained on the C17 column for the two glycoproteins transferred from the Con A column (i.e., previously captured by Con A column); mobile phase used was 20 mM BisTRIS, pH 6.0, at 10% (v/v) for 2 min and then the acetonitrile content was increased to 20% at 2.0 min. Con A column conditions are as follows. Binding mobile phase: 20 mM BisTRIS, pH 6.0, containing 100 mM NaCl, 1 mM $CaCl_2$, 1 mM $MnCl_2$, 1 mM $MgCl_2$; eluting mobile phase: 0.2 M Me-α-D-Man in the binding mobile phase. Pressure drop: 1.0 MPa for both running mobile phase and sample injection. Sample injection: 12 s. Reproduced from Ref. [45], with permission.

monolith by step gradients of an increased percentage of acetonitrile in the mobile phase allowed the separation of ovotransferrin, trypsinogen, and lactalbumin (Fig. 10a), while the two glycoproteins were "parked" on the Con A monolith. In second step, the Con A column was reconnected to the C17 monolithic column and subsequently eluted with the hapten sugar Me-α-D-Man, under which condition the enriched glycoproteins were moved to the C17 monolith. Thereafter, the Con A column was disconnected again and taken out from the capillary cartridge holder, and the C17 monolith was eluted by step gradients, which resulted in separating the two glycoproteins (Fig. 10b). This approach is an example of peak-parking in glycoproteomic analyses. While the ability of interfacing such a scheme to mass spectrometry was not demonstrated, the potential of this approach is undoubtedly clear. Configuring the monolith-based lectin and monolith-based reversed-phase columns into a valving system will allow automation and high-throughput analyses.

Lectin affinity chromatography was recently miniaturized into a microfluidic device using monoliths fabricated by polymerization of GMA and EDMA [46]. The monolith preparation and lectin immobilization was performed as described for the capillaries. The glass chips used were made by the standard photolithography

and wet chemical etching techniques. The cross-section of the microchannel was
70 μm × 20 μm. The layout of this chip is shown in Fig. 11A. In addition to the
required primary reservoirs, the reservoirs 5 and 6 were added for the microchannel washing to achieve a stable and reproducible electroosmotic flow (EOF)
(Fig. 11A). Using EOF as the driving force, the enrichment of three glycoproteins
(turkey ovalbumin, chicken ovalbumin, and ovomucoid) was achieved with this
microfluidic system. All the glycoproteins were successfully separated into several
fractions, with different affinities toward the immobilized *Pisum sativum agglutinin* lectin. The integrated system reduces the time conventionally required for the
lectin affinity chromatography by ∼30 fold, thus allowing a complete analysis to
be performed in 400 s. Also, lectin affinity chromatography is substantially simplified through integration of the different components into one microfluidic
device, so that minimizing sample handling has allowed the enrichment of glycoproteins at 300 pg level [46]. Further improvements in sensitivity are expected
upon coupling of this microfluidic chip to mass spectrometry.

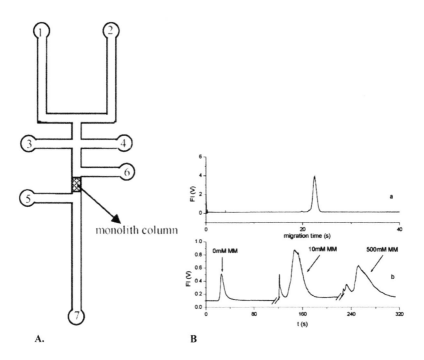

Figure 11. (A) Layout of the microfluidic chip: (i) running buffer reservoir (1); (ii) eluent buffer
reservoir (2); (iii) sample reservoir (3); (iv) sample waste reservoir (4); (v, vi) washing reservoir; (vii)
waste reservoir (7). (B) Electropherograms of ovalbumin by CCZE and the PSA-affinity microfluidic chip-based method. (a) Electropherogram of ovalbumin by CCZE; the experimental conditions
were the same as in Figure 6a. (b) Electropherogram of ovalbumin by PSA- affinity microfluidic
chip. Reproduced from Ref. [46], with permission.

6. Lectins Immobilized to Other Materials

Additional surfaces used for lectin immobilization include magnetic particles [47, 48], colloidal gold [49], and affinity membranes [50]. The immobilization to these different solids has been achieved using different chemical procedures, different lectin materials and demonstrated through analyzing different samples. However, in all cases, minimizing sample handling and reducing analysis times were common primary objectives.

6.1. Magnetic particles

In one approach, 100 µl (3 mg) magnetic porous glass beads with a long-chain alkylamine group (CPG Inc., Lincoln Park, NJ, USA) were transferred to a 1.5-ml tube, and the beads retained by a magnet [47]. The solution was removed, and the beads were resuspended in 200 ml reaction buffer consisting of 50 mM NH_4OAc, pH 7.8, 1 mM $CaCl_2$, 1 mM $MnCl_2$, again retained by a magnet, while the solution was removed. The amine groups of the beads were activated at room temperature under gentle rotation for 1.5 h with 200 ml 5% glutaraldehyde solution prepared in the reaction buffer. The beads were then retained by a magnet and the solution was removed, followed by four times washing with 200 ml reaction buffer. Next, the beads were incubated with the Con A solution, prepared in 100 ml reaction buffer containing 1% $NaCNBH_3$, for 3 h under rotation. After removal of the protein solution, the beads were incubated for 1 h with 200 ml 0.75% glycine, and 1% $NaCNBH_3$ in reaction buffer. Finally, the beads were washed four times with 200 ml reaction buffer before use. These beads were effectively employed for the enrichment of glycopeptides originating from a tryptic digest of ribonuclease B prior to a MALDI-MS analysis.

In another approach, carboxyl-functionalized magnetic beads (Chemicell, Berlin, Germany) were covalently modified with Con A, employing the carbodiimide chemistry [48]. Binding of the samples to Con A magnetic beads was achieved by incubating the sample and the beads for 1 h at room temperature under gentle agitation in the binding buffer. The beads were separated from the supernatant using a magnetic separation device (Bruker Daltonics, Leipzig, Germany). After washing, the bound glycoproteins/peptides were eluted under acidic conditions. The quality and specificity of the Con A beads were demonstrated by capturing different model proteins, i.e., bovine serum albumin (BSA), bovine lactoferrin, ovalbumin, bovine ribonuclease B, and bovine transferrin (1–5 µg, corresponding to 10–300 pmol protein). BSA served as a non-glycosylated "negative" control, while all other proteins were N-glycosylated. Binding of proteins on the Con A beads (20 µL) was performed under neutral conditions for 1 h at RT. After binding, washing, and elution, the supernatants and the eluates were analyzed by MALDI/TOF-mass spectrometry (Fig. 12A) and SDS-PAGE (Fig. 12B). As shown in Fig. 12, a

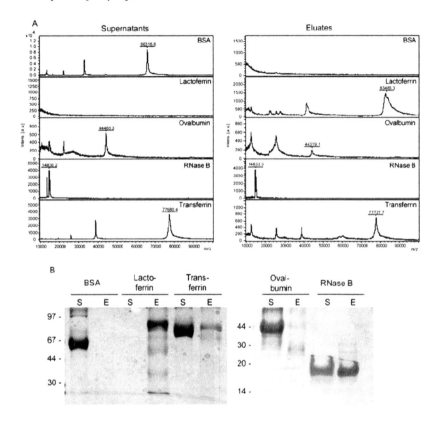

Figure 12. Analysis of supernatants and eluates of model proteins after binding on Con A magnetic beads by MALDI/TOF-MS (A) and SDS-PAGE (B). Samples were prepared with 2,5-DHAP on anchorchip targets for MALDI/TOF-MS analyses and measured on an autoflex II TOF/TOF. Reproduced from Ref. [48], with permission.

non-specific binding of BSA was not observed in either SDS-PAGE or the more sensitive MALDI/TOF-MS, while the other model proteins were bound, demonstrating the efficiency of this procedure. These magnetic beads are available commercially from Bruker Daltonics (Leipzig, Germany).

6.2. Gold foil

A lectin immobilized to gold foil has also been employed to isolate and enrich bacteria prior to MALDI-MS analysis [49]. Con A was immobilized via a self-assembled monolayer. Briefly, gold foil (99.9% pure, 0.1 mm thickness) was cut to approximately the same size as the sample well of a MALDI plate. The foil was extensively washed with 2-propanol which had been dried over the molecular sieves before use. The probes were then immersed in a saturated solution of dithiobis-(succinimidylpropionate) (Pierce, Rockford, IL, USA) in dry 2-propanol for 20 min with a gentle rocking. Next, the probes were repeatedly washed with dry 2-propanol, followed by dry ethanol. The probes were then immersed into a 20 μg/ml solution of the lectin Con A in binding buffer (100 mM potassium phosphate, pH 7.4, containing 100 μM CaCl$_2$) for 20 min with gentle rocking. An excess of lectin was used in order to ensure maximum binding to the activated surface. Following the coupling reaction, the probes were extensively washed with the binding buffer, followed by washes with purified water. The probes were then stored at 5°C. Finally, the probes were affixed to the MALDI plate with a double-sided tape.

Escherichia coli bacteria were prepared in a phosphate-binding buffer and 1 μl aliquot of the sample was applied to the probe. Next, the sample was allowed to incubate in a humid enclosure at room temperature for 2 h. This was performed to ensure optimum binding of bacteria to the lectin. The probe was then washed thoroughly with water prior to the addition of a MALDI matrix. This lectin-derivatized surface allowed samples to be concentrated and readily characterized using ~5,000 cells.

6.3. Membranes

Lectins were immobilized to a membrane surface and utilized in the analysis of a wide variety of microorganisms of importance to human health such as *E. coli* (K-12 and O157:H7) and *Salmonella typhimurium*, as well as the enveloped viruses (Sinbis AR-339) [50]. The probes were constructed using a commercially available affinity membrane (Gelman Ultrabind US450, Pall Gelman, Ann Arbor, MI, USA). The membrane was incubated for 2 h with 1 mg/ml solutions of the lectin (Con A or WGA) in the binding buffer (100 mM sodium acetate pH 7.4, containing 100 μM MnCl$_2$, 100 μM MgCl$_2$, and 100 μM CaCl$_2$). The membrane was washed extensively (3–5 ml, with gentle agitation) with the binding buffer to remove the unbound lectin and then immersed for 1 h in 5 ml of 100 mM TRIS buffer, pH 8.0, to block remaining reactive sites. The membrane was washed again with the binding buffer and water and stored in the binding buffer at 4°C until use. The biocapture surfaces were cut to the approximate size of a sample well of the MALDI sample plate and affixed to the plate using a double-sided tape. One microliter of a bacterial or viral sample, along with an equal volume of the binding buffer, was deposited on each capture surface. The samples were then transferred to a humidity chamber and incubated at room temperature for 2 h to ensure maximum binding. After binding, the samples were washed extensively (2–6 μl) with water to desalt and remove a non-specifically bound material. The bacterial samples were then treated with 1 μl of 5% aqueous trifluoroacetic acid (TFA) to lyse the cells, promoting the appearance of protein biomarkers. The viral samples were treated with 50% acetic acid solution to disrupt the virion, followed by addition of 1 μl of MALDI matrix. Using this approach, the bacteria samples were detected from physiological buffers, urine, milk, and processed chicken samples.

7. Enrichment with Hydrazide-Activated Particles

Enrichment of glycoproteins and glycopeptides employing the hydrazide chemistry, as originally devised by Aebersold and co-workers [51], has recently been used as an interesting alternative to the thus far preferred lectin affinity chromatography. As depicted in Fig. 13, a biological sample containing both non-glycosylated and glycosylated proteins is first subjected to periodate oxidation, where the hydroxyl

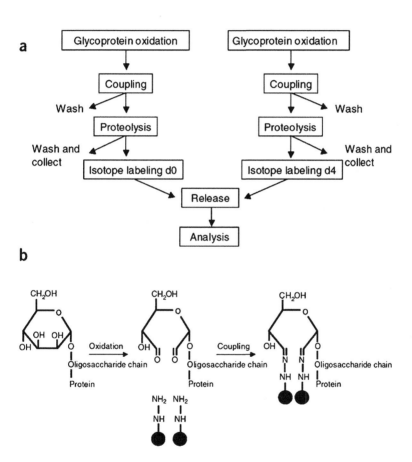

Figure 13. Schematic diagram of quantitative analysis of *N*-linked glycopeptides. (a) Strategy for quantitative analysis of glycopeptides. Proteins from two biological samples are oxidized and coupled to the hydrazide resin. Non-glycosylated peptides are removed by proteolysis and extensive washes. The non-glycosylated peptides are optionally collected and analyzed. The *N*-terminus of glycopeptides are isotope-labeled by succinic anhydride carrying either d0 or d4. The beads are then combined and the isotopically tagged peptides are released by PNGase F and analyzed by MS. (b) Oxidation of a carbohydrate to an aldehyde followed by covalent coupling to the hydrazide resin. Reproduced from Ref. [51], with permission.

groups in glycan moieties attached to the glycoprotein backbone are converted to aldehydes. The oxidized sample is then covalently bound to the hydrazine-modified resin via a Schiff-base chemistry. Next, all unbound sample is washed, while the glycan moities of the bound glycoproteins are released by PNGase F treatment, or subjected to the tryptic digestion, followed by elution of the unbound peptides and an enzymatic release of the bound glycopeptides.

Such hydrazide-modified beads are now commercially available, and if needed, they can be prepared by reacting polyacrylamide resins with hydrazine. This method works exclusively for the amide-carrying matrices, but can also be applied to various types of chromatographic supports after their proper modification.

In comparison to the affinity-based techniques, which are usually solvent-sensitive and essentially require using aqueous buffers, the hydrazide chemistry enrichment has one significant advantage, which is the lack of non-specific binding. Due to the strong covalent bond between glycopeptides and the hydrazine resin, non-glycosylated components can be extensively washed out by pure organic solvents, such as methanol or acetonitrile, eliminating any possibility of non-specific binding. Additionally, the bound glycopeptides can be subjected to a further derivatization prior to the release, such as isotope labeling; consequently, this methodology could become easily applicable to quantitative glycoproteomic studies.

Although this approach appears highly efficient, it still suffers from several drawbacks. First, the periodate oxidation followed by a covalent attachment of the modified glycan moieties to the hydrazide-activated beads results in the loss of any structural information pertaining to the oligosaccharides attached to the glycopeptide backbone. Second, PNGase F releases *N*-glycans only. Although *O*-glycosylated glycoproteins or glycopeptides are also efficiently enriched, they have to be released from the resin differently, thus complicating the analytical approach. Also, regeneration or reuse of the activated resin is not possible, since the *N*- or *O*-glycans remain covalently bound to the surface of the beads after the release of the enriched glycoproteins or glycopeptides.

The hydrazide chemistry has been successively evaluated with the set of standard glycoproteins and led to the identification of novel sites of *N*-glycosylation on α-1-antichymotrypsin [51]. The same approach has also been applied to the enrichment of human blood serum glycoproteins where 145 unique peptides corresponding to 57 unique proteins were identified, and to the enrichment of glycoproteins originating from cell membranes, where 104 unique peptides corresponding to 64 unique proteins were identified.

Recently, the hydrazide chemistry has played a key role in a comprehensive study involving multiple depletion and separation techniques, focusing on the analysis and identification of human blood serum glycoproteome [52]. After the enrichment, glycopeptides were subjected to an ion-exchange fractionation and LC-MS/MS, which resulted in a credible identification of 303 *N*-glycopeptides together with their corresponding *N*-glycosylation sites.

In this study [52], serum glycoproteins suspended in the coupling buffer (100 mM sodium acetate and 150 mM NaCl, pH 5.5) were oxidized by adding 15 mM sodium periodate at room temperature for 1 h. After removal of sodium periodate, the sample was conjugated to the hydrazine resin at room temperature for 10–24 h. Non-glycosylated proteins were then removed by washing the resin six times with the equal volume of 8 M urea in 0.4 M ammonium bicarbonate. After the last wash and removal of the urea solution, the resin was diluted with three bed volumes of water. Trypsin was added at a concentration of 1 μg of trypsin/200 μg of serum protein and digested at 37°C overnight. The peptides were reduced by adding 8 mM tris(2-carboxyethyl) phosphine hydrochloride at room temperature for 30 min and alkylated by adding 10 mM iodoacetamide at room temperature for 30 min. The trypsin-released peptides were removed by washing the resin three times with 1.5 M NaCl, 80% acetonitrile, 0.1% formic acid, then with methanol and with 0.1 M ammonium bicarbonate. *N*-linked glycopeptides were released from the resin by addition of PNGase F (1 μl/40 mg serum proteins) overnight. A summary of this protocol is outlined in Fig. 14.

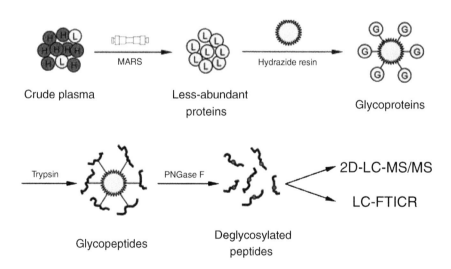

Figure 14. Strategy for plasma *N*-glycoprotein analysis using immunoaffinity subtraction, glycoprotein capture, and mass spectrometry. Crude plasma (red balls marked with "H" represent high abundance proteins; blue balls marked with "L" represent low-abundance proteins) is first subjected to multicomponent depletion using the pre-packed multi-affinity removal column, followed by incubating the flow-through (minus the retained abundant proteins) with hydrazide resin. The captured glycoproteins (blue balls marked with "G") are then washed and digested (on-resin) with trypsin, and the *N*-glycopeptides are specifically released from the resin by PNGase F. The resulting deglycosylated peptides can either be identified by 2-D LC-MS/MS or directly analyzed by LC-FTICR to validate the number of glycosylation sites. Reproduced from Ref. [52], with permission.

8. Conclusions

More than 100 years after their discovery, lectins remain a most remarkable class of proteins to be studied well into the future years. While unique biochemical specificities of new lectins are likely to be discovered in this process, the current studies of immobilized lectins aim at their standardization as useful media for biochemical isolations and bioanalytical studies. To this date, various investigations, which pertain to the immobilization of lectins to either traditional gels or newer materials, such as macroporous silica or monolithic matrices, represent both a very rich chemistry and successful applications to important scientific problems.

This is evident from the exponential increase in the number of applications employing lectin affinity chromatography for glycoprotein enrichment. The number of publications involving the use of lectin affinity chromatography as an analytical tool has rapidly increased from 300 to over 1,000 between 1977 and 1985 (Fig. 15). Since then, the increase in the use of this analytical tool has been steady; however, a sharp increase has been noticed over the past two years. This has mainly been driven by the realization that in proteomic analyses there is a need for the reduction of sample complexity through fractionation. Lectin affinity chromatography has been recently established as a very valuable tool for the fractionation of complex samples. This is due to the fact that samples are fractioned first into glycosylated and non-glycosylated proteins, and then into different fractions depending on the selectivity of a lectin material to certain glycan structures. With

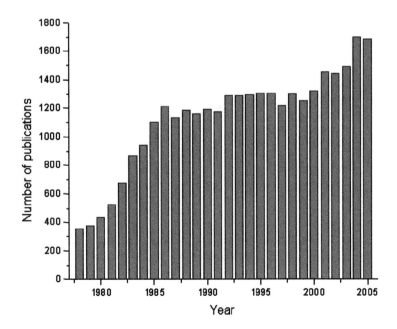

Figure 15. Number of lectin affinity chromatography application over the past 30 years.

the emerging importance of glycoproteomics and functional glycomics to the biomedical field, both the accumulated knowledge on lectins and their behavior, and a search for their better analytical uses will continue with the emphasis on comprehensiveness and high-sensitivity measurements.

References

[1] R. Kellner, F. Lottspeich and H.E. Meyer, Microcharacterization of proteins, Wiley-VCH, Weinheim, 1999.

[2] J.M. Walker, The protein protocols handbook, Humana Press, Totowa, 1996.

[3] U.K. Laemmli, Nature, 227 (1970) 680.

[4] P. Gravel and O. Golaz, The protein protocols handbook, Humana Press Inc., Totowa, NJ, 1996.

[5] J. Charlwood, J.M. Skehel and P. Camilleri, Anal. Biochem., 284 (2000) 49.

[6] C. Borchers and K.B. Tomer, Biochemistry, 38 (1999) 11734.

[7] D. Seimetz, E. Frei, M. Schnolzer, T. Kempf and M. Wiessler, Biosci. Rep., 19 (1999) 115.

[8] M.A. Smith, S.K. Bains, J.C. Betts, E.H.S. Choy and E.D. Zanders, Clin. Diagn. Lab. Immun., 8 (2001) 105.

[9] N.H. Packer and M.J. Harrison, Electrophoresis, 19 (1998) 1872.

[10] R.D. Cummings, Meth. Enzym., 230 (1994) 66.

[11] S.R. Carlsson, Glycobiology: A practical approach, Oxford University Press, Oxford, UK, 1993.

[12] R. Lotan, G. Beattie, W. Hubbell and G.L. Nicolson, Biochemistry, 16 (1977) 1787.

[13] V. Kinzel, D. Kubler, J. Richards and M. Stohr, Science, 192 (1976) 487.

[14] J. Porath, Methods Enzymol., 34 (1974) 13.

[15] R. Axén, J. Porath and S. Ernback, Nature, 214 (1967) 1302.

[16] J. Kohn and M. Wilchek, Chem. Anal., 66 (1983) 599.

[17] M. Wilchek and T. Miron, Appl. Biochem. Biotechnol., 11 (1985) 191.

[18] L. Sundberg and J. Porath, J. Chromatogr. A, 90 (1974) 87.

[19] S.C. March, I. Parikh and P. Cuatreca, Anal. Biochem., 60 (1974) 149.

[20] Y. Mechref, W. Ma, G. Hao and M.V. Novotny, Biochem. Biophys. Res. Commun., 255 (1999) 451.

[21] Y. Mechref, L. Zidek, W. Ma and M.V. Novotny, Glycobiology, 10 (2000) 1.

[22] J. Nawarak, S. Phutrakul and S.-T. Chen, J. Proteome Res., 3 (2004) 383.

[23] Z. Yang and W.S. Hancock, J. Chromatogr. A, 1070 (2005) 57.

[24] P.-O. Larsson, Methods Enzymol., 104 (1984) 212.

[25] R.D. Voyksner, D.C. Chen and H.E. Swaisgood, Anal. Biochem., 188 (1990) 72.

[26] P.-O. Larsson, Methods Enzymol., 104 (1984) 212.

[27] M. Geng, X. Zhang, M. Bina and F. Regnier, J. Chromatogr. B, 752 (2001) 293.

[28] F.N. Xi, J.M. Wu and M.M. Luan, Chinese Chem. Lett., 16 (2005) 1089.

[29] C.A.K. Borrebaeck, J. Soares and B. Mattiasson, J. Chromatogr., 284 (1984) 187.

[30] E.D. Green, R.M. Brodbeck and J.U. Baenziger, Anal. Biochem., 167 (1987) 62.

[31] M. Madera, Y. Mechref, I. Klouckova and M.V. Novotny, J. Chromatogr. B, 845 (2007) 121.

[32] M. Madera, Y. Mechref, I. Klouckova and M.V. Novotny, J. Proteome Res., 5 (2006) 2348.

[33] A. Palm and M.V. Novotny, Anal. Chem., 69 (1997) 4499.

[34] A.K. Palm and M.V. Novotny, Rapid Commun. Mass Spectrom., 18 (2004) 1374.

[35] A.K. Palm and M.V. Novotny, Rapid Commun. Mass Spectrom., 19 (2005) 1730.

[36] A. H. Que and M.V. Novotny, Anal. Bioanal. Chem., 375 (2003) 599.

[37] A.H. Que, Y. Mechref, Y. Huang, J.A. Taraszka, D.E. Clemmer and M.V. Novotny, Anal. Chem., 75 (2003) 1684.

[38] A.H. Que and M.V. Novotny, Anal. Chem., 74 (2002) 5184.

[39] D. Josic and A. Buchacher, J. Biochem. Biophys. Methods, 49 (2001) 153.

[40] A. Jungbauer, A. Buchacher and D. Josic, J. Chromatogr. B, 752 (2001) 191.

[41] F. Svec. In: Z. Deyl and F. Svec (Eds.), Capillary electrochromatography, Elsevier, Amsterdam, 2001, p. 183.

[42] F. Svec, E.C. Peters, D. Sykora and J.M.J. Frechet, J. Chromatogr. A, 887 (2000) 3.

[43] Z. Pan, H. Zou, W. Mo, X. Huang and R. Wu, Anal. Chim. Acta, 466 (2002) 141.

[44] M. Bedair and Z.E. Rassi, J. Chromatogr. B, 1044 (2004) 177.

[45] M. Bedair and Z.E. Rassi, J. Chromatogr A, 1079 (2005) 236.

[46] X. Mao, Y. Luo, Z. Dai, K. Wang, Y. Du and B. Lin, Anal. Chem., 76 (2004) 6941.

[47] T.N. Krogh, T. Berg and P. Hojrup, Anal. Biochem., 247 (1999) 153.

[48] K. Sparbier, S. Koch, I. Kessler, T. Wenzel and M. Kostrzewa, J. Biomol. Tech., 16 (2005) 407.

[49] J.L. Bundy and C. Fenselau, Anal. Chem., 71 (1999) 1460.

[50] J.L. Bundy and C. Fenselau, Anal. Chem., 73 (2001) 751.

[51] H. Zhang, X.-J. Li, D.B. Martin and R. Aebersold, Nat. Biotechnol., 21 (2003) 660.

[52] T. Liu, W.-J. Qian, M.A. Gritsenko, D.G. Camp II, M.E. Monroe, R.J. Moore and R.D. Smith, J. Proteome Res., 4 (2005) 2070.

[53] M. Madera, Y. Mechref and M.V. Novotny, Anal. Chem., 77 (2005) 4081.

[54] K. Kalghatgi, L. Varady and C. Horvath, J. Chromatogr., 458 (1988) 207.

[55] D. Josič, W. Hoffmann, R. Habermann, A. Becker and W. Reutter, J. Chromatogr., 397 (1987) 39.

[56] Z. El Rassi, Y.H. Truei, Y.F. Maa and C. Horvath, Anal. Biochem., 169 (1988) 172.

[57] J. Dakour, A. Lundblad and D. Zopf, Anal. Biochem., 161 (1987) 140.

[58] E.D. Green, R.M. Brodbeck and J.U. Baenziger, J. Biol. Chem., 262 (1987) 12030.

[59] L. Xiong and F.E. Regnier, J. Chromatogr. B, 782 (2002) 405.

[60] D.J. Anderson, E.L. Branum and J.F. Obrien, Clin. Chem., 36 (1990) 240.

[61] D.G. Gonchoroff, E.L. Branum, S.L. Cedel, B.L. Riggs and J.F. Obrien, Clin. Chim. Acta, 199 (1991) 43.

[62] Z. El Rassi, C. Horvath, R.K. Yu and T. Ariga, J. Chromatogr. B, 488 (1989) 229.

[63] B. Zhang, M.M. Palcic, D.C. Schriemer, G. Alvarez-Manilla, M. Pierce and O. Hindsgaul, Anal. Biochem., 299 (2001) 173.

[64] S. Honda, S. Suzuki, T. Nitta and K. Kakehi, J. Chromtogr., 438 (1988) 73.

[65] D. Renauer, F. Oesch, J. Kinkel, K.K. Unger and R.J. Wieser, Anal. Biochem., 151 (1985) 424.

Lectins: Analytical Technologies
C.L. Nilsson (Editor)
299

Chapter 12

Proteomic Techniques for Functional Identification of Bacterial Adhesins

Elisabet Carlsohn[a] and Carol L. Nilsson[b]

[a]Proteomics Core Facility at Göteborg University, Box 435,
SE 405 30 Göteborg, Sweden
[b]National High Magnetic Field Laboratory, Florida State University,
1800 E. Paul Dirac Dr., Tallahassee, FL 32310, USA

1. Introduction

Proteomics includes the identification, characterization and quantification of proteins including isoforms, polymorphism and modifications, and also definition of all protein–protein interactions and intracellular signaling that occurs in a cell under a given time and condition. Proteomics can be described as a multidisciplinary research activity in which separation science, mass spectrometry (MS) and bioinformatics play crucial roles. During comparative proteomic investigations of cell lines or microbial strains, lectins are frequently revealed as important functional modulators of a biological response. In the case of bacteria, surface lectins expressed at the microbial surface provide a means of attachment to host cell surface carbohydrates, the first essential step for colonization and infection. When the microbes' carbohydrate affinities are known, a functional proteomic approach to lectin identification can be employed, thus matching protein identity to carbohydrate ligand. We have previously employed a carbohydrate-containing cross-linking probe to select bacterial surface adhesins for trypsin digestion, MALDI-TOF MS and identification against genome sequence. That strategy was successful in the identification of the low-abundant Lewis b (Le^b)-binding adhesin of *Helicobacter pylori* (*H. pylori*) [1] as well as the identification of a sialic acid-binding adhesin from the same organism [2]. Protein identification was obtained through the enrichment of approximately 300 femtomoles of adhesin from solubilized cells. We present an overview of techniques used in proteomics and describe a functional proteomic approach for identification of a lactoferrin-binding protein of *H. pylori*.

2. Protein Separation Techniques

Proteomics is usually divided into expression proteomics, which includes analysis of protein expression, as well as quantification, functional and structural characterization of proteins in cell lysates and tissues, and cell-map proteomics, which attempts to define all protein–protein interactions and intracellular signaling that occurs in a cell under a given condition. Expression proteomics relies heavily on two-dimensional gel electrophoresis (2D-GE) [3, 4] computer analysis to reveal patterns of protein expression and mass spectrometric analysis of enzymatically cleaved peptides (Fig. 1). Cell-mapping proteomics is performed either by affinity purification and identification of protein complexes by MS or by direct DNA readout by yeast two-hybrid, phage display, ribosome display, and RNA–peptide fusion.

2.1. Analytical 2D-GE

Analytical 2D-GE, introduced 30 years ago by O'Farrell and Klose [3, 4], is a powerful method for separation and visualization of very complex mixtures of proteins. The technique relies on the fact that all proteins in an electric field

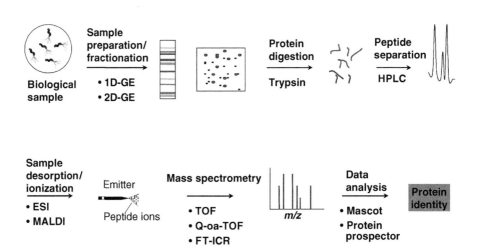

Figure 1. General procedures in mass spectrometry (MS)-based proteomics. The proteins are isolated and separated, usually by 2D- or 1D-GE. The protein spots are visualized by Coomassie, silver or flourescent staining and proteins of interest are cut from the gel and digested with trypsin. The peptide mixture may then be separated and enriched by high-pressure liquid chromatography (HPLC) prior to the desorption/ionization of the peptide molecules by either electrospray ionization (ESI) or matrix-assisted laser desorption/ionization (MALDI). Mass measurement and fragmentation analysis occur in the mass analyzers and the resulting mass spectra are used to search against known protein sequences in databases.

migrate at a speed that is dependent on their respective size and electric charge. Due to the differences in isoelectric point and molecular weight, the separated proteins will appear in different areas of the gel. The strength of analytical 2D-GE is the ability to visualize proteome maps in order to identify differences in protein pattern or expression of different protein isoforms of, for example, healthy versus nonhealthy cells or bacterial strains with different pathogenic properties.

The introduction of differential in-gel electrophoresis (DIGE) [5], which allows separation of two sets of protein mixtures from different sources (e.g., wildtype and mutant) in the same gel, has minimized previous reproducibility difficulties associated with 2D-gels. By labeling the different pools of proteins with fluorescent dyes (Cy2, Cy3 or Cy5), the protein profile can easily be imaged by the fluorescent excitation of the different dyes. The strongest feature of the DIGE technology is the possibility to corun an internal standard with the samples. In order to equalize gel-to-gel variations, spot volumes inside each gel and between different gels can be normalized to the volumes of the internal standard and prevent incorrect conclusions regarding biological variations. However, this approach has some limitations. Due to the corun of up to three different samples in one gel, it is not possible to load as much material on to the gel as when single samples are analyzed, making it difficult to study low-abundance proteins. In addition, labeling with any of the three dyes will add $\sim$500 Da to the molecular weight of the protein, which might cause issues when proteins with low molecular weights are analyzed.

One limitation of 2D-GE is that this technique tends to discriminate certain types of proteins. Hydrophobic proteins, such as outer membrane proteins (OMPs), the group in which most bacterial adhesins can be found, are difficult to solubilize in the IEF sample buffer. Solubilized proteins are prone to precipitation at their isoelectric point, resulting in escape from the transfer between the first to the second dimension. In addition, this technique possesses a limited dynamic range, so detection of proteins with a low copy number in a complex mixture will require either enrichment of the specific proteins or removal of high-abundance proteins (a recent review was written by Ahmed and Rice [6]). Several studies have been focused on improvement of 2D-GE protocols for making this procedure compatible with membrane proteins, but so far, the success has been limited [7–9]. Therefore, in order to achieve comprehensive analysis of all types of proteins, which is the overall goal in proteomics, the limitations of 2D-GE need to be overcome by developing new types of separation methods.

2.2. Alternatives to 2D-GE

One alternative technique to 2D-GE is two-dimensional liquid-phase electrophoresis (2D-LPE) [10], which is based on the same principles as 2D-GE except that the first separation step occurs in solution by use of preparative liquid-phase isoelectric focusing (IEF). By adding ampholytes to the sample, a pH gradient is generated and complex mixtures of proteins can be fractionated by an applied electric field. For a more detailed review of preparative IEF see Ref. [11]. In the second dimension, the proteins are separated by one-dimensional (1D) GE and then eluted directly from the gel into new liquid fractions. Because the proteins remain in solution, they are easily studied by either immunoblotting and/or MS.

Other proteomic approaches gaining popularity include gel-free separations based on online coupling of multi-dimensional liquid chromatography (LC) to MS instruments. Proteins are digested in solution and the resulting peptides are separated by 2D-LC. The first dimension usually separates the peptides according to their charge and in the second dimension they are separated according to hydrophobicity. Recently, several large-scale proteomic studies have used this approach. In a study by Washburn et al., nearly 1,500 proteins, including 131 integral membrane proteins, were identified from crude yeast extracts [12]. Vollmer et al. successfully used 2D-LC tandem MS (MS/MS) for proteomic analysis of E. coli cells [13].

A relatively new combination with high potential in proteomic analysis is the 1D-GE combined with nano-LC coupled to high-resolution MS analysis [14]. In this method, whole cell extracts are separated by 1D GE and the entire sample lane is cut into equally sized pieces, digested with trypsin and analyzed by nano-LC-MS. A major advantage of this approach is the high reproducibility of the 1D-GE, which also allow separation of proteins in a wide molecular mass range. By adding a fractionation step prior to the 1D-GE separations, this approach has been shown to facilitate the identification of low-abundance membrane proteins in H. pylori [15].

2.3. Reduction of sample complexity

The wide concentration range of proteins in biological samples usually complicates proteomic analyses. A common issue is that highly abundant proteins suppress the detection and identification of low-abundance proteins. However, recent studies have demonstrated that the introduction of a depletion step prior to either 2D-GE or LC-MS analysis makes it possible to analyze low-abundance proteins in, for example, human body fluids [16, 17].

Another way to reduce the complexity of the biological samples and enable identification of low-abundance proteins is to study the subcellular proteomes by use of subcellular fractionation. Subcellular fractionation is based on two major steps, disruption of the cellular organization and fractionation of the resulting cell lysate, in order to separate the organelles based on their different physiological properties. Centrifugation is the most efficient method for organelle isolation and can be used for fractionation of the total proteome into cytosolic, cytoskeletal, nuclear and membrane proteins.

In addition, differential detergent extraction methods have been used for enrichment of membrane proteins. It was demonstrated early that the inner and outer membrane of gram-negative bacteria differs in sensitivity to detergents. Sodium lauryl sarcosine (SLS), for example, has been shown to selectively disrupt the inner membrane and enable exclusive purification of the outer membrane [18]. Moreover, the use of TritonX-114 has been shown to facilitate enrichment of integral membrane proteins [19].

The combination of subcellular fractionation with digitonin and nano-LC-MS analysis was used by Nielsen et al. to identify 1,685 plasma membrane proteins from mouse hippocampus, of which more than 60% were membrane proteins [20]. For a recent review on fractionation strategies, see Ref. [21].

3. Biological Mass Spectrometry

By providing accurate mass measurement of low amounts (femtomoles) of peptides from gel-separated samples or complex mixtures, the opportunity to obtain sequence information, and the ability to characterize a number of post-translational modifications, establishes mass spectrometry as an essential tool in proteomic research. The principle of MS is to separate gas phase ions in vacuum and detect them according to their mass-to-charge ratio (m/z).

One could say that the history of MS began April 30, 1897 when Sir J.J. Thomson first announced the discovery of the corpuscles (electrons) to the Royal Institution at the University of Cambridge. Inspired by the work of the German physicist Wilhelm Wien, who in 1898 observed that canal rays (positive ions) could be deflected by strong electric and magnetic fields, Thomson started his own studies in 1905 on ions generated by residual gas molecules in discharge tubes. Wien used superimposed parallel electric and magnetic fields and the ions followed parabolic curves with different m/z ratios. Thomson used a similar setup but constructed a tube with much better vacuum and got sharper parabolas for the ions. In 1906 he received the Nobel Prize in Physics for his "investigations on the conduction by gases". These discoveries then led to the invention of the first mass spectrometer by Thomson and his student Francis Aston in 1912. More sophisticated mass spectrometers were subsequently designed and constructed by Arthur Dempster (1918), who used a 180° magnetic analyzer to focus ions into an electrical collector and introduced electron ionization, while Aston used both electrostatic and magnetic fields to focus ions on a photographic plate in his first generation of mass spectrographs (1919). About two decades later, Alfred Nier introduced sector magnets (60°) and later, together with his student Edgar Johnson, an electrostatic sector (90°) in combination with the sector magnet resulted in double focusing and was the initiation for the development of high-resolution MS of organic molecules. For a more detailed review, see Ref. [22].

3.1. Methods of ionization

In order to analyze biomolecules by MS, they must first be desorbed and ionized. Currently, there are two ionization methods used for mass spectrometric analysis of proteins and peptides. These are matrix-assisted laser desorption/ionization (MALDI) and electrospray ionization (ESI).

3.1.1. MALDI

In the beginning of the 1960s, it was demonstrated that irradiation of low-mass organic compounds with a laser pulse, produces ions that could be successfully mass analyzed. These experiments, however, revealed an upper mass limit. In 1981, experiments by particle-induced desorption methods like fast-atom bombardment (FAB) and the equivalent liquid secondary ion MS (LSIMS) showed that a matrix compound was needed to mediate the ionization [23, 24]. In 1987–1988, Franz Hillenkamp and Michael Karas showed that the use of a matrix compound in laser desorption could be used for desorption and ionization of larger biomolecules [25, 26]. They also found that introduction of the matrix allowed the laser incidence spot to be refreshed between each pulse, thus greatly enhancing shot-to-shot reproducibility. This was the foundation of MALDI. At the same time, Koichi Tanaka et al. [27] demonstrated laser desorption of biomolecules dissolved in glycerol containing a fine powder of cobolt particles using a nitrogen laser. Tanaka shared half of the Nobel Prize in Chemistry 2002 for this invention.

MALDI ionization forms mainly singly charged molecule ions in the gas phase. The analyte is first cocrystallized with a large molar excess of a matrix compound, usually a derivative of cinnamic acid, which absorbs UV light above 300 nm. Ions are produced by irradiating a spot of the sample by short pulses (a few ns) of UV light ($\lambda = 337$ nm) from a nitrogen laser. This allows the matrix to absorb energy, which causes rapid electronic excitation of the matrix molecules. Relaxation proceeds via radiationless states down to vibrationally excited states and finally into heat. This causes the matrix and analyte molecules in the irradiated spot to be desorbed in a miniature gas phase. In this gas phase, collisions between matrix molecules lead to protonation of the matrix. Hydroxyl groups in matrix compounds are much more acidic in the gas phase than in the liquid state. The analyte is then ionized by proton transfer from ionized matrix molecules. The acceleration of the ions into the mass analyzer occurs by applying a high voltage ($+20-30$ kV) giving all ions the same kinetic energy. As a pulsed ionization technique, MALDI is almost exclusively associated with time-of-flight (TOF) analyzers as the laser pulse starts the time measurements and the ionization (MALDI-TOF MS).

3.1.2. ESI

ESI for MS was first reported by Malcolm Dole in the late 1960s, who managed to ionize several high-molecular weight compounds at atmospheric pressure [28]. However, it was not until the mid 1980s, when the group of Fenn and Alexandrov independently used electrospray to generate gas phase ions that the development of the modern-day technique of ESI-MS occurred [29, 30]. Four years later, in 1988, the group of John Fenn et al. demonstrated identification and accurate mass measurement (0.01%) of polypeptides and mid-size proteins and the breakthrough of the ESI ionization technique was a fact [31]. For this invention, John Fenn shared half of the Nobel Prize in Chemistry 2002 with Tanaka.

The unique feature of ESI compared to other ionization techniques is the ability to form both singly and multiply charged ions and that the ionization occurs at atmospheric pressure. To generate ions by electrospray, the analyte, which is dissolved in an aqueous or aqueous/organic solution, is placed in a thin capillary emitter and a high voltage (typically 1–3 kV) is applied between the spray emitter and the inlet of the mass spectrometer. If the applied electrical field is high enough, charged micro-droplets will be formed from the liquid filament, the so-called Taylor cone (Fig. 2). Either application of a nebulizing gas streaming through the ion source in front of the Taylor cone or passage of the micro droplets through a heated capillary will cause the solvent to evaporate and the droplets to shrink. When the charge-to-size ratio reaches the Rayleigh limit (i.e., the point at which the electrostatic repulsion equals the surface tension), the increasing charges in the droplet overcome the surface tension and droplet fission occurs. This process, called Coulombic fission, will be repeated until only very

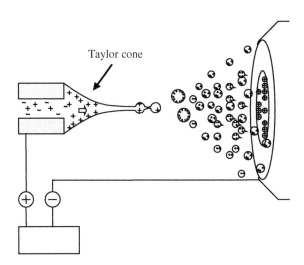

Figure 2. The electrospray process in positive ion mode. The spray is formed by the application of a high-voltage potential between the emitter tip and the inlet of the mass spectrometer.

small droplets with a size of around 10 nm remain. The actual formation of gas-phase ions from the small droplets has been intensively discussed and two different models, the ion evaporation and the charge residue mechanism have been proposed. The ion evaporation model suggests that the increase in charge density of the droplets of around 10 nm radius will force the analyte ion out of the solution before the droplets break up [32].

However, in the charge residue model, also named SID (single ion in droplet) model, Dole et al. suggest that the Coulombic fission will be continued until only very small droplets (less than 10 nm) containing only one ion will remain and evaporation of the solvent from this droplet will result in gas phase ionization of the analyte ion [28]. For small droplets ($R < 1$ nm) both of these theories may be correct, but when the size of the droplets increases (1 nm $< R < 10$ nm) it is most likely that ionization according to the charge residue model occurs. Ionization of multiprotonated proteins is best described by the charge residue mechanism.

The introduction of micro-electrospray [33] and later the nano-electrospray [34] have improved the ionization efficiency and thereby enabled the analysis of low sample volumes. Moreover, because nano-ESI is a softer ionization technique compared to traditional ESI, this ionization process will result in formation of ions with more charges and is therefore more useful in analysis of biological samples. The combination of ionization at atmospheric pressure and the continuous flow of solvent used in ESI allows direct coupling with separation techniques, such as liquid chromatography and capillary electrophoresis.

3.2. Mass analyzers for proteomic applications

The separation and mass measurement of gas phase ions occur under vacuum (10^{-6} Torr or better) in a mass analyzer. There are a number of different types of mass analyzers (e.g., TOF, quadrupole orthogonal acceleration TOF(Q-oa-TOF) and Fourier transform ion cyclotron resonance (FT-ICR)) currently used in proteomic research. Important instrumental properties include mass measurement accuracy, mass resolution, dynamic mass range, sensitivity, and capability of MS/MS.

3.2.1. Time-of-flight (TOF)
The first, in practice working TOF analyzer was introduced by Wiley and McLaren in 1955 [35]. This is the simplest of all mass analyzers, which separate ions according to their velocity and determine their m/z ratio from measurements of the flight time taken to travel through a field-free region (usually 1–2 m in length) between the ion source and the detector. The gas phase ion with charge (q) and mass (m) are affected with the same acceleration voltage (V) and thus obtain almost the same

kinetic energy. Ions of different m/z ratio will obtain different velocities and thus reach the detectors at different flight times. Thus, the flight time of the ions is directly proportional to the square root of their m/z ratio. Because the TOF analyzer is based on pulsed ionization, it is well suited in combination with MALDI and hence this technique became popular with the introduction of MALDI [25, 26].

A linear TOF spectrometer has a limited resolving power (experimental mass resolution $R = m/\Delta m$; where Δm is the full width at half-maximum) due to the spread in flight times of ions with the same m/z ratios. This is caused by (a) ions are not formed at exactly the same time; (b) ion formation is spread in space and (c) different initial kinetic energies. Two techniques are used to compensate for these phenomena. Ions can be focused by (1) delayed extraction [35], and/or (2) a reflectron [36]. In delayed extraction or time-lag energy focusing, the ions are formed in a field-free region and are focused using a pulsed two-grid ion source that was introduced already in 1955 by Wiley and McLaren. It is based upon the principle that ions of different distributions like (a), (b) and (c) above can be focused according to flight time by the use of a time delay (<400ns) before the acceleration voltage is applied as a potential gradient.

The reflectron is a type of ion mirror, which not only increases the flight path but also compensates for differences in kinetic energy for ions with the same m/z. The principal of the reflectron is that ions with a higher energy will penetrate a retarding field (a number of ring electrodes with increasing potential) more deeply. Thus, the slower ions will not go so far into the reflectron and thus catch up with the faster ones at the detector, and all ions with the same m/z will be better time-focused at the detector. A MALDI-TOF spectrometer equipped with both delayed extraction and a reflectron may have a resolution above 15,000. The high sensitivity of the MALDI-TOF instruments is due to multiplex recording, which means that all ions produced from one laser pulse is recorded without scanning and the spectra from 50 to 100 pulses are averaged into one single spectrum.

The first use of MALDI-TOF MS analysis in what we now address as proteomic research was reported by several groups in 1993. Mann and coworkers, for example, demonstrated the usefulness of peptide mass mapping (PMF) in which the molecular mass of peptides enzymatically cleaved from proteins are measured and used to identify proteins in sequence databases [37] and Henzel's group reported successful protein identification from only femtomoles of protein digests from *E. coli* cells separated by 2D-GE analysis [38]. The high sensitivity, wide mass range, ease of handling and automation has since then given MALDI-TOF MS a high impact in the field of protein research. In particular, PMF for identification of proteins from non-complex samples has become one of the most important biological applications and still forms the core for protein identification in proteomic research.

3.2.2. Quadrupole orthogonal acceleration time-of-flight (Q-oa-TOF)
The Q-oa-TOF mass analyzer is a hybrid instrument which combines the high sensitivity of the TOF mass analyzer with the ability to obtain efficient

precursor ion selection by use of the quadrupole mass filter and dissociation in a collision cell [39]. The Qoa-TOF analyzer was constructed for pulsed electron ionization already in 1989 by Dawson and Guilhaus [40] and was commercialized by Micromass (now Waters) with the Q-Tof® spectrometer in 1996. This type of instrument has also become very important for the development of biological MS.

There are several advantages with this hybrid instrument including improved resolution, mass measurement accuracy, and sensitivity compared to earlier instruments. The relatively high resolution and mass accuracy achieved on Q-oa-TOF instruments are mainly due to the narrow beam packet, which is pushed down into the TOF analyzer orthogonally to the transfer ion optics. In MS mode, the quadrupole and the hexapole (RF-only) act as focusing devices by transmitting ions to the TOF analyzer, where they are pushed down and separated according to their m/z. In MS/MS mode, the quadrupole is set to allow only ions within a very narrow m/z range (the normal isotopic cluster) to pass through into the collision cell for low energy collision-induced dissociation (CID). The process of CID includes multiple low-energy collisions of the analyte ion (precursor) with an inert gas, often argon, which converts translational energy into vibrationally excited states with increasing stretching of bond amplitudes. This leads finally to dissociation of the precursor ion into product ions, which are mass measured in the TOF analyzer.

3.2.2.1. Q-oa-TOF in proteomics
The Q-oa-TOF instrument equipped with a nano-electrospray source has greatly improved gas-phase sequencing of peptides. In 1996, Morris et al. reported low-femtomole- and attomole-range sequencing of peptides with relatively high mass accuracy obtained from a Q-oa-TOF prototype [39]. Later publications have verified the possibility of the Q-oa-TOF to sequence low-concentrations of peptides, which in turn has increased the level of confidence for protein identification [41]. In addition, de novo sequencing by this type of instrument has become crucial for the complete elucidation of protein primary structures, especially when the source organism is poorly represented or distantly related to any other organism in the protein- and expressed sequence tag (EST)-databases.

3.2.3. Fourier transform ion cyclotron resonance (FT-ICR)
A full description of FT-ICR MS and ion dissociation modes for this technique is given in Chapter 14 (Adamson and Hakansson). Briefly, this instrument has the highest performance of all mass analyzers available, for a review see Refs. [42, 43]. The instrument possesses extraordinary resolution and mass accuracy. For example, two peptides with a mass difference of less than 0.0005 Da (approximately the mass of an electron) were resolved by use of a home-built FT-ICR instrument equipped with a 9.4 T magnet [44]. The mass accuracy is typically in

the low ppm or sub-ppm level. The high performance is due to the fact that ions are detected by their cyclotron resonance frequencies, which can be measured extremely accurately. The ICR analyzer is a type of ion trap (a Penning trap) consisting of a cell under high vacuum (10^{-10} Torr) located inside a superconducting magnet (3–14.5 T). The stability of the magnetic field is essential and several parameters including resolution, mass accuracy, signal-to-noise ratio and dynamic mass range improve with increasing magnetic field.

Due to their high performance, FT-ICR instruments are very suitable for analysis of complex biological samples and thus their introduction has been an important complement to the MALDI-TOF and ESI-Q-oa-TOF instruments in proteomic research (for a review see Refs. [45, 46]). In addition, FT-ICR MS provides complementary fragmentation techniques for characterization of post-translational modifications such as glycosylation and phosphorylation [47–49].

The development of a new generation of hybrid FT-ICR instruments such as the hybrid linear ion trap FT-ICR mass analyzer (Fig. 3) have made FT-ICR instruments more user friendly. Recent studies [14, 50–53] by hybrid linear ion trap-FT, which combine high accuracy mass measurement of ions in the ICR cell with efficient fragmentation and detection in the linear ion trap, has shown its high potential for proteomic analysis. FT-ICR MS is now intensively used in both bottom-up (peptide levels) and top-down (intact protein) analysis in research laboratories. The approximate number of installed instruments worldwide is currently over 700, as compared to 200 instruments ten years ago.

3.3. Peptide fragmentation

One of the most powerful tools in MS is the fragmentation technique, which combines two stages of mass analysis (MS/MS). This is used when measurement of the

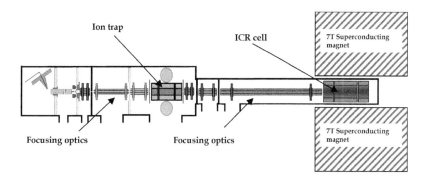

Figure 3. Schematic drawing of the hybrid linear ion trap FT-ICR instrument (LTQ-FT, Thermo Finnigan) equipped with a 7 T magnet.

molecular mass is not sufficient for identification, for example, identifying proteins from complex mixtures. The most common method of fragmentation is CID, also called collisionally activated decomposition (CAD).

3.3.1. Collision-induced dissociation (CID)

The sequencing of peptides by MS/MS analysis depends upon the formation of a complete set of sequence ions via random cleavage of each of the peptide bonds. Because this is the weakest bond in a peptide this type of fragmentation usually results in formation of *b*- and *y*-ions with charge on the N- and C-terminus, respectively (Fig. 4). Each peptide is usually cleaved just once. In a typical CID experiment, the precursor ions enter the collision cell containing high pressure of an energised, chemically inert collision gas – Ar, He, N_2, CO_2, etc. The precursor ion will undergo repeated collisions with the collision gas, building up potential energy in the molecule, until the fragmentation threshold is reached and the product ions are formed. The product ions are then analyzed in a second stage of mass analysis. Ideally, each of the product ions would be formed in equal abundance, allowing the peptide sequence to be read out from the MS/MS spectrum. This is, however, not the case because some sequence ions are not observed and some side reactions complicate the mass spectrum.

The simplest CID spectra to assign are those obtained from doubly charge tryptic peptides. There are several reasons why trypsin is preferable in proteolytic digestion of proteins. First, useful CID data is usually only obtained from peptides less than 2–3 kDa in size, and trypsin generally produces peptides from 500 to 3,000 Da. The second reason has to do with the desirability of placing basic residues, notably arginine, at the C-terminus of a peptide. It is a general observation in low energy CID that the presence of arginine in the middle of a peptide will

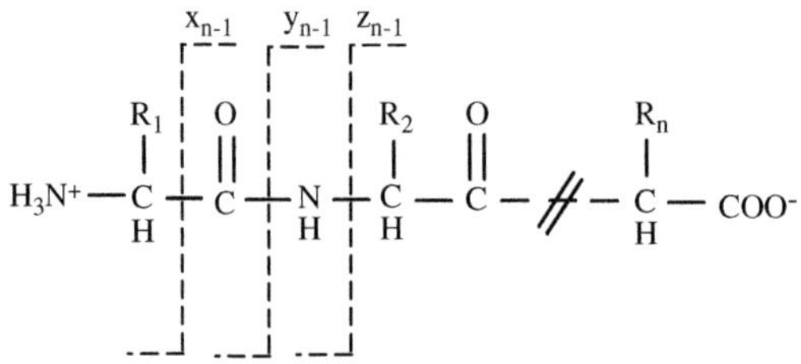

Figure 4. Nomenclature of peptide fragmentation. For protonated peptides the cleavage of the amide bond in the peptide backbone will generate different series of fragmentation ions [54]. In the formation of *a*, *b*, *c* ions the charge is retained at the N-terminus and the major N-terminus-containing ion series is the *b*-ion series. For *x*, *y* and *z* ions the charge is retained on the C-terminus and the major C-terminus-containing ion series is the *y*-ion series.

often result in the absence of fragmentations of peptide bonds adjacent to the arginine. Trypsin cleaves on the C-terminal side of arginine and lysine and by putting the basic residues at the C-terminus, tryptic peptides dissociate mainly by formation of a series of y-ions throughout the length of the peptide.

4. Bacterial Adhesins and Virulence

4.1. Helicobacter pylori

The first observation of this spiral-shaped bacterium in the human stomach was made in the late 18th century when Walery Jaworski discovered spiral bacteria in gastric sediments from humans. But it was not until 1983 when the Australian physician Barry Marshall and the pathologist Robert Warren [55, 56], managed to culture bacteria, which they had isolated from the human mucosal surface, that the bacterium was identified. Later, by drinking a cocktail of the bacteria and then treating the resulting gastritis with antibiotics, Marshall proved that these bacteria caused gastric inflammation. Warren and Marshall received the 2005 Nobel Prize in Medicine for their discovery. The bacterium was initially named *Campylobacter pyloridis* but was renamed in 1989 to *Helicobacter pylori* as specific morphologic, structural and genetic features indicated that it should be placed in a new genus.

This human/primate-specific microorganism is now known to colonize a majority of the global population [57]. In most infected individuals, the colonization is asymptomatic; however, *H. pylori* infection is known to be the major cause of chronic active gastric inflammation and peptic ulcer. Moreover, *H. pylori* has been shown to be a risk factor for the development of gastric adenocarcinoma and primate-specific B-cell mucosa-associated lymphoid tissue lymphoma [58–60]. In 1994, the WHO International Agency for Cancer Research declared *H. pylori* a human carcinogen of the highest class. The severity of infection is related to the degree of host response to the bacteria and to the polymorphism of the bacteria. The prevalence of *H. pylori* infection is increased in developing countries (up to 90%) where the great majority is colonized early in life [61–63]. Factors found to be associated with infection include overcrowding during childhood and poor hygiene [64]. The transmission procedure is undetermined but may result from a faecal–oral, an oral–oral [65] or a gastro–oral route [64, 66, 67]. Food and water have been suggested as vehicles of transmission [68]. The treatment of *H. pylori* infection is complicated, requiring a minimum of two different antibiotics together with a proton-pump inhibitor [69, 70]. Due to both the increased rate of antibiotic resistance among *H. pylori* strains and the high rate of reinfection in developing countries, for instance, 73% within 8 months in Peru [71], the development of a vaccine against *H. pylori* is urgently needed.

4.2. Virulence factors in H. pylori

Why will only a minor number of *H. pylori*-infected individuals develop gastro-duodenal disease? The answer probably lies in a combination of several factors including microbial, host-genetic and environmental ones. In order to establish colonization and maintain infection, *H. pylori,* in common with other pathogens express a variety of virulence factors. These include the urease enzyme, which neutralizes the acidic human gastric secretions in order to create a favorable environment for the bacteria, the flagella, which provide motility and a number of adhesins, which possess adhesion properties that allow the bacteria to attach tightly to gastric epithelial cells (described in detail under "*H. pylori* outer membrane proteins"). Furthermore, *H. pylori* produces a number of factors harmful to the host. These factors include the CagA pathogenicity island (*Cag PAI*), *H. pylori* neutrophil-activating protein (HP-NAP) and the VacA cytotoxin.

4.3. H. pylori outer membrane proteins (OMPs)

OMPs are a class of proteins present at the surface of all gram-negative bacteria, including *H. pylori*. These proteins are key molecules that interface the bacteria with the environment and they are involved in a number of important processes, including molecular transport across the membrane and interaction with the host carbohydrates. The genome sequence of the two sequenced *H. pylori* strains 26695 and J99 contains one major outer membrane family of 33 (32 in J99) genes, which encodes *H. pylori* OMPs [72, 73]. Sequence analysis of this OMP family revealed identification of several highly conserved motifs which divided the family into two different closely related subfamilies, named *H. pylori* OMPs (Hop) and Hor (Hop-related), see Fig. 5. There are 21 members of the Hop family, which contain highly conserved N-terminal and C-terminal domains. The Hor family is composed of 12 (11 in J99) members that display an overall sequence similarity to the Hop proteins but does not contain the conserved N-terminal motif. All Hop and Hor proteins contain a conserved C-terminal region with alternating hydrophobic and hydrophilic amino acid residues, which have been predicted to form transmembrane β-strands [72, 74, 75]. Interestingly, the surface-exposed regions of the OMPs display a high rate of sequence diversity, which might reflect the different properties of the OMPs.

4.3.1. Adhesins
The majority of *H. pylori* bacteria found in the human stomach are motile and only a minor part of the bacterial population is found to be attached to the epithelial cell surfaces. However, adherence to the gastric epithelium is essential for *H. pylori* colonization, survival and infection. In order to establish this connection to the

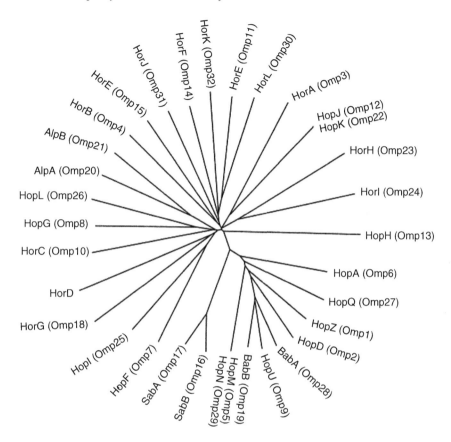

Figure 5. Phylogenetic tree of the two major outer membrane protein (OMP) families, Hop and Hor. The Hop family consists of twenty-one members including all identified adherence-related proteins (BabA, SabA, AlpA, AlpB and HopZ). The Hor family consists of twelve proteins with unknown function. The phylogenic tree is modified from Alm et al. [75].

host, *H. pylori* expresses a number of adhesins on its cell surface. These adhesins bind mainly to specific carbohydrate-based receptor structures exposed on the epithelial surface of the host, and protein identity can be matched to carbohydrate affinity through a functional proteomic approach.

One of the most well-characterized adhesins is the blood-group antigen binding adhesin BabA, which is responsible for binding to the Le^b antigen on human cells [76]. Expression of BabA is not found in all *H. pylori* isolates [76, 77] and is not a prerequisite for *H. pylori* colonization. However, it has been shown that presence of Le^b in the gastric epithelium is essential for attachment of *H. pylori* to gastric epithelial cells and prerequisite for chronic gastritis [78].

The sialic acid-binding adhesin SabA, which binds to sialylated Lewis x (sLe^x) antigens expressed during chronic inflammation was recently identified by

Mahdavi et al. [2]. The authors found that less than 50% of the examined *H. pylori* isolates bound to sLe^x.

Results from adhesion experiments of the closely related OMPs AlpA and AlpB indicate that both these proteins are involved in the specific binding of *H. pylori* to human epithelial cells [79, 80]. In addition a recent study by de Jonge et al. demonstrated that AlpB is required for colonization of the guinea pig stomach [81]. No human receptor for these two proteins has yet been identified.

Another *H. pylori* OMP, the HopZ protein has also been suggested to be involved in adhesion. Peck et al. showed that *H. pylori* strains expressing the HopZ protein were able to adhere to epithelial cells, whereas strains lacking HopZ showed reduction in the number of adhesive bacteria [82]. However, in contrast to these results, de Jonge et al. found no correlation between presence of HopZ and colonization efficiency [81].

It should be noted that all five of these potential adhesins are members of the Hop family and that SabA, BabA and HopZ are located in the same cluster in the phylogenic tree of the Hop and the Hor families [75].

The *H. pylori* adhesin A (HpaA) was first identified as a sialic acid-binding adhesin [83], but was subsequently identified as a lipoprotein [84]. Several studies have demonstrated the surface location of the HpaA protein [83, 85, 86]; however, HpaA was also found located in the bacterial inner membrane, in the cytoplasm [84] and later also associated with the flagellar sheet [87]. The expression of HpaA has been found to be highly conserved among *H. pylori* isolates [88, 89] and HpaA is recognized by antibodies from *H. pylori* infected individuals [88–90]. A recent study did also show that HpaA is an important colonization factor essential for *H. pylori* colonization in mice [15]. Taken together the data make this protein an interesting candidate in the search for potential vaccine antigens that can be used to prevent *H. pylori* infection.

5. Proteomic Analysis of *H. pylori*

Since the completion of the genome sequence of the two *H. pylori* strains 26695 and J99 in 1997 and 1999, respectively [72, 73], a large number of proteomic analyses have been applied to this pathogen. This has made the *H. pylori* proteome one of the best characterized microbial proteomes.

The first proteomic investigation of *H. pylori* aimed towards identification of diagnostic and vaccine candidates [91]. By use of 2D-GE and MALDI-TOF analysis, McAtee and coauthors identified twenty proteins including urease, flagellin, and AlpA, which were found to be reactive with sera from *H. pylori* infected patients. Two years later, the group of Jungblut et al. presented a comparative proteome analysis of three different *H. pylori* strains. They used MALDI-TOF MS to identify 126 proteins from strain 26695. Several virulence factors, including urease and HpaA, were detected, but no OMPs were identified. Their main finding was the high proteomic variability between the strains, which most likely depends on shifts in the amino acid composition of certain proteins [92].

In 2002, Sarabath et al. published a new proteomic approach in which intact *H. pylori* cells were biotinylated followed by affinity purification of membrane proteins using streptavidin. Several virulence factors, including two OMPs (HefA and HP1564) were found among the eighteen identified proteins [93]. The same year, Jungblut and coworkers used immunoproteomics for identification of *H. pylori* antigens. They reported that a number of antigens, including some surface proteins, were recognized differently by sera from patients with different clinical outcomes, and thereby demonstrated the potential to use certain proteins as candidate indicators for clinical manifestations [94]. This group also published a study in which they performed proteomic analysis for characterization of the *H. pylori* secretome [95].

Later, Hynes et al. used a protein chip technology for the comparison of OMP profiles between *H. pylori* strains and found alterations in the protein profile between culture collection strains and clinical isolates with low numbers of passages [96]. In 2004, Lee et al. presented a proteomic analysis of a ferric uptake regulator *H. pylori* mutant [97] and Baik et al. used subcellular fractionation in combination with 2D-GE analysis to identify sixteen OMPs expressed by *H. pylori* strain 26695. Four OMPs (Omp11, Omp14, Omp20 and Omp21) were found to be immunoreactive [98]. Recently, a subproteomic study resulted in identification of numerous virulence factors including some OMPs [99]. These authors are now aiming towards the establishment of a dynamic 2D-GE reference database with multiple subproteomes of *H. pylori*.

Standard proteomic approaches can be useful for mapping protein expression, but cannot be used easily to assign protein functions to their identity. For adhesins, knowledge of a receptor saccharide should be possible to use for the functional identification of the microbial protein, provided that the genome sequence of the microbe is available, using a proteomics approach combined with affinity tagging.

6. Functional Identification a Lactoferrin-Binding Protein of *H. pylori*

Human lactoferrin (hLf) is an iron-transporting glycoprotein of the transferrin family, synthesized by various exocrine glands [100] and neutrophils [101] and released from the latter during inflammation. Lf has a number of properties, which give it an important role in the nonspecific host defense system, such as liberation of bactericidal peptides and iron binding characteristics [102–104]. Because iron is an essential growth factor for several microorganisms, the iron-chelating activity of Lf can inhibit their metabolic activity [105–107]. To overcome iron limitation, the microbes have developed several iron-uptake mechanisms. One of them is expression of surface-bound iron-binding proteins, for example, lactoferrin binding proteins (adhesins), which may be involved in the adherence of the bacteria to, for example, carbohydrates attached to the surface of hLf. Human Lf contains three possible N-glycosylation sites, Asn[138] in the N-lobe and Asn[479] and Asn[624] in the C-lobe [108]. Spik et al. [109] have shown that hLf contains two glycan chains of the sialyl N-acetyl-lactosaminic type and have suggested that N-glycosylation at Asn[624] does not occur.

Surface expressed adhesins, which usually are expressed in low amounts, can be very attractive to study in the development of new therapeutic drugs or vaccine.

In this study we performed affinity cross-linking on the human gastric pathogen *H. pylori* in order to identify a lactoferrin-binding protein. We also performed 2D DIGE in order to study differences in protein expression in *H. pylori* strains with different capacity of lactoferrin binding.

6.1. Materials and methods

6.1.1. Bacterial strains and growth condition
The *H. pylori* wild-type strains 26695, 17874 and 17875 were obtained from the Culture Collection, Göteborg University. The cells were stored at $-80°C$ in soy broth containing 20% glycerol by volume, and were grown on Luria broth plates containing 1.5% agar and ampicillin 100 µg/ml at 37°C. After 24 h, the cells were scraped off and washed three times in phosphate-buffered saline (PBS).

6.1.2. Labeling of receptor conjugate
Human Lf (Sigma #L0520) was conjugated to the trifunctional reagent sulfo-SBED (sulfosuccinimidyl[2-6-(biotinamido)-2-(*p*-azidobanzamido)-hexanoamido]ethyl-1-3′-dithiopropionate, Pierce, USA) following the instruction of the manufacturer. Briefly, 5 µl sulfo-SBED (10 µg/µl) in dimethylsulfoxide (Merck) was added to 100 µl receptor conjugate (1 µg/µl) in 0.1 M potassium phosphate buffer, pH 7.2. The mixture was incubated at room temperature in the dark for 1 h. The product was desalted on a HiTrap column (Amersham Pharmacia Biotech, Sweden) in a 1 ml fraction.

6.1.3. Tagging of H. pylori lactoferrin-binding protein
The protein was tagged using a procedure previously described [1]. In short, freshly harvested bacteria from approximately one-third of a cultivation dish were incubated for 30 min at ambient temperatures with labeled lactoferrin and the photoreactive aryl azide cross-linker group was activated by ultraviolet radiation. The disulfide bond on the linker was reduced by the addition of dithiothreitol (DTT) to give a final concentration of 50 mM. Finally, the tagged bacteria were washed three times in PBS, pH 7.4, and frozen.

6.1.4. Enrichment of biotinylated bacterial proteins with
streptavidin-coated beads

Tagged bacterial pellets were dissolved in 2% SDS containing 25 mM Tris, pH 8.0, followed by dilution to 0.5% with 25 mM Tris and 30 U endonuclease (Sigma) and incubated at 37°C for 30 min. 200 µl Bio Mag streptavidin-coated beads (Polysciences Inc., Warrington, USA) were washed three times in PBS. Bacterial extracts were incubated with beads for 16 h at 4°C and unbound material was washed away from the beads using 0.5% SDS. The beads were finally heated to 95°C in NuPAGE sample buffer (0.14 M Tris, 0.1 M Tris-HCl, 1.1 M glycerol, 70 mM lithium dodecyl sulfate (LDS), 0.5 mM, ethylenediaminetetraacetic acid (EDTA) , 0.2 mM Serva Blue G250, 0.2 mM Phenol Red, 50 mM DTT, pH 8.5) for 15 min.

6.1.5. Electrophoresis and detection of biotinylated proteins

SDS PAGE was carried out with Novex NuPAGE system (Novex, San Diego, CA) according to manufacturer's instructions. Briefly, samples (5 or 25 µl, respectively) extracted from beads were applied on a homogeneous gel of 10%. The gel was either stained with GELCODE® Blue Stain Reagent (Pierce, USA) or electroblotted to a PVDF (0.2 µm) membrane according to the manuals. The PVDF membrane was preincubated in blocking solution (3% BSA, 50 mM HEPES-NaOH, 100 mM NaCl, pH 7.3 for 1.5 h and washed with 0.05% Tween-20, 50 mM HEPES-NaOH, 100 mM NaCl, pH 7.3 (washing buffer). The membrane was then incubated with horseradish peroxidase-conjugated streptavidin (HRP-NeutrAvidin™, Pierce, USA) in washing buffer with 1% BSA added. After 1–1.5 h, the membrane was washed five times in washing buffer, and biotin-containing protein bands were developed with 0.1% H_2O_2 and 0.02% DAB (3,3'-diaminobenzidine hydrochloride, Pierce, USA) in washing buffer. The membranes were then washed with water and dried at room temperature.

6.1.6. Enzymatic digestion

Selected protein band from the Coomassie-stained gel was cut out, placed in separate siliconized tubes and gently macerated with a hand-held mixer. The dye was removed by adding 85 µl of 25 mM NH_4HCO_3 in 50% CH_3CN and the tube was vortexed for 30 min before the supernatant was removed. This procedure was repeated twice.

Destained gel pieces were dried for 40 min in a vacuum concentrator (Speed Vac®, Savant Instruments, USA). 15 µl trypsin (porcine, sequencing grade, modified, Promega, USA) in 25 mM NH_4HCO_3, pH 8, was added. The tubes were incubated for 16 h at 37°C. Peptides were extracted from the gel using 37.5% CH_3CN/2.5% CF_3COOH. The samples were then vortexed for 30 min and centrifuged for 2 min at 5,000 g.

6.1.7. Mass spectrometry

Mass spectra were obtained using a TofSpecE MALDI-TOF mass spectrometer (Micromass, Manchester, UK) equipped with a reflectron. Aliquots of 0.5 µl matrix solution (α-cyano-4-hydroxy-cinnamic acid 10 mg/mL in acetonitrile/H_2O, 1:1) were mixed with 0.5 µl of the samples. All mass spectra were acquired in the positive reflectron mode at an accelerating voltage of 20 kV. The instrument was calibrated externally using the peptide standards ACTH (MW 2465.20) and Angiotensin II (MW 1046.54). A mass from the autodegradation of trypsin (2211.105) was used as internal lock mass during post-processing of the mass spectra. Resulting values for the twenty most intense monoisotopic peaks were used to identify known proteins through database searches using MS-Fit (http://prospector.ucsf.edu). All searches were performed with the following parameters: protein molecular mass range 1,000–120,000 Da, all species allowed and one missed cleavage for tryptic digests. A mass deviation of ±200 ppm was tolerated.

6.1.8. ESI-qTOFMS/MS

Samples for which no confident identification could be made with peptide mapping were analyzed using nanoflow electrospray MS/MS in a Q-TOF (Waters, UK). Samples were enriched using C_{18} Zip Tips (Millipore, Bedford, MA, USA), eluted with acetonitrile:H_2O containing 0.1% formic acid, and sprayed from gold-coated glass capillaries (Waters). Doubly and triply protonated peptides were fragmented using argon as a collision gas. Instrument calibration was performed using fragment ions from Glu-fibrinopeptide B and a fourth-order polynomial fit. Fragment ion spectra were post-processed using MaxEnt3 in the Masslynx software, saved in Sequest-compatible format and used to search the entire NCBI nr database using MASCOT (http://www.matrixscience.com). A mass deviation of ±200 ppm was tolerated.

6.1.9. Sample preparation for 2D DIGE

Total protein concentration was determined by the Bio-Rad, Dc Protein Assay as described by the manufacturer's Bio-Rad, and similar protein concentrations were used for three dye 2D DIGE analysis. Approximately 100 µg of each *H. pylori* substrain (26695 wildtype), 26695 V (increased capacity of lactoferrin binding) and 26695⁻ (HP0486 knock-out strain) suspended in lysis buffer (30 mM Tris, 5 mM magnesium acetate, 8 M urea, 4% CHAPS) was labeled with Cy Dyes according to the manufacturer's instructions (Amersham Bioscience). Cyanine dyes were reconstituted in dimethylformamide (DMF) and added to labeling reactions in ratio of 100 pmol CyDye:50 µg protein. Protein labeling was achieved by incubation on ice in the dark for 30 min. The reaction was quenched by addition of 10 mM lysine followed by incubation on ice for a further 10 min.

6.1.10. 2D GE

Prior to IEF, labeled samples to be separated in the same gel were mixed and added to an equal volume of rehydration buffer (450 µl) (4% CHAPS, 8M urea, 1% IPG buffer 3–10, 13 mM DTT) before being applied to 24 cm DryStrips (pH 3–10 NL) in a strip holder for rehydration at 30 V for 12 h. The first dimension separation was carried out using an IPGphor IEF system (Amersham Bioscience). The total focusing time was 60 kVh and the strips were equilibrated in 8 M urea, 30% glycerol, 50 mM Tris, pH 8.8, 2% SDS and 40 mM DTT. After equilibration second-dimension electrophoresis was performed in an Ettan Dalt (Amersham Bioscience) at 15 mA/gel for 15 h using 12% polyacrylamide gels cast in 20 × 24 cm glass plates, of which one inner side was precoated with (bind-)silane. After 2-DE, gels were scanned on the 2920 Master Imager (Amersham Bioscience) using excitation/emission wavelengths specific for the different fluorescent dyes. Differentially expressed proteins were identified with the differential in-gel analysis (DIA) module of DeCyder (Amersham Bioscience).

6.2. Results and discussion

The identity of the lactoferrin binding protein (OMP, Fig. 6) was obtained by peptide mapping of the enzymatically digested protein in an MALDI-TOF. 10 out of 17 peptides were matched to HP0486 within a mass error of 200 ppm. MS/MS

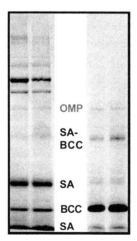

Figure 6. Gel/Blot from lactoferrin affinity cross-linking of *H. pylori* strain 26695. SDS-PAGE of proteins extracted from magnetic beads after binding of biotin-tagged *H. pylori* proteins. To the left, a Coomassie stained gel and right, PVDF blot stained with streptavidin-HRP/DAB. Bands indicated with abbreviations was identified as SA, streptavidin monomer/dimer, BCC, *H. pylori* biotin carboxyl carrier protein, SA-BCC, streptavidin-BCC complex and OMP, *H. pylori* OMP HP 0486.

analysis of two doubly protonated peptides, *m/z* 957.4, ALEYGIGmYLDYQFSK (m = oxidized methionine) and *m/z* 770.4, GIYPTETFVNLTGK was performed in a Q-TOF. The extensive series of *b*- and *y*-ions obtained confirmed the sequence of both peptides and the identity of the protein (Fig. 7).

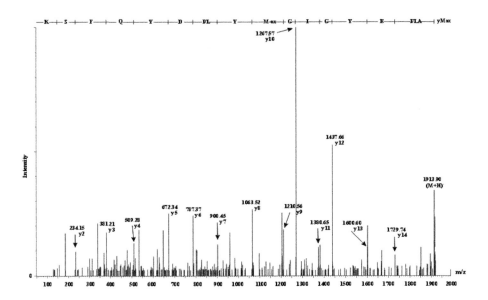

Figure 7. Product-ion spectrum of a doubly protonated tryptic peptide at *m/z* 957.4 confirmed the sequence ALEYGIGmYLDYQFSK, where m = oxidized methionine, previously matched with MALDI-MS data.

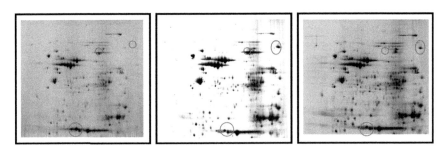

Figure 8. Comparison of expressed protein pattern in different *H. pylori* substrains. Left: HP0486 knockout, Middle: 26695 colony with increased lactoferrin binding capacity, Right: 26695 wildtype strain.

Variation of the lactoferrin binding capacity within different colonies in *H. pylori* strain 26695 was studied by radioimmunoassay assay (RIA, data not shown). In order to visualize alterations in protein pattern between a colony with high binding affinity (26695:V) to lactoferrin compared to the wildtype strain 26695 and to a 26695 HP0486 knockout strain, we performed 2D DIGE. Several major differences in the protein expression were observed (Fig. 8).

To reveal knowledge about *H. pylori* lactoferrin binding properties, work will be continued with identification of altered proteins. We will also study alteration in lactoferrin binding properties in clinical strains isolated from adenocarcinoma, duodenal ulcer patients and asymptomatic carriers in order to determine the role of lactoferrin binding in pathogenicity.

Acknowledgments

The support of the NSF National High Field FT-ICR Mass Spectrometry Facility (DMR 0084173) and Knut and Alice Wallenberg Foundation (Stockholm, Sweden) is gratefully acknowledged.

References

[1] T. Larsson, J. Bergstrom, C. Nilsson and K.A. Karlsson, FEBS Lett., 469 (2000) 155–158.

[2] J. Mahdavi, B. Sonden, M. Hurtig, F.O. Olfat, L. Forsberg, N. Roche, J. Angstrom, T. Larsson, S. Teneberg, K.A. Karlsson, S. Altraja, T. Wadstrom, D. Kersulyte, D.E. Berg, A. Dubois, C. Petersson, K.E. Magnusson, T. Norberg, F. Lindh, B.B. Lundskog, A. Arnqvist, L. Hammarstrom and T. Boren, Science 297 (2002) 573–578.

[3] J. Klose, Humangenetik, 26 (1975) 231–243.

[4] P.H. O'Farrell, J. Biol. Chem., 250 (1975) 4007–4021.

[5] M. Unlu, M.E. Morgan and J.S. Minden, Electrophoresis, 18 (1997) 2071–2077.

[6] N. Ahmed and G.E. Rice, J. Chromatogr. B Anal. Technol. Biomed. Life Sci., 815 (2005) 39–50.

[7] S.H. Bae, A.G. Harris, P.G. Hains, H. Chen, D.E. Garfin, S.L. Hazell, Y.K. Paik, B.J. Walsh and S.J. Cordwell, Proteomics, 3 (2003) 569–579.

[8] K. Bunai and K. Yamane, J. Chromatogr. B Anal. Technol. Biomed. Life Sci., 815 (2005) 227–236.

[9] T.A. Rhomberg, O. Karlberg, T. Mini, U. Zimny-Arndt, U. Wickenberg, M. Rottgen, P.R. Jungblut, P. Jeno, S.G. Andersson and C. Dehio, Proteomics, 4 (2004) 3021–3033.

[10] C.L. Nilsson, T. Larsson, E. Gustafsson, K.A. Karlsson and P. Davidsson, Anal. Chem., 72 (2000) 2148–2153.

[11] P.G. Righetti, A. Castagna, B. Herbert, F. Reymond and J.S. Rossier, Proteomics, 3 (2003) 1397–1407.

[12] M.P. Washburn, D. Wolters and J.R. Yates,Nat. Biotechnol., 19 (3) (2001) 242–247.

[13] M. Vollmer, E. Nagele and P. Horth, J. Biomol. Tech., 14 (2003) 128–135.

[14] J.V. Olsen and M. Mann, Proc. Natl. Acad. Sci. USA, 101 (2004) 13417–13422.

[15] E. Carlsohn, J. Nystrom, I. Bolin, C.L. Nilsson and A.M. Svennerholm, Infect. Immun., 74 (2006) 920–926.

[16] M. Ramstrom, C. Hagman, J.K. Mitchell, P.J. Derrick, P. Hakansson and J. Bergquist, J. Proteome Res., 4 (2005) 410–416.

[17] C. Sihlbom, P. Davidsson and C.L. Nilsson, J. Proteome Research, 4 (2005) 2294–2301.

[18] C. Filip, G. Fletcher, J.L. Wulff and C.F. Earhart, J. Bacteriol., 115 (1973) 717–722.

[19] I. Fialka, C. Pasquali, R. Kurzbauer, F. Lottspeich and L.A. Huber, Electrophoresis, 20 (1999) 331–343.

[20] P.A. Nielsen, J.V. Olsen, A.V. Podtelejnikov, J.R. Andersen, M. Mann and J.R. Wisniewski, Mol. Cell. Proteomics, 4 (2005) 402–408.

[21] T. Stasyk and L.A. Huber, Proteomics, 4 (2004) 3704–3716.

[22] M.A. Grayson, Philadelphia: Chemical Heritage Foundation (2002).

[23] M. Barber, R.S. Bordoli, R.D. Sedgewick and A.N. Tyler, J. Chem. Soc. Chem. Commun., (1981) 325–327.

[24] M. Barber, R.S. Bordoli, R.D. Sedgewick and A.N. Tyler, Nature, 293 (1981) 270–275.

[25] M. Karas, U. Bachmann, U. Bahr and F. Hillenkamp, Int. J. Mass spectrom. Ion Processes, 78 (1987) 53–68.

[26] M. Karas and F. Hillenkamp, Anal. Chem., 60 (1988) 2299–2301.

[27] K. Tanaka, H. Waki, Y. Ido, S. Akita and Y. Yoshida, Rapid. Commun. Mass Spectrom., 2 (1988) 151–153.

[28] M. Dole, L.L. Mach, R.L. Hines, R.C. Mobley, R.C. Ferguson and M.B. Alice, J. Chem. Phys., 49 (1968) 2240–2247.

[29] M.L. Aleksandrov, L.N. Gall, V.N. Krasnov, V.I. Nikolaev, V.A. Pavlenko and V.A. Shkurov, Dokl Akad Nauk SSSR. 277 (1984) 379–383.

[30] M. Yamashita and J.B. Fenn., J. Phys. Chem., 88 (1984) 4451–4459.

[31] J.B. Fenn, M. Mann, C.K. Meng, S.F. Wong and C.M. Whitehouse, Science, 246 (1989) 64–71.

[32] B.A. Thomson and J.V. Iribarne, J. Chem. Phys., 71 (1979) 4451–4463.

[33] M.R. Emmett and R.M. Caprioli, J. Am. Soc. Mass Spectrom., 5 (1994) 605–613.

[34] M. Wilm and M. Mann, Anal. Chem., 68 (1996) 1–8.

[35] W.C. Wiley and I.H. McLaren, Rev. Sci. Instrum., 26 (1955) 1150–1157.

[36] B.A. Mamyrin, V.I. Karataev, D.V. Shmikk and V.A. Zagulin, Sov. Phys. JETP, 37 (1973) 45.

[37] M. Mann, P. Hojrup and P. Roepstorff, Biol. Mass Spectrom., 22 (1993) 338–345.

[38] W.J. Henzel, T.M. Billeci, J.T. Stults, S.C. Wong, C. Grimley and C. Watanabe, Proc. Natl. Acad. Sci. USA, 90 (1993) 5011–5015.

[39] H.R. Morris, T. Paxton, A. Dell, J. Langhorne, M. Berg, R.S. Bordoli, J. Hoyes and R.H. Bateman, Rapid Commun. Mass Spectrom., 10 (1996) 889–896.

[40] J.H.J. Dawson and M. Guilhaus, Rapid Commun. Mass Spectrom., 3 (1989) 155–159.
[41] D.B. Kristensen, K. Imamura, Y. Miyamoto and K. Yoshizato, Electrophoresis, 21 (2000) 430–439.
[42] A.G. Marshall, Int. J. Mass Spectrom., 200 (2000) 331–356.
[43] A.G. Marshall, C.L. Hendrickson and G.S. Jackson, Mass Spectrom. Rev., 17 (1998) 1–35.
[44] F. He, C.L. Hendrickson and A.G. Marshall, Anal. Chem., 73 (2001) 647–650.
[45] J. Bergquist, Curr. Opin. Mol. Ther., 5 (2003) 310–314.
[46] B. Bogdanov and R.D. Smith, Mass Spectrom. Rev., 24 (2005) 168–200.
[47] K. Håkansson, H.J. Cooper, R.R. Hudgins and C.L. Nilsson, Curr. Org. Chem., 7 (2003) 1503–1525.
[48] K. Håkansson, H.J. Cooper, M.R. Emmett, C.E. Costello, A.G. Marshall and C.L. Nilsson, Anal. Chem., 73 (2001) 4530–4536.
[49] M.J. Chalmers, K. Hakansson, R. Johnson, R. Smith, J. Shen, M.R. Emmett and A.G. Marshall, Proteomics, 4 (2004) 970–981.
[50] J.S. Andersen, Y.W. Lam, A.K. Leung, S.E. Ong, C.E. Lyon, A.I. Lamond and M. Mann, Nature, 433 (2005) 77–83.
[51] M.L. Nielsen, M.M. Savitski and R.A. Zubarev, Mol. Cell. Proteomics, 4 (2005) 835–845.
[52] J.V. Olsen, S.E. Ong and M. Mann, Mol. Cell. Proteomics, 3 (2004) 608–614.
[53] M.M. Savitski, M.L. Nielsen and R.A. Zubarev, Mol. Cell. Proteomics, 4 (2005) 1180–1188.
[54] P. Roepstorff and J. Fohlman, Biomed. Mass Spectrom., 11 (1984) 601.
[55] B.J. Marshall and J.B. Warren, Lancet, 1 (1984) 1311–1315.
[56] J.B. Warren and B.J. Marshall, Lancet, 1 (1983) 1273–1275.
[57] J. Parsonnet, Infect. Dis. Clin. North Am., 12 (1998) 185–197.
[58] R.P. Logan, Lancet, 344 (1994) 1078–1079.
[59] D. Forman, Scand. J. Gastroenterol Suppl., 215 (1996) 48–51.
[60] E. Bayerdorffer, A. Neubauer, B. Rudolph, C. Thiede, N. Lehn, S. Eidt and M. Stolte, Lancet, 345 (1995) 1591–1594.
[61] P.K. Bardhan, Clin. Infect. Dis., 25 (1997) 973–978.
[62] B.E. Dunn, H. Cohen and M.J. Blaser, Clin. Microbiol. Rev., 10 (1997) 720–741.
[63] M.F. Go, Aliment. Pharmacol. Ther., 16 (Suppl. 1) (2002) 3–15.
[64] K.J. Goodman and P. Correa, Int. J. Epidemiol., 24 (1995) 875–887.
[65] A. Lee, Scand. J. Gastroenterol. Suppl., 187 (1991) 9–22.
[66] F. Megraud, Aliment Pharmacol. Ther., 9 (Suppl 2) (1995) 85–91.
[67] G. Cammarota, A. Tursi, M. Montalto, A. Papa, G. Veneto, S. Bernardi, A. Boari, V. Colizzi, G. Fedeli and G. Gasbarrini, J. Clin. Gastroenterol, 22 (1996) 174–177.
[68] P.D. Klein, D.Y. Graham, A. Gaillour, A.R. Opekun and E.O. Smith, Lancet, 337 (1991) 1503–1506.
[69] F. Bazzoli, P. Pozzato and T. Rokkas, Helicobacter, 7 (Suppl. 1) (2002) 43–49.
[70] P. Unge, Curr. Top. Microbiol. Immunol., 241 (1999) 261–300.

[71] A. Ramirez-Ramos, R.H. Gilman, R. Leon-Barua, S. Recavarren-Arce, J. Watanabe, G. Salazar, W. Checkley, J. McDonald, Y. Valdez, L. Cordero and J. Carrazco, Clin. Infect. Dis., 25 (1997) 1027–1031.

[72] J.F. Tomb, O. White, A.R. Kerlavage, R.A. Clayton, G.G. Sutton, R.D. Fleischmann, K.A. Ketchum, H.P. Klenk, S. Gill, B.A. Dougherty, K. Nelson, J. Quackenbush, L. Zhou, E.F. Kirkness, S. Peterson, B. Loftus, D. Richardson, R. Dodson, H.G. Khalak, A. Glodek, K. McKenney, L.M. Fitzegerald, N. Lee, M.D. Adams, E.K. Hickey, D.E. Berg, J.D. Gocayne, T.R. Utterback, J.D. Peterson, J.M. Kelley, M.D. Cotton, J.M. Weidman, C. Fujii, C. Bowman, L. Watthey, E. Wallin, W.S. Hayes, M. Borodovsky, P.D. Karp, H.O. Smith, C.M. Fraser, and J.C. Venter, Nature, 388 (1997) 539–547.

[73] R.A. Alm, L.S. Ling, D.T. Moir, B.L. King, E.D. Brown, P.C. Doig, D.R. Smith, B. Noonan, B.C. Guild, B.L. deJonge, G. Carmel, P.J. Tummino, A. Caruso, M. Uria Nickelsen, D.M. Mills, C. Ives, R. Gibson, D. Merberg, S.D. Mills, Q. Jiang, D.E. Taylor, G.F. Vovis and T.J. Trust, Nature, 397 (1999) 176–180.

[74] J. Bina, M. Bains and R.E. Hancock, J. Bacteriol., 182 (2000) 2370–2375.

[75] R.A. Alm, J. Bina, B.M. Andrews, P. Doig, R.E. Hancock and T.J. Trust, Infect. Immun., 68 (2000) 4155–4168.

[76] D. Ilver, A. Arnqvist, J. Ogren, I.M. Frick, D. Kersulyte, E.T. Incecik, D.E. Berg, A. Covacci, L. Engstrand and T. Boren, Science, 279 (1998) 373–377.

[77] E.E. Hennig, R. Mernaugh, J. Edl, P. Cao and T.L. Cover, Infect. Immun., 72 (2004) 3429–3435.

[78] P.G. Falk, L. Bry, J. Holgersson and J.I. Gordon, Proc. Natl. Acad. Sci. USA, 92 (1995) 1515–1519.

[79] S. Odenbreit, M. Till, D. Hofreuter, G. Faller and R. Haas, Mol. Microbiol., 31 (1999) 1537–1548.

[80] S. Odenbreit, G. Faller and R. Haas, Int. J. Med. Microbiol., 292 (2002) 247–256.

[81] R. de Jonge, Z. Durrani, S.G. Rijpkema, E.J. Kuipers, A.H. van Vliet and J.G. Kusters, J. Med. Microbiol., 53 (2004) 375–379.

[82] B. Peck, M. Ortkamp, K.D. Diehl, E. Hundt and B. Knapp, Nucleic Acids Res., 27 (1999) 3325–3333.

[83] D.G. Evans, D.J. Evans, Jr., J.J. Moulds and D.Y. Graham, Infect. Immun., 56 (1988) 2896–2906.

[84] P.W. O'Toole, L. Janzon, P. Doig, J. Huang, M. Kostrzynska and T.J. Trust, J. Bacteriol., 177 (1995) 6049–6057.

[85] K. Blom, B.S. Lundin, I. Bolin and A. Svennerholm, FEMS Immunol. Med. Microbiol., 30 (2001) 173–179.

[86] A.M. Lundstrom, K. Blom, V. Sundaeus and I. Bolin, Microb. Pathog., 31 (2001) 243–253.

[87] A.C. Jones, R.P. Logan, S. Foynes, A. Cockayne, B.W. Wren and C.W. Penn, J. Bacteriol., 179 (1997) 5643–5647.

[88] I. Bolin, H. Lonroth and A.M. Svennerholm, J. Clin. Microbiol., 33 (1995) 381–384.

[89] J. Yan, Y.F. Mao and Z.X. Shao, World J. Gastroenterol., 11 (2005) 421–425.

[90] A. Mattsson, A. Tinnert, A. Hamlet, H. Lonroth, I. Bolin and A.M. Svennerholm, Clin. Diagn. Lab. Immunol., 5 (1998) 288–293.

[91] C.P. McAtee, K.E. Fry and D.E. Berg, Helicobacter, 3 (1998) 163–169.

[92] P.R. Jungblut, D. Bumann, G. Haas, U. Zimny-Arndt, P. Holland, S. Lamer, F. Siejak, A. Aebischer and T.F. Meyer, Mol. Microbiol., 36 (2000) 710–725.

[93] N. Sabarth, S. Lamer, U. Zimny-Arndt, P.R. Jungblut, T.F. Meyer and D. Bumann, J. Biol. Chem., 277 (2002) 27896–27902.

[94] G. Haas, G. Karaali, K. Ebermayer, W.G. Metzger, S. Lamer, U. Zimny-Arndt, S. Dierscher, U.B. Goebel, K. Vogt, A.B. Roznowski, B.J. Wiedenmann, T.F. Meyer, T. Aebischer and P.R. Jungblut, Proteomics, 2 (2002) 313–324.

[95] D. Bumann, S. Aksu, M. Wendland, K. Janek, U. Zimny-Arndt, N. Sabarth, T.F. Meyer and P.R. Jungblut, Infect. Immun., 70 (2002) 3396–3403.

[96] S.O. Hynes, J. McGuire, T. Falt and T. Wadstrom, Proteomics, 3 (2003) 273–278.

[97] H.W. Lee, Y.H. Choe, D.K. Kim, S.Y. Jung and N.G. Lee, Proteomics, 4 (2004) 2014–2027.

[98] S.C. Baik, K.M. Kim, S.M. Song, D.S. Kim, J.S. Jun, S.G. Lee, J.Y. Song, J.U. Park, H.L. Kang, W.K. Lee, M.J. Cho, H.S. Youn, G.H. Ko and K.H. Rhee, J. Bacteriol., 186 (2004) 949–955.

[99] S. Backert, T. Kwok, M. Schmid, M. Selbach, S. Moese, R.M. Peek, Jr., W. Konig, T.F. Meyer and P.R. Jungblut, Proteomics, 5 (2005) 1331–1345.

[100] P.L. Masson, J.F. Heremans and C. Dive, Clin. Chem. Acta, 14 (1966) 735–739.

[101] M. Baggiolini, C. de Duve, P.L. Masson and J.F. Heremans, J. Exp. Med., 131 (1970) 559–570.

[102] R.R. Arnold, M.F. Cole and J.R. McGhee, Science, (1977) 197.

[103] J. Oram and B. Reiter, Biochim. Biophys. Acta, (1968) 170.

[104] B. Reiter and J. Oram, Nature, 216 (1967) 328–330.

[105] P. Rainard, Vet. Microbiol., 11 (1986) 103–115.

[106] B. Reiter, J.H. Brock and E.D. Steel, Immunology, 28 (1975) 83–95.

[107] G. Spik, A. Cheron, J. Montreuil and J.M. Dolby, Immunology, 35 (1978) 663–671.

[108] M.W. Rey, S.L. Woloschuk, H.A. deBoer and F.R. Pieper, Nucl. Acids. Res., 18 (1990) 5288.

[109] G. Spik, B. Coddeville and J. Montreuil, Biochimie, 70 (1988) 1459–1469.

Lectins: Analytical Technologies
C.L. Nilsson (Editor)
© 2007 Elsevier B.V. All rights reserved.

Chapter 13

The Use of Lectins in Bioaffinity MALDI Probes

Crystal Kirmiz, Caroline S. Chu and Carlito B. Lebrilla

Chemistry Department, University of California, Davis, One Shields Avenue, Davis, CA 95616, USA

1. Introduction and Background

1.1. What are lectins and what are they used for?

Lectins are proteins or glycoproteins that are present in a variety of organisms. They possess the ability to weakly bind glycans with high specificity. According to Rudiger and Gabius, 2001, there are three characteristics a protein or glycoprotein must possess to define itself as a lectin. These characteristics are as follows: (1) A lectin is a protein that binds a glycan. (2) Immunoglobulins are not lectins, that is, lectins do not exclusively exist as a result of an immune response; they can be present as a result of stress or change in environment as well. (3) Lectins do not change the structure and characteristics of a glycan they bind to [1].

While lectins are present in most organisms, the majority of lectins used in research today are isolated from components of plants. Although lectins can be found in many components of plants, the richest abundance of lectins is found in the seeds. As yet, the biological functions of lectins are not fully known, although there are several theories that have been formulated for them. Despite the lack of understanding for the biological roles of lectins, they have been exploited for several years in many applications. Commonly, lectins have been utilized to isolate glycoproteins from biological fluids, such as serum and plasma [1]. The use of lectins has advanced many areas of medical research, such as aging, cancer prevention, as well as other clinical problems [1]. Desribed below is the work conducted in the Lebrilla Laboratory utilizing lectin biology to isolate glycans present in the egg jelly of the frog *Xenopus laevis*.

2. A General Method for Producing Bioaffinity MALDI Probes

2.1. Introduction

Prior to the work of developing bioaffinity matrix-assisted laser desorption/ionization (MALDI) lectin probes, we examined the general use of bioaffinity probes for oligosaccharide analysis. The methods developed in this phase of the project were ultimately used with the lectins to extract glycans from complex and heterogenous mixtures. The following is an overview of some of the bioaffinity experiments conducted.

Biologically active probes for MALDI can be highly useful for the rapid isolation of a compound from a mixture, thus allowing researchers to complete rapid analysis using mass spectrometry (MS). Hutchens suggested the use of probe surfaces that are designed to extract specific molecules from unfractionated biological fluids and extracts [2]. More pertinent to this study is the report by Li and co-workers in which avidin bound to agarose beads was used to extract biotinylated peptides and proteins [3].

The key is to find a surface that can bind proteins rapidly and easily, while maintaining the function of the proteins. If most of the protein can be immobilized on such a surface, the immobilized protein would then provide a binding site for other substrates [4], and this method would provide a general procedure for the rapid production of bioaffinity probes.

2.2. Development of the bioaffinity probe

The concept of the bioaffinity probe specifically for oligosaccharides was first examined with biotinylated oligosaccharides using avidin immobilized on the MALDI probe surface. To establish that oligosaccharides can be examined in this way, standard compounds were used. For example, maltohexaose, a linear hexasaccharide was derivatized with biotin at the reducing end using standard methods [5]. MS analysis of the resulting compounds showed that biotinylation was complete (100% conversion). MALDI mass spectra were obtained to determine the difference in ionization characteristics between native maltohexaose and the derivatized compounds. Interestingly, the biotinylated maltohexaose yielded stronger signals than that of the underivatized compound when the compounds were analyzed by MALDI (Fig. 1a). This behavior is likely due to the presence of the basic amine group in the derivatized compound. To confirm the structure and to show that structural information can be obtained from the biotinylated oligosaccharide, collision induced dissociation (CID) was performed. The CID spectrum of the biotinylated maltohexaose showed fragmentation occurring at every glycosidic bond to produce a series of Y-type ions (Fig. 1b). To produce a bioaffinty probe, we searched for a substrate that can readily bind proteins without chemical modifications. We wanted to take advantage of the natural ability of proteins to adhere to hydrophobic surfaces. A number of surfaces and materials were examined, including polyester, micropolyethylene, and polyvinylidene fluoride, but we found simple transparency film (Canon NP transparency film type E) to be the most effective.

Avidin, in the form of neutravidin, was immobilized on the membrane source by placing a solution of the avidin on it and letting it dry. This produced an

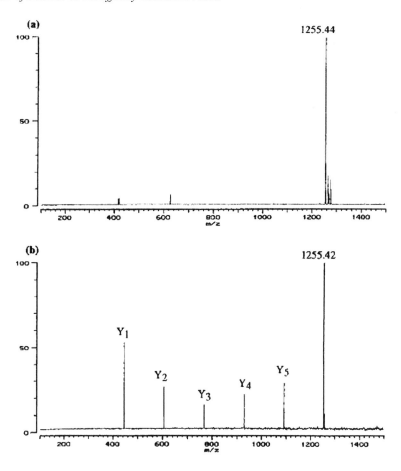

Figure 1. (a) MALDI-FT-ICR MS spectrum of biotinylated maltohexaose. (b) CID of biotinylated maltohexaose.

avidin-immobilized surface that could not be readily washed away and was used directly for MALDI-MS analysis.

The avidin-immobilized surface was tested by attaching it via adhesive to the metal MALDI probe surface. A mixture containing native and biotinylated oligosaccharides was applied to the probe and allowed to incubate for a few minutes. This was followed by a washing step to remove unbound material. To enhance the Na^+-coordinated quasimolecular ion, a solution of NaCl as well as matrix (2,5-dihydroxybenzoic acid) was used. The entire process took less than 20 min. After the sample was allowed to dry, it was analyzed by MALDI-FT-ICR MS. As a control, the same sample mixture was placed on a film with no avidin. The results are illustrated in Fig. 2a for an oligosaccharide from human milk LNDFH-1. In the avidin-treated surface, only the biotinylated compounds were observed (Fig. 2b), while in the control surface both biotinylated (*m/z* 1,264.49) and native LNDFH-I (*m/z* 1,022.37) were present in nearly equal abundances.

Experiments with mixtures of oligosaccharides indicated that the affinity surface behaved generally for other oligosaccharides. In order to determine if the bioaffinity probe would differentiate between glycans with different structures, experiments with a mixture of glycans composed of linear and branched structures were conducted. This was done to observe whether the bioaffinity probe shows a preference for linear or branched structures. The glycans used for this experiment were two hexasaccharides, maltohexaose (a linear glycan), and

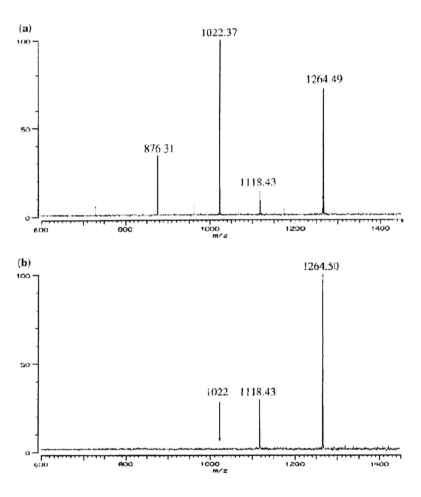

Figure 2. (a) MALDI FT-ICR MS spectrum of a mixture of biotinylated (*m/z* 1,264.49) and native (*m/z* 1,022.37)) LNDFH-I from a control probe surface. (b) Competition experiment involving the mixture in (a) on an avidin-treated surface. Only the biotinylated compound is observed in the mass spectrum. The native compound (arrow) is absent.

LNDFH-I (a branched glycan). Fig. 3a shows a spectrum of a mixture of the biotinylated and native glycans.

The biotinylated maltohexaose (*m/z* 1,255.44) and LNDFH-I (*m/z* 1,264.48), as well as the native maltohexaose (*m/z* 1,013.32) and LNDFH-I (*m/z* 1,022.36) are present. Fig. 3b shows only the biotinylated glycans after the washing step is completed. The glycans are in relatively equal abundance showing that the probe does not prefer one form of structure over another. These experiments proved that bioaffinity technology could be a powerful tool in the analysis of mixtures of glycans. Applications to oligosaccharides were developed in this set of experiments,

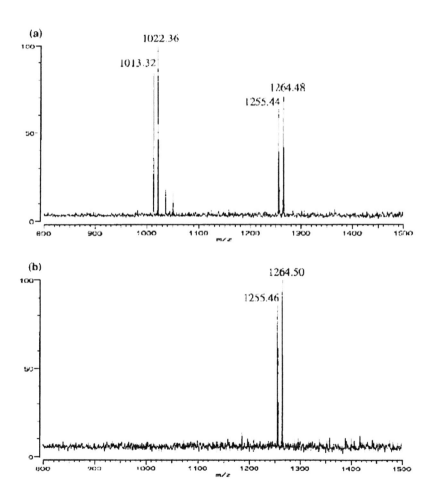

Figure 3. (a) MALDI FT-ICR MS spectrum of a mixture of biotinylated maltohexaose (*m/z* 1,255.44) and LNDFH-I (*m/z* 1,264.48) with their native analogues (*m/z* 1,013.32 and 1,022.36, respectively). (b) MALDI FT-ICR MS of the same mixture on a bioaffinity MALDI probe. Only the biotinylated species are observed.

although this method will work with any biotinylated compound. Other biopolymers, such as lectins, can be analyzed in the same manner. The advantage of this
process is that the probe can be prepared in a few minutes and with high reliability
because it does not require the formation of covalent bonds.

3. Identification and Structure Elucidation of Lectin-Binding Oligosaccharides by Bioaffinity MALDI FT-ICR MS

3.1. Introduction

The extracellular matrix surrounding most animal eggs plays important roles in
reproduction. One of those, mediating sperm–egg binding, is critical for fertilization.
However, the block of further sperm penetration (polyspermy) after fertilization is
similarly important for creating viable offspring. Cortical granule lectin (CGL) is
released by the egg of the South African frog *X. laevis* on fertilization to prevent
secondary sperm penetration [6–8]. The extracellular matrix of *X. laevis* contains
a vitelline envelope and a jelly coat composed of three morphologically distinct
regions designated J1, J2, and J3. The lectin interacts with the innermost layer (J1)
of the jelly coat to form a fertilization layer, which acts as an impenetrable barrier
for additional sperm penetration. Sialic acid and sulfate residues are known con-
stituents of the glycan moieties, but most, if not all, sulfate sugar esters are
believed to reside in the J1 layer. Indeed, binding studies indicate that the lectin
attaches specifically to two highly anionic glycoproteins, with apparent molecular
masses of 630 (gp630) and 450 kDa (gp450) [9].

Identification of the oligosaccharide components that bind to the CGL is important for elucidating the nature of the interaction. However, traditional methods for
studying lectin–carbohydrate binding demand considerable effort in the isolation and
identification of the substrates, requiring a larger amount of material than can be
obtained from biological sources. For this set of experiments bioaffinity probe composed of immobilized CGL in conjunction with MALDI FT-ICR MS was used to isolate oligosaccharides that have affinity for the lectin. CID was also used to prove
structural information on strongly binding and weakly binding oligosaccharides.

Only one other report has been published using a lectin-based bioaffinity probe.
Bundy and Fenselau employed these probes to immobilize an oligosaccharide and
identify a binding epitope in a bacterium [10]. The goal for these experiments was
to use CGL along with bioaffinity technology to rapidly separate glycans bound to
the lectin from a mixture, as well as identify these glycans.

3.2. The use of bioaffinity technology and lectins to isolate and characterize frog glycans

In the previous section neutravidin was immobilized on a polymer film surface and
the affinity of the treated surface for biotinylated oligosaccharides was illustrated [5].

Because of the high binding affinity of biotin for avidin – it is the strongest noncovalent interaction – the native oligosaccharides were readily washed away so that only biotinylated oligosaccharides were observed in the MALDI-MS spectra. Lectin–oligosaccharide interactions exhibit neither the same strength of binding nor the specificity. For this reason, the lectin probes were not expected to be highly selective. The researchers studied several model lectins to determine whether selectivity would be detected in the MALDI-MS spectra.

The lectin *Ulex europaeus* agglutinin (UEA) preferentially retains α-L-fucose. A mixture containing the trisaccharides Fucα1-2Galβ1-4Glu (**1**) and GlcNAcβ1-4GlcNAcβ1-4GlcNAc (**2**) was examined to determine whether the fucosylated compound **1** would be preferentially retained. Approximately equal amounts were placed on the polymer probe (without lectin) and a spectrum was obtained to observe the ionization behavior of the two compounds (Fig. 4a). Fucosylated compound **1** is ~50% of the nonfucosylated compound, the base peak, suggesting a slightly better ionization efficiency for the nonfucosylated compound **2**. A polymer film was then treated with the UEA as described in the previous section pertaining to bioaffinity technology to produce the bioaffinity probe. The same sample mixture was placed on the lectin-immobilized probe. The resulting spectrum is shown in Fig. 4b. Note that the relative intensities are reversed with the nonfucosylated substrate **2** now 60% of the fucosylated compound **1**, the base peak, suggesting that indeed **1** was preferentially retained. The polymer film without lectin was pretreated and used in the same manner as that outlined in the previous experiment as a control. The spectrum for this set of experiments is shown in Fig. 4c, confirming the binding of the oligosaccharides to the lectin.

As a further test of the efficacy of the lectin probe, wheat germ agglutinin (WGA), which specifically retains β-GlcNAc, was examined by employing a mixture of three GlcNAc oligomers (GlcNAcβ1-4GlcNAc (**3**), GlcNAcβ1-4GlcNAcβ1-4GlcNAc (**4**), and GlcNAcβ1-4GlcNAcβ1-GlcNAcβ1-4GlcNAc (**5**)). Binding studies in solution show that the preferences for the oligomers followed the order $5 > 4 > 3$ [11, 12]. Equimolar mixtures of the compound applied and analyzed on the bare polymer film show that the tetramer of the trimer have similar signal responses, while the dimer has significantly less (Fig. 5a). It has previously been shown that small oligomers, particularly disaccharides, bind sodium cations more weakly than larger oligomers causing a decrease in abundance of these species [13]. Furthermore, the instrument used for these experiments was optimized for masses corresponding to the tri- and tetrasaccharides rather than the disaccharides. The same mixture applied to the lectin-immobilized polymer probe yielded the spectrum shown in Fig. 5b. The disaccharide **3** is totally absent, while the trisaccharide **4** is diminished relative to the tetrasaccharide **5**.

By comparing MALDI-MS spectra of the mixture on the untreated polymer probe and the lectin-immobilized polymer probe, the natural affinity of the lectin is duplicated in the relative abundances in the mass spectra. This indicates that the lectin remained active on the probe surface, and the comparison of the relative

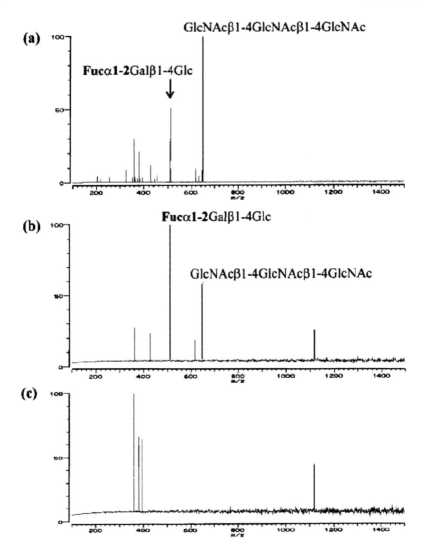

Figure 4. MALDI FT-ICR MS spectra of a mixture containing approximately equimolar concentration of Fucα1-2Galβ1-4Glc and GlcNAcβ1-4GlcNAcβ1-4GlcNAc on (a) polymer film probe, (b) a polymer film probe treated with lectin UEA followed with washings to remove unbound oligosaccharides, and (c) a polymer film probe not treated with lectin followed with washings. The lectin is specific for α-fucose.

intensity of the quasimolecular ion in both bare and lectin-immobilized films provided a method for observing the affinity of the lectin.

The oligosaccharides obtained from gp630 contain neutral, sialylated, sulfated, and doubly sulfated oligosaccharides. The neutral and anionic components were released and determined using standard MALDI-MS procedures. The results are

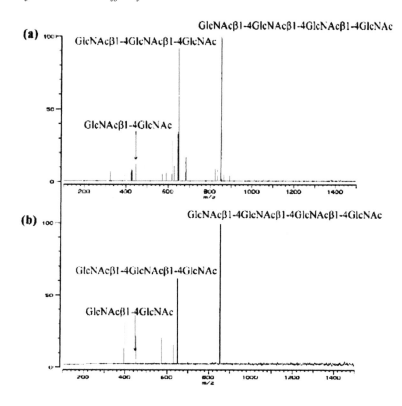

Figure 5. MALDI FT-ICR MS spectra of a mixture containing GlcNAc dimer, trimer, and tetramer in approximately equal concentrations from (a) a polymer film probe and (b) a polymer film probe treated with the lectin WGA followed with washings to remove unbound oligosaccharides. The binding of this lectin favors tetramer > trimer > dimer.

summarized in Tables 1–3 along with the composition of the oligosaccharides according to their *m/z*. Although the neutral components of the whole egg jelly have been characterized [14, 15], there is relatively little structural information on the anionic components. A small number of sialylated oligosaccharides have been elucidated, and there are a few reported structures of sulfated oligosaccharides [16]. From the results in the tables, it can be concluded that sulfated oligosaccharides make up a larger component of the anionic oligosaccharides. They are also the major reason for the anionic character of the protein. The presence of sulfate esters has long been known. The sulfated oligosaccharides are particularly important in the J1 layer, the origin of the 630 protein [17]. Unfortunately, the limited amount of material precluded the lectin-binding experiments with the pure glycoprotein. The lectin-binding experiments were performed with the whole egg jelly of which the oligosaccharides in gp630 formed a subset.

Figs. 6a and 7a show representative MALDI-MS spectra, in the negative mode, of the whole egg jelly mixture on the polymer film probe.

Table 1: Neutral Oligosaccharides from Gp630 Determined by Positive Ion Mass Spectometry

No.	Neutral oligosaccharides in positive mode [M + Na]⁺ theor *m/z*	exp *m/z*	Hex	Fuc	HexNAc
1	611.23	611.22	1	0	2
2	716.26	716.27	2	1	1
3	757.28	757.27	1	1	2
4	773.28	773.29	2	0	2
5	862.32	863.23	2	2	1
6	903.34	903.35	1	2	2
7	919.34	919.35	2	1	2
8	935.33	935.34	3	0	2
9	960.36	960.38	1	1	3
10	1065.40	1065.40	2	2	2
11	1227.45	1227.46	3	2	2
12	1373.51	1373.53	3	3	2
13	1389.50	1389.53	4	2	2

Table.2: Anionic Oligosaccharides with One Sulfate Ester from Gp630 Determined by Negative ion Mass Spectrometry

No.	Singly sulfated [M−H]⁻ theor *m/z*	exp *m/z*	Hex	Fuc	HexNAc	sulfate
1	505.13	505.13	0	0	2	1
2	667.19	667.18	1	0	2	1
3	813.24	813.23	1	1	2	1
4	829.24	829.23	2	0	2	1
5	975.30	975.29	2	1	2	1
6	1032.32	1032.31	2	0	3	1
7	1073.35	1073.32	1	0	4	1
8	1219.40	1219.38	1	1	4	1
9	1295.40	1235.48	2	0	4	1
10	1381.46	1381.43	2	1	4	1
11	1486.49	1486.46	3	2	3	1
12	1527.51	1527.48	2	2	4	1
13	1673.57	1373.54	2	3	4	1
14	1933.61	1933.60	2	2	6	1

Table 3: Anionic Oligosaccharides with two Sulfate Esters from Gp630 Determined by negative Ion Mass Spectrometry

		Doubly sulfated $[M+Na-H]^-$					
No.	theor *m/z*	exp *m/z*	Hex	Fuc	HexNac	sulfate	Na^+
1	1175.41	1175.37	1	0	4	2	1
2	1321.34	1321.39	1	1	4	2	1
3	1483.40	1483.36	2	1	4	2	1

The spectra in the same mass ranges of the mixture of the affinity MALDI probes are shown in Figs. 6b and 7b. While the signal-to-noise ratios varied between spectra, the relative intensities of the peaks remained essentially constant with less than 10% variations. A comparison of Fig. 6a and b and Fig. 7a and b shows specific intensities that are attenuated and signals that are enhanced. For example, *m/z* 1,381 is diminished in the CGL probe while *m/z* 1,219 is enhanced (Fig. 6b). Similarly, in Fig. 7b, *m/z* 1,934 is enhanced in the CGL probe. Some minor components disappeared while others appeared. For example, *m/z* 1,073, with composition corresponding to one hexose (Hex), four *N*-acetylhexoses, and one sulfate group, was not observed in the CGL probe. The majority of the signals in the CGL probe correspond to sulfated oligosaccharides. Similarly, there are signals observed in the CGL bound probe that are not found in the untreated polymer film probe; some including *m/z* 1,129 and 1,291 correspond to sialylated oligosaccharides. Signals of sialylated oligosaccharides are typically suppressed by sulfated oligosaccharides during MALDI-MS and are therefore not strongly observed in the whole egg mixtures with the polymer film probe. Other signals observed only in the lectin probe correspond to neutral oligosaccharides. For example, *m/z* 1,041 and 1,203 in the negative mode correspond to *m/z* 1,065 ($M+Na^+$) and 1,227, respectively, in the positive mode. The structures of these compounds are known and both have galactose in the nonreducing ends [15].

The most noticeable changes in intensities between the polymer film probe and the lectin-treated probe in Figs. 6 and 7 occurred for three abundant ions, *m/z* 1,219, 1,381, and 1,934. The ion *m/z* 1,381 was clearly not retained as were ions *m/z* 1,219 and 1,934. These species and few others were further investigated using CID.

The CID spectrum of *m/z* 1,219, a retained oligosaccharide shown in Fig. 8, is consistent with a pure compound whose proposed structure is shown (**6**, Chart 1). The most abundant peak corresponds to the loss of fucose. Fucose loss was a common fragmentation of oligosaccharides into the positive and negative ion modes. The ions of two *N*-acetylhexoses (HexNAc) from the quasimolecular ion to give *m/z* 813 indicated that the two HexNAc were at the nonreducing end. The loss of

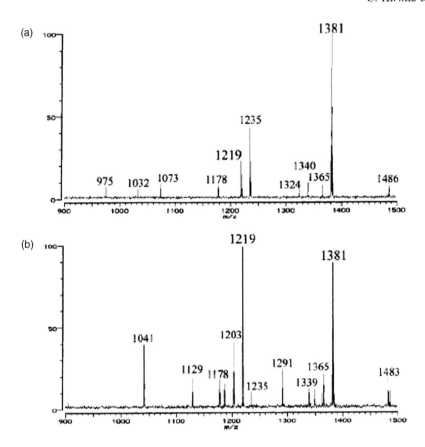

Figure 6. MALDI FT-ICR MS spectra in the negative mode (*m/z* 900–1,500) of an *O*-linked oligosaccharide mixture released from *X. laevis* egg jelly glycoproteins on (a) a polymer film probe and (b) a bioaffinity probe with immobilized CGL followed with washings. The species corresponding to *m/z* 1,381 is not as strongly retained as that corresponding to *m/z* 1,219.

fucose from *m/z* 813 to yield *m/z* 667 and the subsequent loss of a Hex determined the position of the fucose. The loss of HexNAc-ol, the alditol-converted reducing end, and the presence of *m/z* 282 corresponding to a sulfated *N*-acetylhexose (HexNAc-SO$_3^-$) yielded the biantennary nature of the compound. Moreover, the position of the sulfate was on one of the nonreducing termini.

Two other retained species *m/z* 1,365 and 1,483 were examined in a similar manner. Based on the CID spectrum, the proposed structure of *m/z* 1,365 is provided (**7**). The species *m/z* 1,483 yielded a composition that contained two sulfate ester groups. CID of *m/z* 1,483 indicated that two adjacent residues on the reducing end contain the sulfate ester **8**. It could not be determined which of the two residues were on the terminal position.

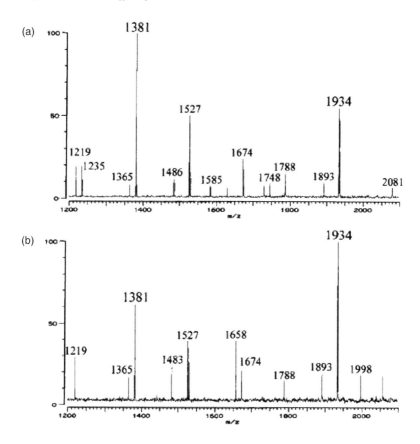

Figure 7. MALDI FT-ICR MS spectra in the negative mode (*m/z* 1,200–2,100) of an *O*-linked oligosaccharide mixture released from *X. laevis* egg jelly glycoproteins on (a) a polymer film probe and (b) a bioaffinity probe with immobilized CGL followed with washings. The species corresponding to *m/z* 1,934 is strongly retained.

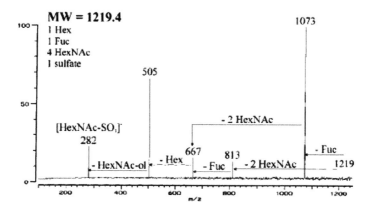

Figure 8. Tandem MS (MS/MS) spectrum of *m/z* 1,219, a species retained on the CGL probe. The fragmentation pattern is consistent with the structure proposed in Chart 1.

Chart 1: Chart of glycan structures that bound to CGL.

The CID of *m/z* 1,381 yielded a mixture of compounds. HPLC separation on an amine column was performed, resulting in two peaks. One peak yielded the CID spectrum in Fig. 9a. The fragmentation pattern was consistent with a single dominant species whose structure is proposed as **9**. The important structural feature here is that the sulfated residue is not on the nonreducing end.

The second HPLC peak produced the CID presented in Fig. 9b. The presence of *m/z* 241 is consistent with a sulfated Hex unit, while *m/z* 505 is consistent with two

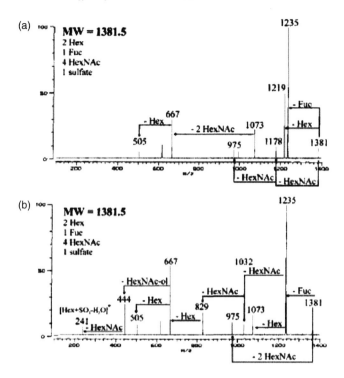

Figure 9. Tandem MS (MS/MS) spectra of two species of *m/z* 1,381. (a) Mass spectrum of a single component. (b) Mass spectrum of a two-component mixture. These compounds are not strongly retained on the CGL probe. The fragmentation patterns are consistent with the structures proposed in Chart 1.

HexNAc, one of which is sulfated. Thus, this sample was likely composed of two species with the sulfate on different residues. Fig. 9b is a spectrum of a mixture with structures proposed as **10** and **11**, based on the similar analyses performed on Fig. 9a. The important feature of these two oligosaccharides is that the sulfate is again on an internal residue. A similar analysis of another unretained oligosaccharide, *m/z* 1,235, yielded another sulfated oligosaccharide with the sulfate ester also in the internal residue **12**.

The use of bioaffinity MALDI probes to identify specific lectin substrates provides a rapid tool for screening protein–substrate binding. The oligosaccharides that are preferentially bound are separated from the other oligosaccharides in Chart 1. The structures of seven sulfated oligosaccharides were determined using CID.

The results obtained in these experiments were consistent with the notion that the nonreducing ends of the oligosaccharide moieties represent the recognition components of the glycoprotein. The oligosaccharides that were preferentially retained all had sulfate esters on the terminal (nonreducing) position. The oligosaccharides that were not preferentially retained had sulfate esters in the internal residues. The position of the sulfate esters is therefore important and sulfate esters must be intimately involved in the binding of CGL to the egg jelly. The lectins themselves are highly glycosylated so that the interactions between the lectin and the egg jelly may occur through oligosaccharide–oligosaccharide interactions via the sulfate groups. This notion is supported by the behavior of oligosaccharides, which are known to be strong chelators of sulfate groups [12]. The binding of the lectin to oligosaccharide may also involve ionic interactions because the binding is calcium dependent [18]. This method provided a powerful tool to determine lectin–glycan interactions which can lead to several beneficial applications.

References

[1] H. Rudiger and H.J. Gabius, Glycoconj. J., 18 (2001) 589–613.
[2] M. Wilchek and E.A. Bayer, Anal. Biochem., 171 (1988) 1–32.
[3] K.P. Henrikson, S.H.G. Allen and W.L. Maloy, Anal. Biochem., 94 (1979) 366–370.
[4] R.M. Whittal and L. Li, Anal. Chem., 67 (1995) 1950–1954.
[5] H. Wang, K. Tseng and C.B. Lebrilla, Anal. Chem., 71 (1999) 2014–2020.
[6] R.E. Wyrick, T. Nishihar and J.L. Hedrick, Proc. Natl. Acad. Sci. U.S.A., 71 (1974) 2067–2071.
[7] L.C. Greve and J.L. Hedrick, Gamete Res., 1 (1978) 13–18.
[8] R.D. Grey, D.P. Wolf and J.L. Hedrick, Dev. Biol., 36 (1974) 44–61.
[9] B.S. Bonnell, D. Reinhart and D.E. Chandler, Dev. Biol., 174 (1996) 32–42.
[10] J. Bundy and C. Fenselau, Anal. Chem., 71 (1999) 1460–1463.
[11] A.K. Allen and A. Neuberge, Biochem. J., 135 (1973) 307–314.
[12] A.W. Wong, M.T. Cancilla, L.R. Voss and C.B. Lebrilla, Anal. Chem., 71 (1999) 205–211.
[13] M.T. Cancilla, S.G. Penn, J.A. Carroll and C.B. Lebrilla, J. Am. Chem. Soc., 118 (1996) 6736–6745.
[14] G. Strecker, J.M. Wieruszeski, Y. Plancke and B. Boilly, Glycobiology, 5 (1995) 137–146.
[15] K. Tseng, J.L. Hedrick and C.B. Lebrilla, Anal. Chem., 71 (1999) 3747–3754.
[16] Y. Plancke, J.M. Wieruszeski, C. Alonso, B. Boilly and G. Strecker, Eur. J. Biochem., 231 (1995) 434–439.
[17] J.L. Hedrick, A.J. Smith, E.C. Yurewicz, G. Oliphant, D.P. Wolf, Biol. Reprod., 11 (1974) 534–542.
[18] T.A. Quill and J.L. Hedrick, Arch. Biochem. Biophys., 333 (1996) 326–332.

Lectins: Analytical Technologies
C.L. Nilsson (Editor)
© 2007 Elsevier B.V. All rights reserved.

Chapter 14

Electrospray Ionization Fourier Transform Ion Cyclotron Resonance Mass Spectrometry for Lectin Analysis

Julie T. Adamson and Kristina Håkansson

Department of Chemistry, University of Michigan, 930 North University Avenue, Ann Arbor, MI 48109-1055, USA

1. Introduction

Since the introduction of electrospray ionization (ESI) [1, 2] and matrix-assisted laser desorption/ionization (MALDI) [3, 4], these "soft" ionization techniques have dramatically changed the field of mass spectrometry. ESI and MALDI allow for the ionization of large non-volatile molecules, while yielding little or no fragmentation. They are applicable to a wide variety of molecules, ranging from peptides, proteins, oligonucleotides, oligosaccharides, and polymers, to lipids. These techniques are also suitable for extremely large molecules, rendering mass determination of species larger than 100,000 Da feasible. Currently, ESI and MALDI are the most commonly used ionization techniques for mass spectrometric analysis of biomolecules.

Fourier transform ion cyclotron resonance mass spectrometry (FT-ICR MS) was introduced in 1974 by Comisarow and Marshall [5], and since its inception has proven itself to be a valuable tool for the characterization of biomolecules. The employment of FT-ICR MS for biomolecular structural characterization offers several advantages, including high mass resolution, ultrahigh mass accuracy, and a wide mass range. FT-ICR MS is also an extremely versatile technique, and is capable of performing several varieties of tandem mass spectrometric methods (see Section 6) [6–11].

The most commonly used ionization technique combined with FT-ICR MS for biomolecule analysis is ESI [12, 13]. The first successful implementation of ESI as a mass spectrometric ionization technique was shown simultaneously by Yamashita and Fenn [14] and by Aleksandrov et al. [2]. One of the major advantages of ESI is its ability to generate multiply charged ions. Multiple charging, which generates ions with lower *m/z* values, is beneficial because FT-ICR MS resolving power decreases linearly with increasing *m/z* [15]. Furthermore, mass resolving power, mass accuracy, and limits of detection are also enhanced in the low *m/z* region. Contrary to ESI, MALDI typically produces singly charged ions, resulting in larger *m/z* values. MALDI can also be coupled to FT-ICR MS, and constitutes a powerful technique for the characterization of moderately sized bio-molecules (<3–5 kDa) [16–22].

There are several advantages gained when ESI is coupled to mass spectrometry. Due to the "soft" nature of the ESI process, non-covalent complexes may be preserved and examined in the gas phase. For example, the non-covalent interactions between concanavalin A monomers and between concanavalin A complexes and carbohydrates have been examined extensively using ESI [23–25]. The ionization and accurate molecular weight characterization of extremely large biomolecules is also feasible when ESI is combined with mass spectrometry. Presently with ESI FT-ICR MS, it is possible to achieve unit (isotopic) mass resolution for proteins over 100 kDa [26]. This possibility renders "top-down" characterization with FT-ICR MS an attractive choice of analyses. Another advantage of coupling ESI to FT-ICR MS is that on-line liquid chromatography (LC) may be readily implemented. However, due to the typically high flow rates used during liquid chromatographic separations, great care must be taken to avoid jeopardizing the vacuum of the mass analyzer. Several methods are available to effectively couple LC to mass spectrometry [27–30]. For example, flow splitters are often utilized for the coupling of LC to ESI-MS. In addition, pneumatically assisted electrospray is also used to overcome the flow-rate limitations of conventional ESI [31]. Although FT-ICR MS is capable of examining complicated mixtures, it is still desirable to couple it to a chromatographic interface. For highly complex mixtures, too many ions in the FT-ICR analyzer cell result in "space-charge" effects, which will reduce instrument performance. In addition, coupling of chromatographic separation to FT-ICR MS is advantageous for reducing ion suppression effects during ESI.

2. ESI FT-ICR MS Instrumentation

There are several key features common to all ESI FT-ICR instruments, including an electrospray source, ultrahigh vacuum system, magnet, analyzer cell (ICR cell), and data acquisition system [32]. We will briefly describe each of the key components of ESI FT-ICR MS, along with the operating principles behind this technique (see Section 3 on operating principles).

Interest in the electrospray process dates back to the beginning of this century [33, 34], and is the subject of several reviews [1, 35–38]. Multiple variations of ESI sources exist, and may be as simplistic as a metal capillary at an elevated voltage relative to a counter electrode. Liquid is passed through the capillary where it is nebulized at the tip due to a strong applied electric field, on the order of 10^6 V m^{-1} [39], which induces charge accumulation in the liquid surface at the end of the capillary. This excess of charge results in a dispersion of charged droplets due to Coulomb repulsion overcoming the surface tension. Following the production of charged droplets, and subsequent free ions, an efficient method of sampling and transferring ions from atmospheric pressure into the mass analyzer is required. Several strategies have been employed for initial droplet desolvation and transport. Fenn and co-workers developed an interface using a countercurrent flow of dry gas (typically N_2) combined with a glass capillary [1]. Alternative approaches involve the use of heated metal [40] or glass capillaries [35], or a heated chamber [41]. The combination of increased temperature and drop in pressure (as will be discussed) both aid in the desolvation of charged droplets.

An ultrahigh vacuum system is necessary for any FT-ICR instrument because ion detection needs to occur in an extremely low-pressure environment, in order to achieve characteristic FT-ICR MS high resolution. Typical pressures are in the range of 10^{-9} to 10^{-10} Torr [15, 32]. However, these low-pressure requirements are only applicable to the mass analyzer cell, therefore atmospheric pressure ionization techniques are still compatible with FT-ICR MS. Henry et al. were the first to report the coupling of ESI to FT-ICR MS in 1989 [42]. The difficulty in establishing this union is due to the extreme pressure difference between the ESI source and mass analyzer. The coupling of ESI to FT-ICR MS requires that several stages of differential pumping separate the ion source from the ICR cell. Focusing lenses with small orifices and focusing multipole lenses in combination with high capacity (e.g., turbomolecular or cryogenic) pumps can be utilized to generate the required pressure gradient.

Ion motion within the ICR cell, which is located within the magnet, is the governing principle behind FT-ICR MS and will be discussed shortly in the next section. The type of magnet in an FT-ICR instrument may vary, but currently superconducting magnets of 3–14.5 T are utilized. Spatial homogeneity of the magnetic field, typically of a few ppm over a 10-mm sphere, is crucial for high performance [43]. Unlike the magnets used for nuclear magnetic resonance (NMR) spectroscopy, those in FT-ICR MS are solenoidal magnets with wide bores. Increasing magnetic field strength offers several improved parameters, including mass resolving power, upper mass limit, mass accuracy, and signal-to-noise ratio [44]. However, increasing cost is the obvious drawback.

The ESI source is almost always located outside the magnetic field, with very few exceptions [45, 46]. This configuration renders it necessary for ions to be transported along the magnetic field lines into the ICR cell. As they progress, analyte ions must pass through the so-called magnetic mirror region of the magnetic field. As the magnetic field strength increases, the radial velocity of analyte ions also increases [47]. In order to keep kinetic energy constant, the axial velocity must therefore decrease. To efficiently pass through the magnetic mirror region, suitable ion guides are necessary. Such ion focusing may be accomplished using several methods, including the use of multipole ion guides [48–50] or DC wire ion

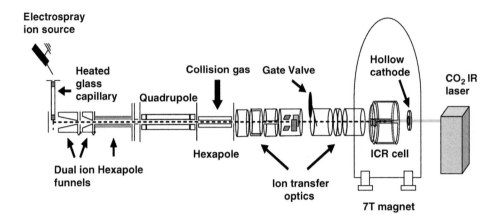

Figure 1. A schematic diagram of an APEX-Q FT-ICR mass spectrometer from Bruker Daltonics (Billerica, MA, USA), located at the University of Michigan in Ann Arbor, MI, USA.

guides [51]. The purpose of such techniques is to focus ions along the central axis of the magnet and thereby allow them to overcome the repulsive forces of the magnetic mirror [52, 53], which may slow down or even deflect analyte ions away from the ICR cell. Alternatively, ions can be accelerated through the magnetic fringe field into the ICR cell [54].

As an example of a typical ESI FT-ICR MS setup, a schematic diagram of an instrument located at the University of Michigan in Ann Arbor, MI, is shown in Fig. 1. The instrument is equipped with a microelectrospray ion source and a dual ion funnel inlet. The ion funnels consist of a series of ring electrodes with applied RF and DC electric potentials. Smith et al. first introduced an ESI ion funnel in 1998 as a means to more effectively focus and transmit ions from regions of higher pressure to lower pressure [55]. This instrument is also equipped with two hexapoles for external ion accumulation of electrosprayed ions [56]. The quadrupole and second hexapole allow for mass-selective ion accumulation in addition to external collision activated dissociation (CAD) (see Section 6 for a description of CAD). The dynamic range and sensitivity of an FT-ICR mass spectrometer is generally enhanced when external ion selection is utilized prior to ion accumulation in the ICR cell [57]. Furthermore, ESI is a continuous ion source, whereas FT-ICR MS is a pulsed detector. External ion accumulation is utilized to keep the duty cycle high. To transport ions into the ICR cell and overcome the magnetic mirror effect, the instrument contains a series of ion transfer optics. The ICR cell in this instrument is cylindrical, which is a relatively common geometry. However, several other varieties are also possible [58]. Located behind the ICR cell is a hollow cathode and CO_2 laser, which are used to fragment ions by electron capture dissociation (ECD), electron detachment dissociation (EDD), and infrared multiphoton dissociation (IRMPD) (see Section 6 for a description of these techniques).

3. FT-ICR MS Operating Principles

Under the influence of a spatially homogenous magnetic field, $\mathbf{B} = -B_0\mathbf{z}$ where B_0 is magnetic field strength, an ion experiences a force that is perpendicular to both the direction of its velocity and the magnetic field:

$$\boldsymbol{F} = \text{mass} \times \text{acceleration} = q\boldsymbol{v} \times \mathbf{B} \tag{1}$$

where q is charge and $\boldsymbol{v}$ is velocity. This force, called the Lorentz force, bends ions into a circular orbit which is perpendicular to the magnetic field axis (see Fig. 2). If v_{xy} denotes ion speed in the xy plane (perpendicular to the magnetic field axis) and because angular acceleration is equal to v_{xy}^2/r, Eq. (1) can be re-written as:

$$\frac{mv_{xy}^2}{r} = qv_{xy}B_0 \tag{2}$$

Here, m is ionic mass and r is the radius of the circular orbit. If these ions do not change speed (no collisions), Eq. (2) can be further re-written as the basic cyclotron equation:

$$\omega = \frac{qB_0}{m}(\text{SI units}) \tag{3}$$

where ω is angular velocity. Referring to Eq. (3), it is apparent that an ion's angular velocity is only dependent on ionic mass, charge, and magnetic field strength (where B_0 is constant). By dividing Eq. (3) by 2π, an ion's cyclotron frequency, v (in Hz), can be determined. The unique feature of an ion's cyclotron frequency in FT-ICR MS is that it is independent of ion velocity and kinetic energy. Unlike many other types of mass spectrometers, "focusing" of ions with a spread in kinetic energy is unnecessary [15, 32]. It is clear that ion cyclotron frequency is inversely proportional to m/z (in which z is the number of elementary charges), thus clarifying why multiple charging from ESI is beneficial for FT-ICR MS. In addition, it is important to maintain very low pressure in the analyzer cell of an FT-ICR mass spectrometer, because collisions between ions and neutrals decrease the radius of the ion cyclotron orbit and cause ions to drift away from the center of the analyzer cell. Once these ions collide with the analyzer cell walls, they are neutralized and lost.

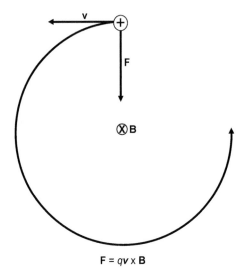

Figure 2. Ion cyclotron motion. The Lorentz force bends ions into a circular orbit which is perpendicular to the magnetic field axis. The magnetic field is directed into the plane of the figure. Negatively charged ions orbit in the opposite direction of positively charged ions.

The magnetic field in an FT-ICR instrument traps ions radially in the ICR cell, but does not prevent them from leaving the cell along the axial direction (along the magnetic field axis). In order to trap ions axially, trapping plates are mounted perpendicular to the magnetic field with an applied potential, typically 1–2 V. These applied voltages create a potential well in the analyzer cell, causing ions to oscillate between the two endplates. However, while the applied electric field traps ions along the magnetic field axis, it also repels ions away from the center of the analyzer cell in a radially outward direction. This outward-directed force, which opposes the inward directed Lorentz force, results in an additional type of ion motion within the analyzer cell, called magnetron motion. This force causes the center of an ion's cyclotron motion to orbit around the central axis of the cell, as shown in Fig. 3.

Magnetron motion effectively reduces the magnetic field strength, thus reducing the ion cyclotron frequency described in Eq. (2) [15, 32]. Magnetron motion may also cause radial ion diffusion, mass shifts, and sidebands. Fortunately, the frequencies of magnetron motion and trapping frequencies (due to ion oscillation) are much lower than the cyclotron frequencies, and are generally only detected as sidebands when the analyzer cell is misaligned or ion motional amplitudes reach the dimensions of the cell [59–61]. Ideally, in the absence of the electric field and additional ion motion, only one ion of known mass could be used for frequency to *m/z* calibration.

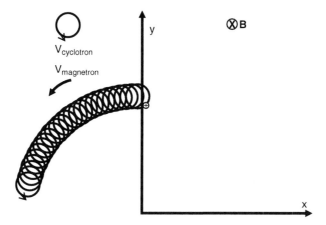

Figure 3. Ion trajectory showing ion cyclotron motion and ion magnetron motion. The latter motion causes the center of an ion's cyclotron motion to orbit about the center of the analyzer cell. The magnetic field is directed into the plane of the figure.

However, the resulting perturbation of ion cyclotron frequency due to the applied trapping potential necessitates that two ions of known mass are required for calibration [62]. Such calibration can be accomplished with "internal" or "external" calibration, where calibrant ions are either present along with analyte ions in a sample or are analyzed separately.

Once ions have been trapped in the analyzer cell, an effective means of detection is required. However, the initial ion cyclotron radius in the analyzer cell is small (<1 mm) compared to the analyzer cell dimensions (on the order of 10 cm), thus direct detection of an ion's cyclotron frequency would be relatively difficult. Therefore, ion excitation is necessary in order to accelerate ions to a larger detectable cyclotron radius. Such acceleration is accomplished by applying a spatially uniform electric field containing a range of frequencies, which vary depending on the *m/z* range being examined. The application of differential RF voltages to a pair of opposing excitation plates in the analyzer cell generates the required electric field (see Fig. 4). Broadband ion cyclotron excitation is commonly achieved by a frequency-sweep ("chirp") excitation, which uses a constant-amplitude sinusoidal waveform [63, 64].

Those ions which are in resonance with the frequencies of the applied RF voltage gain kinetic energy and spiral outwards to a larger cyclotron radius. All ions are excited coherently as a tight packet, which is commonly referred to as an ion cloud. The orbiting ion cloud continually passes by the other set of opposing plates in the analyzer cell, called the detection plates (see Fig. 4). This set of plates has parallel resistance and capacitance components and connects to a detection preamplifier [15] (see Fig. 4). The orbiting ion cloud induces an alternating current

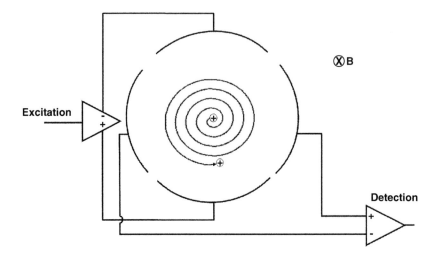

Figure 4. Ion excitation and detection in an ICR analyzer cell. A sinusoidal voltage is applied to a pair of opposing excitation plates. Ions in resonance with the applied frequencies spiral outwards to a larger cyclotron radius. An image current is induced in an opposing pair of detection plates, which are connected to an amplifier.

on the two opposing detection plates, which is called the image current. The image current is measured as a time-domain signal and is composed of all the frequencies of the ion clouds in the analyzer cell. By applying a Fourier transform to the time domain signal, the frequency components of the signal can be determined. A modified version of the original cyclotron equation (correcting for ion motion due to the electric field) can now be used to calculate m/z ratios. The inherent advantages of ion detection in an FT-ICR mass spectrometer are that it is non-destructive and a wide range of m/z ratios can be detected simultaneously (multi-channel advantage). Furthermore, ICR signal increases linearly with ion charge, demonstrating once more the advantage of using ESI as an ionization source.

4. Performance of FT-ICR MS

FT-ICR MS is capable of providing extremely high mass accuracy. This useful characteristic is possible because the frequencies of ion clouds in the analyzer cell can be measured with extreme accuracy. With external calibration, FT-ICR MS is typically capable of achieving low ppm mass accuracy, but only when standards and samples are analyzed under identical instrumental conditions. Internal calibration, where samples and standards of known masses are mixed together and analyzed under the same conditions, has been shown to improve mass accuracy by an order of magnitude compared to external calibration [15, 65]. With ESI FT-ICR MS, internal calibration is usually accomplished by electrospraying analytes and internal standards simultaneously using a single electrospray needle, or delivering them separately using a dual electrospray source. Several groups have focused on developing the latter to obtain more accurate mass measurements [66–70]. One of the greatest limiting factors to ultra-high mass accuracy in FT-ICR MS is the space-charge effect [71], which causes frequency shifts due to Coulombic interactions between ions in the analyzer cell. Reducing the number of ions in the cell and/or operating at a higher magnetic field strength limits space-charge effects. To account for space-charge effects and improve mass accuracy, a number of mass calibration procedures and techniques have been developed for FT-ICR MS [65]. Alternatively, certain commercially available FT-ICR instruments provide automatic gain control (AGC) as a method to regulate the number of ions entering the analyzer cell. AGC guarantees that samples and external calibrants are detected under identical instrument conditions, therefore resulting in low ppm mass accuracy without the use of internal calibrants [72, 73].

Another advantage offered by FT-ICR MS is its tremendous resolving power. Several factors explain why this technique is capable of such superior resolving power. One reason is that the detection signal in the analyzer cell typically lasts for several seconds, which results in numerous cycles of cyclotron motion. Under low-pressure conditions, FT-ICR MS resolving power is proportional to the length of the time domain signal [74]. Another reason is that cyclotron frequency is only dependent on m, q, and B_0, therefore FT-ICR MS performance is not affected by a spread in kinetic energy. With FT-ICR MS, resolving power in excess of 10^6 has been obtained for large biomolecules. As previously mentioned, Kelleher et al. have achieved isotopic resolution of a 112 kDa protein with unit mass resolution [26]. This example constitutes the highest mass molecule for which unit mass resolution has been obtained to date. Unit mass resolution is attained when there is a separation of the isotopomers differing in the number of heavy stable isotopes, such as ^{13}C, ^{15}N, ^{17}O, ^{18}O, ^{2}H, ^{33}S, ^{34}S, and ^{36}S. With unit mass resolution, the monoisotopic mass (i.e., the mass of a molecule containing only ^{12}C, ^{14}N, ^{16}O, ^{1}H, and ^{32}S) of moderately sized biomolecules, such as peptides and small proteins, can be determined. In addition, isotopic resolution allows for the determination of the charge state of a molecule, based upon the m/z spacing between isotopic peaks [75]. For example, if the m/z difference between adjacent isotopomers differing by 1 Da (such as ^{12}C and ^{13}C) is 0.5 in the mass spectrum, the charge state of the molecule is 1/0.5, i.e., 2.

One of the major factors governing the detection limits in FT-ICR MS is that typically at least 100 charges of a given m/z ratio are necessary in order to generate a measurable signal. Although ESI is advantageous because it generates multiply charged ions, ESI is also a concentration sensitive technique [76]. Despite these limitations, detection limits with ESI FT-ICR MS can still be extremely low. For samples with low levels of analyte, it is beneficial to couple on-line separation, such as capillary electrophoresis (CE) or nano-LC to FT-ICR MS [77].

Nano-LC in particular, with its ability to pre-concentrate the sample and desalt online, greatly improves sensitivity [78]. Desalting samples is desirable because salts greatly interfere with the analyte signal, and cause background interference. Reproducible detection of several hundred attomoles of peptides loaded onto a nano-LC column coupled to FT-ICR MS has been demonstrated by Quenzer et al. [79]. The coupling of CE to FT-ICR MS has permitted the detection of human hemoglobin from a single red blood cell containing several hundred attomoles of the protein [80], and allowed for the detection of low attomole levels of carbonic anhydrase from a single human red blood cell [81]. McLafferty and co-workers have reported the lowest FT-ICR detection limit for whole proteins to date, which was 0.7–3.0 attomoles for proteins ranging in size from 8 to 29 kDa [81].

5. Top-Down and Bottom-Up Proteomics

The high mass accuracy provided by FT-ICR MS allows for the accurate molecular weight determinations of both peptides and proteins. The traditional mass spectrometry-based approach to proteomics, the "bottom-up approach," begins with a one- or two-dimensional (2D) electrophoretic or chromatographic separation followed by proteolysis and mass spectrometric analysis [82–84]. The results of this analysis provide an accurate "mass fingerprint" which can be used to identify the protein of interest [85–89]. A newer approach for protein identification has emerged over the past several years, which relies on the high mass accuracy provided by FT-ICR MS. This approach uses on-line LC combined with accurate-mass measurement of peptides from enzymatic digests of complex mixtures of proteins, followed by searching of protein databases [90, 91]. High mass accuracy measurements of peptides greatly reduce the chances of false-positives during database searching.

A "top-down" approach to protein characterization with ESI FT-ICR MS has emerged as an attractive alternative to the "bottom-up" approach [92, 93]. McLafferty and co-workers have recently applied such methodology for the characterization of proteins in excess of 200 kDa, which to date are the largest proteins which have been analyzed with tandem mass spectrometry [94]. Characterization of intact proteins is made possible by the production of multiply charged protein ions from ESI, along with the high mass accuracy and resolving power of FT-ICR MS. This combination has allowed for the characterization and identification of intact proteins in complex mixtures [95–97]. One of the major proteomic applications of intact protein analysis is for the determination of posttranslational modifications and sequence errors [92, 93, 98–100]. A comparison of the calculated molecular weight of a protein (predicted by the DNA sequence) and the experimentally determined weight reveals sequence variations and the presence of modifications. Tandem mass spectrometry can indicate their locations. High mass accuracy measurements of intact proteins are also useful for the determination of

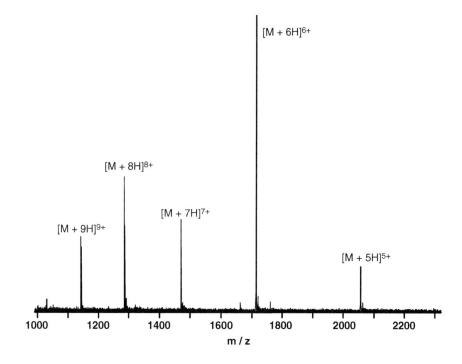

Figure 5. Positive ion mode ESI FT-ICR mass spectrum (10 scans) of a mushroom lectin from *L. decastes*.

disulfide bridges (mass difference of 2 Da), deamination (mass difference of 1 Da), metal oxidation states, and adduct formation. Recently in our laboratory, we have utilized ESI FT-ICR MS for the top-down and bottom-up characterization of a mushroom lectin from *Lyophyllum decastes*. Fig. 5 shows the broadband mass spectrum of the purified protein. In positive ion mode, several charge states of the intact protein are observed. From these results, an accurate molecular weight of the intact protein could be determined.

6. Tandem Mass Spectrometry in ESI FT-ICR MS

Tandem mass spectrometry (MS/MS or MS^n) involves the activation of a primary ion of interest ("precursor" or "parent" ion), followed by dissociation or reaction, and analysis of the secondary ions ("product" ions). The purpose of fragmenting biomolecular ions is to obtain information regarding their primary structure. If cleavage occurs throughout the biomolecule between individual building blocks, sequence information can be obtained by examining differences in mass between product ions. This information aids in the identification of proteins and determination of present modifications. A description of precursor ion isolation and ion activation techniques in FT-ICR MS will follow. Each ion activation method will be described in detail, as the means of activation is fundamental to ion fragmentation and defines what product ions are produced. There are a wide variety of fragmentation techniques available in FT-ICR MS. These include sustained off-resonance irradiation collision activated dissociation (SORI-CAD) [6], IRMPD [7, 8], blackbody infrared radiative dissociation (BIRD) [9, 10], surface induced dissociation (SID) [101–103], ECD [104], and EDD [105]. Several applications of these techniques for the characterization of lectins will also be described.

6.1. Precursor ion isolation

Isolation of precursor ions for tandem mass spectrometry in FT-ICR MS can be accomplished by the application of specific excitation waveforms to the cell plates. Similar to the technique described for ion excitation, a specific waveform is utilized that contains the resonant frequencies of all ions in the cell except the precursor ion of interest. The excitation signal amplitude must be large enough that unwanted ions are excited to a cyclotron radius that causes them to collide with the analyzer cell walls. Due to the application of this waveform, all ions resonant with the applied frequencies are radially ejected from the ICR cell while precursor ions of interest remain in the center of the analyzer cell. Following precursor ion isolation, the next step is ion activation and dissociation followed by product ion detection.

There are several varieties of waveforms which can be used for ion excitation and precursor ion selection in FT-ICR MS. Arbitrary waveform generators can be used to produce tailored excitation fields for ion ejection. One of the better-known waveforms used for ion excitation and isolation is stored waveform inverse Fourier transform (SWIFT) [106]. First, the *m/z* ratios of ions to be ejected from the analyzer cell are determined and the corresponding frequencies are calculated to determine the required frequency domain function. This function is then inverse Fourier-transformed into the time domain, and this stored waveform is used to adjust the potential applied to the excitation plates of the analyzer cell.

Generally, precursor ion isolation and tandem mass spectrometry with FT-ICR MS is temporal, as opposed to spatial. However, as shown in the instrument schematic in Fig. 1, external ion isolation using a quadrupole is also possible in hybrid-type instruments. As previously discussed, the dynamic range and sensitivity of FT-ICR MS is generally enhanced when external ion selection is utilized prior to ion accumulation in the ICR analyzer cell. In addition to mass selective external accumulation with a quadrupole, Wang et al. have shown that mass selection can be achieved using an external linear octopole [107].

6.2. Sustained off-resonance irradiation collision-activated dissociation

CAD is the most common ion activation method in modern day mass spectrometers [108–111]. The CAD process involves two steps: ion activation and ion dissociation. Ions are activated through inelastic collisions with neutral gas molecules, which results in a conversion of translational energy into internal energy. Internal energy is redistributed throughout the ion, and if it exceeds its threshold energy dissociation occurs. There are two categories of CAD: low-energy and high-energy. When referring to CAD in FT-ICR MS, only low-energy CAD has been implemented thus far.

To increase the translational energy of ions in the FT-ICR analyzer cell, resonance excitation or off-resonance excitation can be used. By applying a resonant frequency at the precursor's ion cyclotron frequency, the ion radius increases rapidly and through collisions with neutral gas molecules (which are pulsed into the ICR cell) ions increase their internal energy and dissociate. However, there are several issues associated with this approach. One is that product ions are not formed in the center of the analyzer cell, which negatively affects resolving power and renders further fragmentation (MS^n) more difficult [112]. In addition, the magnetic field and size limitations of the analyzer cell restrict the translational energy that can be imparted into the precursor ions. Finally, resonance excitation typically results in a more complicated spectrum because of extensive product ion fragmentation [113].

A "softer" approach is to use SORI-CAD, which is the preferred method of CAD in FT-ICR MS. In SORI, an excitation frequency is applied which is slightly off-resonance from the precursor ion's cyclotron frequency, resulting in an alternating increase and decrease in ion cyclotron radius [114, 115]. The alternating period is equal to the difference between the ion cyclotron frequency and applied off-resonance excitation frequency. The offset resonance is typically 500–2,000 Hz. Prior to off-resonance excitation, an inert gas is pulsed into the analyzer cell to act as a collision target. During the SORI excitation event, precursor ions continually collide with the background gas and their internal energy thereby increases. Typically, product ions formed through SORI-CAD have cyclotron frequencies which are different from the applied off-resonance frequency. However, in cases when the frequencies do match ("blind spots"), it is necessary to use a different excitation frequency (for example, offset frequencies of $+1,000$ Hz instead of $-1,000$ Hz). There are several advantages of using off-resonance excitation, as opposed to resonance excitation. The time-scale of off-resonance excitation is longer, which results in more collisions, because the ion cloud radius continually increases and decreases without ions leaving the analyzer cell. Another advantage is that product ions form relatively close to the center of the analyzer cell. The latter feature simplifies subsequent excitation and detection of product ions. Following the SORI-CAD event, a pump down delay is incorporated into the experimental sequence to lower the pressure of the analyzer cell and thereby restore high-resolution measurements.

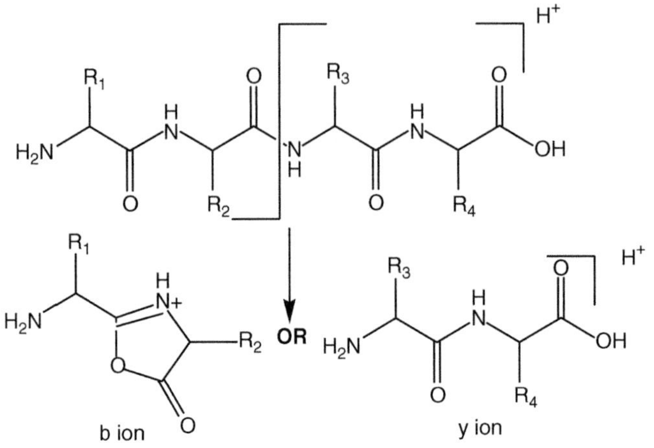

Figure 6. Fragmentation of the peptide amide bond from "slow-heating" MS/MS techniques available for FT-ICR MS. Such fragmentation techniques yield *b* and *y* type ions.

SORI-CAD is considered a "slow-heating" technique, because fragmentation occurs through multiple low-energy collisions and results in dissociation through the lowest energy pathways [116]. This fragmentation technique is extensively applied to peptides and proteins for obtaining structural information. When peptides and proteins are fragmented with SORI-CAD, neutral molecule losses (such as water and ammonia) in addition to *b* and *y* type product ions are observed (Fig. 6). Because dissociation occurs through the lowest energy pathways, labile modifications (such as glycosylation and phosphorylation) are often cleaved in SORI-CAD.

6.3. Infrared multiphoton dissociation

Another activation method available in FT-ICR MS is IRMPD, which as the name implies is a photodissociation technique. Dissociation of biomolecules with IRMPD is achieved by irradiating precursor ions with infrared photons. Similar to SORI-CAD, IRMPD is a "slow-heating" technique. Isolated precursor ions are slowly heated to their dissociation threshold, and fragment through the lowest energy pathways. IRMPD is a low-energy fragmentation technique, and requires the absorption of IR radiation from an extensive number of photons before dissociation occurs. This characteristic requires that an IR active mode is present in the ion, which is true for both peptides and proteins. Typically, 10.6 μm CO_2 lasers are used for IRMPD. The absorption of one photon corresponds to 0.117 eV of energy [111]. The IR laser is fired through a window, typically made of BaF_2 or

ZnSe, through the center of the analyzer cell; although off-axis IRMPD is also applicable depending on instrumental design [117]. The length of the IR pulse required for fragmentation depends on ion structure, but Little et al. have shown that optimum irradiation times for peptides and proteins are on the order of several hundred milliseconds for typical laser powers under 10 W [8].

One of the greatest advantages of IRMPD compared to SORI-CAD is that there is no need to pulse a collision gas into the analyzer cell; therefore, there is no degradation of the high vacuum and no need for an additional pump delay. Dissociation experiments with IRMPD are typically much faster compared to their SORI-CAD counterparts. While SORI-CAD requires off-resonance excitation, which may be problematic in congested spectra, these issues are not present with IRMPD. Also, IRMPD does not suffer from the "blind spots" that can disadvantage SORI-CAD. IRMPD is straightforward to implement and offers several levels of control, including control over the laser firing pulse length and laser power. In addition, because precursor ions are not moved off-axis during the IRMPD event, product ions are formed on-axis. If product ions were to form off axis, there would be a loss in resolution and it would be necessary to axialize these ions prior to excitation and detection. Another advantage of on-axis product ion formation is that further stages of tandem mass spectrometry (MS^n) are simplified. One possible disadvantage of on-axis product ion formation is that secondary fragmentation often occurs. Product ions are still on-axis and in the path of the IR laser beam, therefore further fragmentation of both precursor and product ions occurs. Such secondary fragmentation complicates spectral interpretation, but on the other hand provides more extensive structural information. Amide bond cleavage is typically more extensive with IRMPD compared to SORI-CAD, which renders IRMPD useful for peptide sequencing.

As discussed previously, ion activation techniques such as SORI-CAD and IRMPD typically generate b and y type product ions for peptides and proteins. However, when labile modifications such as glycosylation and phosphorylation are present, these modifications are often preferably cleaved. As an example of the type of fragmentation observed following IRMPD of a modified peptide, the IRMPD fragmentation pattern of a glycopeptide from a tryptic digest of an *Erythrina cristagalli* lectin is shown in Fig. 7. The dotted lines in the figure indicate cleavage sites. In this example, only cleavages within the glycan structure are observed. IRMPD of N-glycosylated peptides is understood to selectively cleave glycosidic bonds rather than the peptide backbone bonds for both xylose type (shown here) and complex type glycopeptides [117–119], whereas a mixture of peptide and glycan cleavage has been observed for high-mannose type glycopeptides [120]. The fragmentation pattern observed in Fig. 7 provides valuable information regarding the glycan composition of this glycopeptide; however, no information is obtained regarding the peptide amino acid sequence or modification location. Such fragmentation behavior can be problematic if the sequence and structure of this peptide were unknown.

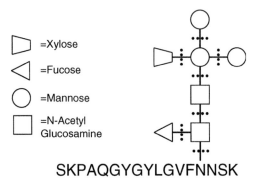

□ =Xylose

◁ =Fucose

○ =Mannose

□ =N-Acetyl
   Glucosamine

SKPAQGYGYLGVFNNSK

Figure 7. Fragmentation pattern observed following positive ion mode IRMPD of a glycopeptide from an *E. cristagalli* lectin obtained with ESI FT-ICR MS. Only glycosidic cleavages are observed following IRMPD, which is typical for certain types of N-linked glycopeptides.

6.4. Blackbody infrared radiative dissociation

Blackbody infrared radiative dissociation is achieved in FT-ICR MS by heating the vacuum chamber surrounding the analyzer cell. Such heating results in the emission of blackbody infrared photons from the walls of the analyzer cell, which are absorbed by trapped ions. The absorption of thermal, infrared radiation increases the temperature of ions trapped in the ICR cell, resulting in subsequent fragmentation. There are two essential requirements for the successful implementation of BIRD [111, 121]. One requirement is that the ambient pressure is relatively low (below 10^{-6} Torr), such that photon absorption and emission significantly competes with collisional activation. In addition, because BIRD is a relatively slow process, trapped ions must be stored for longer periods of time in order to obtain sufficient fragmentation. For peptides and proteins, a BIRD spectrum can take between 10 and 1,000 s to obtain [43]. Due to these trapping requirements, BIRD is typically only performed in FT-ICR instruments.

Generally, BIRD is not the preferred fragmentation technique if only MS/MS of peptides and proteins for protein identification is required. There are several disadvantages associated with this type of ion activation. Due to the time requirements necessary to equilibrate the temperature of the analyzer cell and obtain a BIRD spectrum, long time periods are necessary to complete an experiment. This time-scale makes BIRD incompatible with on-line separations. In addition, McLafferty and co-workers have shown that fragmentation efficiencies of proteins with BIRD are typically lower compared to other "slow-heating" techniques such as CAD and IRMPD [122]. Similar to CAD and IRMPD, BIRD of peptides and proteins results in *b* and *y* type ions. Also, extensive water losses are characteristic of this technique.

Despite these disadvantages, BIRD is a unique fragmentation technique because it permits the determination of kinetic dissociation parameters of biomolecules, thereby providing valuable information regarding gas-phase stability. In order to obtain this information, the abundance of a precursor ion and its product ions are monitored over time and at several different temperatures (typically below 250°C) [121]. From this information, first-order dissociation rate constants can be determined. When multiple product ions are observed, the kinetics of each individual reaction can be monitored.

One valuable application of BIRD is for the examination of protein complexes in the gas phase. Klassen and co-workers have utilized BIRD for the characterization of a complex of a single chain variable fragment (scFv) and its native trisaccharide ligand [123]. With BIRD, these authors were able to map the location and strength of hydrogen bonds between the protein and ligand. To date, very few studies have been published on BIRD applied to protein-carbohydrate complexes, but Klassen and co-workers have demonstrated the potential of this technique for the examination of lectin–carbohydrate binding.

6.5. Surface induced dissociation

SID is similar to CAD, except that a solid surface is used as a collision target instead of neutral gas molecules. An advantage of SID compared to CAD is that more internal energy is deposited into precursor ions, which renders the dissociation of high mass biomolecules more feasible. Also, the internal energy distribution of excited ions is much narrower than that in CAD, and energy transfer can be readily controlled by varying the impact energy. Similar to IRMPD and BIRD, there is no need to introduce a collision gas into the analyzer cell. In addition, because SID does not require any RF excitation, there are no "blind spots." SID of peptides and proteins generally produces a similar fragmentation pattern as SORI-CAD (predominantly *b* and *y* type ions). However, unlike the activation techniques discussed thus far (SORI-CAD, IRMPD, and BIRD), SID results in an almost instantaneous energy deposition. Shattering transitions, which are due to the instantaneous decomposition of ions upon collision with the target surface, can be observed at high collision energies [124]. Shattering of peptides produces several additional types of product ions which are not typically observed for "slow-heating" activation techniques (including additional internal fragments and immonium ions) [125]. SID is a powerful technique for probing peptide structure and examining fragmentation energetics and mechanisms. However, despite these advantages, SID is rarely applied for proteomic applications in FT-ICR MS.

6.6. Electron capture dissociation

ECD is a relatively recent ion activation technique, and is based upon the dissociative recombination of protonated polypeptide molecules with low-energy electrons (<0.2 eV) [104, 126–128]. ECD is implemented in FT-ICR MS by

irradiating trapped ions with low-energy electrons, typically from a heated filament electron gun or indirectly heated dispenser cathode. ECD is difficult to
implement in other mass spectrometers because of the trapping requirements;
ions and electrons must be trapped simultaneously for dissociation to occur. To
observe product ions following ECD, the precursor ion must carry at least two
positive charges; electron capture reduces the total charge by one. Moreover, the
electron capture cross-section has been shown to increase as the square of ion
charge; therefore, precursor ions with high charge states are preferred [129].
Fortunately, if ESI is coupled to FT-ICR MS, multiply charged ions are easily
generated for peptides and proteins.

Ion activation techniques which rely on vibrational excitation, such as SORI-
CAD, IRMPD, and BIRD tend to favor the lowest energy fragmentation pathways.
For peptides and proteins those pathways correspond to backbone amide bond
cleavage, which generates b and y type ions (see above). Furthermore, site-specific
fragmentation adjacent to particular amino acids tends to occur with these techniques. For example, with peptides and proteins, the N-terminal side of proline is
a preferred cleavage site [130]. Unlike the previously described ion activation
techniques, ECD has been proposed to be non-ergodic, with fragmentation occurring before energy randomization [104] although this claim has been challenged
based on both theory [131] and experiment [132]. Following electron capture, the
major product ion is typically the charge-reduced radical species. This ion has captured an electron but has not further dissociated. This species may undergo further
fragmentation, resulting in $N–C_\alpha$ bond cleavage yielding N-terminal c' fragments
and C-terminal radical z fragments (see Fig. 8). ECD cleavage tends to be more
random than other ion activation methods, and more extensive sequence coverage
is usually achieved. The only cleavage site that is not typically observed in ECD is
the N-terminal side of proline, due to its cyclic structure. These differences in fragmentation patterns render ECD a valuable tool for providing fragmentation information which is complementary to that obtained from vibrational excitation
techniques.

There are several features of ECD which make it an attractive technique for
the structural characterization of peptides and proteins [126–128]. ECD preferentially cleaves disulfide bonds, which can enhance sequence information gathered
from MS/MS. In addition, ECD does not cleave post-translational modifications
(PTMs) such as phosphorylation and glycosylation. Given that ECD retains labile
modifications to a much higher degree than traditional MS/MS, it greatly facilitates their localization [128]. An example of this behavior is demonstrated in
Fig. 9, which shows the fragmentation pattern of a glycopeptide from a tryptic
digest of a lectin from *E. cristagalli*. Following ECD, extensive peptide backbone
cleavage is observed (c' and z type ions). In fact, there is cleavage between every
amino acid residue (except the N-terminal side of proline, as mentioned above).
The glycosylation site is left intact, allowing for its localization. For comparison,
the IRMPD fragmentation pattern of this same glycopeptide was shown in Fig. 7.

2H+
R1
R3
H2N
OH
R2
R4
H+
OH
NH
OR
H+
R3
HC
or
H+
NH2
z• ion
c ion

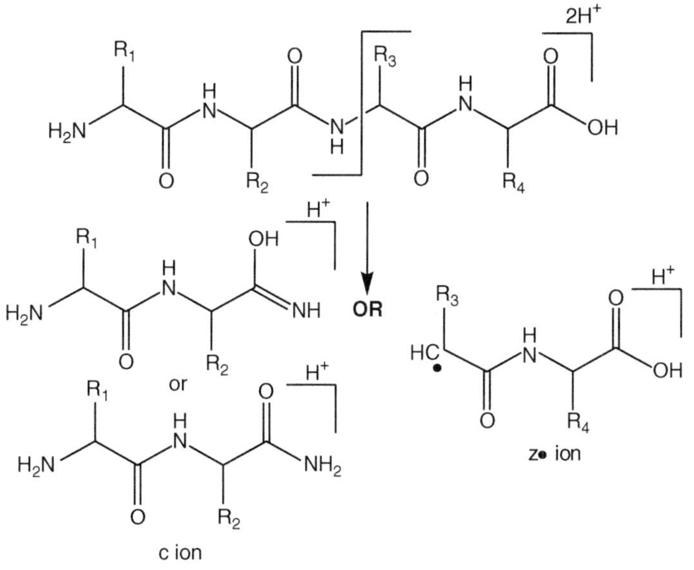

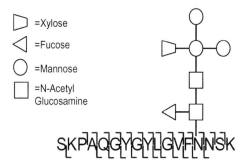

Figure 8. Fragmentation of a peptide backbone amine bond to produce *c'* and radical *z* ions is shown. This fragmentation pattern is typical of ECD and ETD.

=Xylose
=Fucose
=Mannose
=N-Acetyl
Glucosamine
SKPAQGYGYLGVFNNSK

Figure 9. Fragmentation pattern observed following ECD of a glycopeptide from an *E. cristagalli* lectin obtained with ESI FT-ICR MS. Solid lines between amino acids indicate cleavage sites.

Following IRMPD, only glycosidic cleavage is observed. The combination of ECD and IRMPD allows for the sequencing of the peptide and glycan, and also permits the localization of the modification within this peptide. The first demonstration which showed that complementary fragmentation information can be gathered from IRMPD and ECD of a glycopeptide was also demonstrated for a lectin peptide, of *E. corallodendron* [118]. Similarly, the combination of electron transfer dissociation (ETD) and CAD has also shown complementary fragmentation information for a lectin glycopeptide in a 3D-quadrupole ion trap [133]. ETD is not utilized in FT-ICR MS thus far, and has only been implemented in quadrupole ion traps. Similar to ECD, ETD results in c and z type product ions and has also been shown to retain PTMs such as glycosylation. However, the mass accuracy and resolution of FT-ICR instruments surpass those of quadrupole ion traps. The ETD process will not be discussed here, a description of this technique can be found elsewhere [134].

One chief disadvantage of ECD compared to other ion activation techniques is that fragmentation efficiency is relatively low, with typically only 5–20% of precursor ions being converted to product ions [43]. As a consequence, a high signal-to-noise (S/N) ratio is required for precursor ions. Another potential disadvantage of this method is that ECD of proteins larger then 20 kDa typically results in poor fragmentation [135]. This phenomenon has been explained as due to the presence of intramolecular non-covalent interactions. While ECD does indeed cleave protein amine bonds, it does not disrupt non-covalent interactions. To overcome the challenge imposed by the presence of non-covalent interactions, "activated-ion ECD" was developed [135]. In this technique, precursor ions are first vibrationally excited (through collisions or IR-heating) to disrupt tertiary structure and non-covalent interactions, which allows more efficient fragmentation when ions are irradiated with electrons. Although activated-ion ECD was originally proposed to facilitate ECD fragmentation of proteins, this technique is also useful for large peptides. For example, Hakansson et al. have utilized both ECD and activated-ion ECD for fragmenting a glycopeptide from a lectin of *E. corallodendron* [117]. These results are shown in Fig. 10. This glycopeptide is approximately 5 kDa, whereas the glycopeptide shown in Fig. 9 is approximately 3 kDa. Without ion activation, only the charge-reduced species is observed following ECD of the larger glycopeptide (Fig. 10a). These results indicate that intramolecular non-covalent bonds are present, hindering the separation of ECD product ions. However, by isolating the charge-reduced precursor ion and irradiating it with an IR laser, extensive backbone cleavage is observed (Fig. 10b).

7. Negative Ion Mode Tandem Mass Spectrometry

The ion activation techniques discussed thus far for FT-ICR MS are applicable to either positively or negatively charged ions, with the exception of ECD. Negative ion mode mass spectrometry is practical when the peptide or protein of interest contains acidic amino acid residues (glutamic and aspartic acids), or acidic modifications (such as phosphorylation and sulfation). Although negative ion mode mass spectrometry is used in certain cases for the accurate mass determination of peptides and proteins, negative ion mode tandem mass spectrometry of peptides and proteins for obtaining structural information is rare [136].

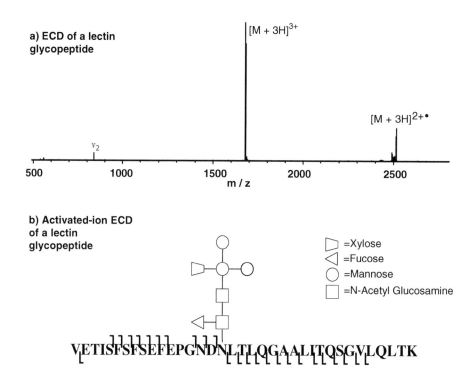

Figure 10. Fragmentation observed following (a) ECD and (b) activated-ion ECD of a glycopeptide from an *E. corallodendron* lectin obtained with ESI FT-ICR MS. Solid lines between amino acids indicate cleavage sites. Following ECD, no peptide backbone cleavage was observed, presumably due to intramolecular non-covalent interactions. Following activated-ion ECD, extensive peptide backbone cleavage was observed (*c'* and *z* type ions).

One disadvantage of MS/MS in negative ion mode is that CAD (and vibrational excitation techniques in general) of deprotonated peptides often results in relatively complex fragmentation spectra. The mechanisms of negative ion mode fragmentation are quite different from positive ion mode. Although several product ions observed in negative ion mode are also observed in positive ion mode CAD, several do not have positive ion mode counterparts [136, 137]. These properties render the interpretation of negative ion mode fragmentation spectra somewhat more difficult compared to positive ion mode.

One recently introduced ion activation technique, EDD, is *only* achieved in negative ion mode [105]. EDD is induced by electron irradiation of polypeptide anions with >10 eV electrons, and yields predominantly N-terminal radical *a* ions and C-terminal *x* ions due to C_α–C bond cleavage. Fragment ions generated by EDD have been shown to retain modifications such as sulfation and phosphorylation, which indicates that EDD also shows the potential for peptide sequencing and for mapping PTMs [105, 138, 139]. Similar to ECD, EDD requires that precursor ions contain at least two charges, indicating that ESI is the most suitable ionization technique. One key disadvantage of EDD is that fragmentation efficiency is much lower compared to other activation techniques. However, this technique shows promise for the characterization of acidic modified peptides. In addition, EDD has also been shown to be a valuable tool for the characterization of nucleic acid structures and folding [140].

In 2005, we utilized negative ion mode tandem mass spectrometry for the characterization of a glycopeptide of a lectin from *E. cristagalli* [141]. This same peptide was examined in positive ion mode, as shown in Figs. 7 and 9. IRMPD of this glycopeptide in negative ion mode resulted in a mixture of peptide and glycan cleavages. In addition to *b* and *y* type ions, several internal cleavage product ions were observed (due to cleavage on both the N- and C-terminal sides of the peptide). EDD of this glycopeptide was also examined. Unlike the fragmentation pattern observed following ECD, only glycan cleavage was seen following EDD. Although the combination of negative ion mode IRMPD and EDD provided complementary fragmentation information, IRMPD data was more difficult to interpret and EDD fragmentation efficiency was quite poor. Our results for this peptide agree with the general consensus that positive ion mode tandem mass spectrometry provides improved peptide and glycan sequence information, and better aid in modification localization.

8. Metal-Assisted Tandem Mass Spectrometry

In addition to protonated and deprotonated molecular ions, metal ion adducts of analytes can also be produced during the ESI process. For certain types of molecules, such as oligosaccharides, fragmentation of metal adducted species is often more useful for obtaining structural information compared to protonated species [142–144]. Metal ions play several important roles in biological processes, ranging from protein structure stabilization, oxygen binding, and catalysis [145]. Typically, metal ion interactions with peptides and proteins are examined in the solution phase. However, in the interior of proteins, aqueous solvation plays a limited role in metal binding (if any role at all). Intrinsic interactions between metal ions and proteins can effectively be examined under solvent-free conditions, in the gas phase [146]. This approach has been utilized extensively for the characterization of metal ions with peptides and proteins [147, 148]. There are several advantages of using mass spectrometry to examine non-covalent interactions in the gas phase, including specificity, speed, and sensitivity [148]. Generally, ESI is used to examine such complexes. However, MALDI has also been applied in certain examples [149–155].

Recently, Lehmann and co-workers have examined the interactions of peptides from human galectin-3 with metal ions in the gas phase using a hybrid ESI FT-ICR mass spectrometer [156]. This instrument is composed of both a linear ion trap and an FT-ICR mass analyzer, and is a commercially available setup. Their results demonstrate that certain galectin-3 peptides have a tendency to form abundant doubly charged adduct ions with Ca^{2+}, as opposed to adduction with alkali metals. They confirmed this behavior by use of the high-resolving power of FT-ICR MS. In these measurements, the isobaric ion pairs $[M + {}^{40}Ca]^{2+}$ and $[M + H + {}^{39}K]^{2+}$ could be distinguished. These two species only differ by 0.009 Da; therefore, high resolving power was necessary to determine which species was present in higher abundance. Many of the peptides that formed adducts with Ca^{2+} were from the N-terminal region of the peptide, which is a collagenase-sensitive region. The authors suggest that this metal-binding preference may also be relevant to in-solution conditions, which may be pertinent considering the role calcium ions play in cellular signaling cascades [157].

9. Summary

The employment of FT-ICR MS for biomolecular structural characterization offers several advantages, including high mass resolution, ultrahigh mass accuracy, and a wide mass range. ESI is the most widely used ionization technique coupled to FT-ICR MS and offers several advantageous characteristics, including multiple charging and the preservation of non-covalent interactions. ESI FT-ICR MS allows for the accurate molecular weight determination of both peptides and proteins, thereby allowing for both a "bottom-up" and "top-down" approach to proteomics. FT-ICR MS offers several tandem mass spectrometric techniques, which have been described in detail in this chapter. Several of these ion activation techniques have been utilized for the characterization of lectins with ESI FT-ICR MS, including IRMPD, ECD, and EDD. In particular, ECD has been shown to be a powerful tool for biomolecular characterization, due to its retention of labile modifications and extensive peptide backbone fragmentation. These numerous examples demonstrate how ESI FT-ICR MS is of great value for lectin analysis and would be an optimal analytical technique if high resolution, high mass accuracy, and detailed structural characterization are required.

Acknowledgments

This work was supported by the Searle Scholars Program, an Analytical Chemistry award from Eli Lilly & Company, a Dow Corning Assistant Professorship, an Eastman Summer Fellowship, and the University of Michigan.

References

[1] J.B. Fenn, M. Mann, C.K. Meng, S.F. Wong and C.M. Whitehouse, Science, 246 (1989) 64.

[2] M.L. Aleksandrov, L.M. Gall, M.V. Krasnov, V.I. Nikolaiev and V.A. Shkurov, Dokl. Akad. Nauk. SSSR, 277 (1984) 379.

[3] M. Karas and F. Hillenkamp, Anal. Chem., 60 (1988) 2299.

[4] K. Tanaka, H. Waki, Y. Ido, S. Akita, Y. Yoshida and T. Yoshida, Rapid Commun. Mass Spectrom., 2 (1988) 151.

[5] M.B. Comisarow and A.G. Marshall, Chem. Phys. Lett., 25 (1974) 282.

[6] J.W. Gauthier, T.R. Trautman and D.B. Jacobson, Anal. Chim. Acta, 246 (1991) 211.

[7] R.L. Woodlin, D.S. Bomse and J.L. Beauchamp, J. Am. Chem. Soc., 100 (1978) 3248.

[8] D.P. Little, J.P. Speir, M.W. Senko, P.B. O'Connor and F.W. McLafferty, Anal. Chem., 66 (1994) 2809.

[9] W.D. Price, P.D. Schnier and E.R. Williams, Anal. Chem., 68 (1996) 859.

[10] R.C. Dunbar and T.B. McMahon, Science, 279 (1998) 194.

[11] R.A. Chorush, D.P. Little, S.C. Beu, T.D. Wood and F.W. McLafferty, Anal. Chem., 67 (1995) 1042.

[12] C.L. Hendrickson and M.R. Emmett, Annu. Rev. Phys. Chem., 50 (1999) 517.

[13] S.A. Lorenz, E.P.I. Maziarz and T.D. Wood, Appl. Spectrosc., 53 (1999) 18A.

[14] M. Yamashita and J.B. Fenn, J. Phys. Chem., 88 (1984) 4451.

[15] A.G. Marshall, C.L. Hendrickson and G.S. Jackson, Mass Spectrom. Rev., 17 (1998) 1.

[16] J.S. Sampson, A.M. Hawkridge and D.C. Muddiman, J. Am. Soc. Mass Spectrom., 17 (2006) 1712.

[17] J.A. Castro, C. Koster and C.L. Wilkins, Rapid Commun. Mass Spectrom., 6 (1992) 239.

[18] J.C. Dunphy, K.L. Busch, R.L. Hettich and M.V. Buchanan, Anal. Chem., 65 (1993) 1329.

[19] R.T. McIver, Jr., Y. Li and R.L. Hunter, Proc. Natl. Acad. Sci. U.S.A., 91 (1994) 4801.

[20] T. Solouki, J.A. Marto, F.M. White, S. Guan and A.G. Marshall, Anal. Chem., 67 (1995) 4139.

[21] A. Brock, D.M. Horn, E.C. Peters, C.M. Shaw, C. Ericson, Q.T. Phung and A.R. Salomon, Anal. Chem., 75 (2003) 3419.

[22] H.J. An, T.R. Peavy, J.L. Hedrick and C.B. Lebrilla, Anal. Chem., 75 (2003) 5628.

[23] K.H. Light-Wahl, B.E. Winger and R.D. Smith, J. Am. Chem. Soc., 115 (1993) 5869.

[24] J.A. Loo, R.R. Ogorzalek Loo and P.C. Andrews, Org. Mass Spectrom., 28 (1993) 1640.

[25] W.D. van Dongen and A.J.R. Heck, Analyst, 125 (2000) 583.

[26] N.L. Kelleher, M.W. Senko, M.M. Siegel and F.W. McLafferty, J. Am. Soc. Mass Spectrom., 8 (1997) 380.

[27] B. Ardrey, Liquid chromatography-mass spectrometry, VCH, New York, 1993.

[28] W.M.A. Niessen, Liquid chromatography-mass spectrometry (chromatographic science), 2nd edition, Marcel Dekker, New York, 1998.

[29] R.E. Ardrey, LC-MS: An introduction, VCH, New York, 1999.

[30] J. Abian, J. Mass Spectrom., 34 (1999) 157.

[31] A.P. Bruins, T.R. Covey and J.D. Henion, Anal. Chem., 59 (1987) 2642.

[32] I.J. Amster, J. Mass Spectrom., 31 (1996) 1325.

[33] J. Zeleny, J. Phys. Rev., 10 (1917) 1.

[34] M. Dole, L.L. Mack, R.L. Hines, R.C. Mobley, L.D. Ferguson and M.B. Alice, J. Chem. Phys., 49 (1968) 2240.

[35] J.B. Fenn, M. Mann, C.K. Meng and S.F. Wong, Mass Spectrom. Rev., 9 (1990) 37.

[36] R.D. Smith, J.A. Loo, R.R. Ogorzalek Loo, M. Busman and H.R. Udseth, Mass Spectrom. Rev., 10 (1991) 359.

[37] R.B. Cole, J. Mass Spectrom., 35 (2000) 763.

[38] N.B. Cech and C.G. Enke, Mass Spectrom. Rev., 20 (2001) 362.

[39] E. De Hoffmann and V. Stroobant, Mass spectrometry: Principles and applications, Wiley, New York, 2001.

[40] S.K. Chowdhury, V. Katta and B.T. Chait, Rapid Commun. Mass Spectrom., 4 (1990) 81.

[41] M.H. Allen and M.L. Vestal, J. Am. Soc. Mass Spectrom., 3 (1992) 18.

[42] K.D. Henry, E.R. Williams, B.-H. Wang, F.W. McLafferty, J. Shabanowitz and D.F. Hunt, Proc. Natl. Acad. Sci. U.S.A., 86 (1989) 9075.

[43] K. Hakansson, H.J. Cooper, R.R. Hudgins and C.L. Nilsson, Curr. Org. Chem., 7 (2003) 1503.

[44] A.G. Marshall and S. Guan, Rapid Commun. Mass Spectrom., 10 (1996) 1819.

[45] S.A. Hofstadler and D.A. Laude, Jr., J. Am. Soc. Mass Spectrom., 3 (1992) 615.

[46] Z. Guan, V.L. Campbell, J.J. Drader, C.L. Hendrickson and D.A. Laude Jr., Rev. Sci. Instr., 66 (1995) 4507.

[47] D.A. Laude Jr., E. Stevenson and J.M. Robinson. In: R.B. Cole (Ed.), Electrospray ionization mass spectrometry: Fundamentals, instrumentation & applications, Wiley, New York, 1997, p. 291.

[48] R.T. McIver, Jr., R.L. Hunter and W.D. Bowers, Int. J. Mass Spectrom. Ion Processes, 64 (1985) 67.

[49] S.A. Campbell, E.M. Marzluff, M.T. Rodgers, J.L. Beauchamp, M.E. Rempe, K.F. Schwinck and D.L. Lichtenberger, J. Am. Chem. Soc., 116 (1994) 5257.

[50] S.C. Beu, M.W. Senko, J.P. Quinn, F.M. Wampler and F.W. McLafferty, J. Am. Soc. Mass Spectrom., 4 (1993) 557.

[51] P.A. Limbach, A.G. Marshall and M. Wang, Int. J. Mass Spectrom., 125 (1993) 135.

[52] F.F. Chen, Introduction to plasma physics and controlled fusion, Plenum Press, New York, 1984.

[53] R.T. McIver, Jr., Int. J. Mass Spectrom. Ion Processes, 98 (1990) 35.

[54] P. Kofel, M. Allemann, H. Kellerhals and K.P. Wancek, Int. J. Mass Spectrom. Ion Processes, 65 (1985) 97.

[55] S.A. Shaffer, D.C. Prior, G.A. Anderson, H.R. Udseth and R.D. Smith, Anal. Chem., 70 (1998) 4111.

[56] M.W. Senko, C.L. Hendrickson, M.R. Emmett, S.D.-H. Shi and A.G. Marshall, J. Am. Soc. Mass Spectrom., 8 (1997) 970.

[57] M.E. Belov, E.N. Nikolaev, G.A. Anderson, H.R. Udseth, T.P. Conrads, T.D. Veenstra, C.D. Masselon, M.V. Gorshkov and R.D. Smith, Anal. Chem., 73 (2001) 253.

[58] S. Guan and A.G. Marshall, Int. J. Mass Spectrom. Ion Processes, 146/147 (1995) 261.

[59] M. Allemann, H.-P. Kellerhals and K.-P. Wanczek, Chem. Phys. Lett., 84 (1981) 547.

[60] D. Mitchell, S. Delong, D. Cherniak and M. Harrison, Int. J. Mass Spectrom. Ion Processes, 91 (1989) 273.

[61] A.G. Marshall and P.B. Grosshans, Anal. Chem., 63 (1991) 215A.

[62] E.B. Ledford, Jr., D.L. Rempel and M.L. Gross, Anal. Chem., 56 (1984) 2744.

[63] M.B. Comisarow and A.G. Marshall, Chem. Phys. Lett., 26 (1974) 489.

[64] A.G. Marshall and D.C. Roe, J. Chem. Phys., 73 (1980) 1581.

[65] L.-K. Zhang, D.L. Rempel, B.N. Pramanik and M.L. Gross, Mass Spectrom. Rev., 24 (2005) 286.

[66] J.C. Hannis and D.C. Muddiman, J. Am. Soc. Mass Spectrom., 11 (2000) 876.

[67] J.W. Flora, J.C. Hannis and D.C. Muddiman, Anal. Chem., 73 (2001) 1247.

[68] K. Tang, A.V. Tolmachev, E. Nikolaev, R. Zhang, M.E. Belov, H.R. Udseth and R.D. Smith, Anal. Chem., 74 (2002) 5431.

[69] A.I. Nepomuceno, D.C. Muddiman, H.R.I. Bergen, J.R. Craighead, J.J. Burke, P.E. Caskey and J.A. Allan, Anal. Chem., 75 (2003) 3411.

[70] M.J. Chalmers, J.P. Quinn, G.T. Blakney, M.R. Emmett, H. Mischak, S.J. Gaskell and A.G. Marshall, J. Proteome Res., 2 (2003) 373.

[71] J.B. Jeffries, S.E. Barlow and G.H. Dunn, Int. J. Mass Spectrom. Ion Processes, 54 (1983) 169.

[72] S.M. Peterman, C.P. Dufresne and S. Horning, J. Biomol. Tech., 16 (2005) 112.

[73] J. Griep-Raming, W. Metelmann-Strupat, S. Horning, H. Muenster, M. Baumert and J. Henion, 51st ASMS Conference on Mass Spectrometry and Allied Topics: Montreal, Quebec, Canada, 2003.

[74] A.G. Marshall, M.B. Comisarow and G. Parisod, J. Chem. Phys., 71 (1979) 4434.

[75] K.D. Henry and F.W. McLafferty, Org. Mass Spectrom., 25 (1990) 490.

[76] P. Kebarle and L. Tang, Anal. Chem., 65 (1993) 972A.

[77] R.T. Kennedy and J.W. Jorgenson, Anal. Chem., 61 (1989) 1128.

[78] D.C. Moon and J.A. Kelley, Biomed. Mass Spectrom., 17 (1988) 229.

[79] T.L. Quenzer, M.R. Emmett, C.L. Hendrickson, P.H. Kelly and A.G. Marshall, Anal. Chem., 73 (2000) 1721.

[80] S.A. Hofstadler, J.C. Severs, R.D. Smith, F.D. Swanek and A.G. Ewing, Rapid Commun. Mass Spectrom., 10 (1996) 919.

[81] G.A. Valaskovic, N.L. Kelleher and F.W. McLafferty, Science, 273 (1996) 1199.

[82] S.D. Patterson and R. Aebersold, Electrophoresis, 16 (1995) 1791.

[83] J.R. Yates, Annu. Rev. Biophys. Biomol. Struct., 33 (2004) 297.

[84] K. Zhu, F.R. Miller, T.J. Barder and D.M. Lubman, J. Mass Spectrom., 39 (2004) 770.

[85] W.J. Henzel, T.M. Billeci, J.T. Stults, S.C. Wong, C. Grimley and C. Watanabe, Proc. Natl. Acad. Sci. U.S.A., 90 (1993) 5011.

[86] P. James, M. Quadroni, E. Carafoli and G. Gonnet, Biochem. Biophys. Res. Commun., 195 (1993) 58.

[87] M. Mann, P. Hojrup and P. Roepstorff, Biol. Mass Spectrom., 22 (1993) 338.

[88] D.J.C. Pappin, P. Hojrup and A.J. Bleasby, Curr. Biol., 3 (1993) 327.

[89] J.R. Yates, S. Speicher, P.R. Griffin and T. Hunkapiller, Anal. Biochem., 214 (1993) 397.

[90] T.P. Conrads, G.A. Anderson, T.D. Veenstra, L. Pasa-Tolic and R.D. Smith, Anal. Chem., 72 (2000) 3349.

[91] M.S. Lipton, L. Pasa-Tolic, G.A. Anderson, D.J. Anderson, D.L. Auberry, J.R. Battista, M.J. Daly, J. Fredrickson, K.K. Hixson, H. Kostandarithes, C. Masselon, L.M. Markillie, R.J. Moore, M.F. Romine, Y.F. Shen, E. Stritmatter, N. Tolic, H.R. Udseth, A. Venkateswaran, L.K. Wong, R. Zhao and R.D. Smith, Proc. Natl. Acad. Sci. U.S.A., 99 (2002) 11049–11054.

[92] N.L. Kelleher, H.Y. Lin, G.A. Valaskovic, D.J. Aaserud, E.K. Fridriksson and F.W. McLafferty, J. Am. Chem. Soc., 121 (1999) 806.

[93] F.W. McLafferty, E.K. Fridriksson, D.M. Horn, M.A. Lewis and R.A. Zubarev, Science, 284 (1999) 1289.

[94] X. Han, M. Jin, K. Breuker and F.W. McLafferty, Science, 314 (2006) 109.

[95] W. Li, C.L. Hendrickson, M.R. Emmett and A.G. Marshall, Anal. Chem., 71 (1999) 4397.

[96] P.K. Jensen, L. Pasa-Tolic, K.K. Peden, S. Martinovic, M.S. Lipton, G.A. Anderson, N. Tolic, K.-K. Wong and R.D. Smith, Electrophoresis, 21 (2000) 1372.

[97] S.-W. Lee, S.J. Berger, S. Martinovic, L. Pasa-Tolic, G.A. Anderson, Y. Shen, R. Zhao and R.D. Smith, Proc. Natl. Acad. Sci. U.S.A., 99 (2002) 5942.

[98] J.A. Loo, C.G. Edmonds and R.D. Smith, Anal. Chem., 63 (1991) 2488.

[99] N.L. Kelleher, Anal. Chem., 76 (2004) 197A.

[100] Y. Ge, B.G. Lawhorn, M. ElNaggar, E. Strauss, J.-H. Park, T.P. Begley and F.W. McLafferty, J. Am. Chem. Soc., 124 (2002) 672.

[101] M.A. Mabud, M.J. Dekrey and R.G. Cooks, Int. J. Mass Spectrom. Ion Processes, 67 (1985) 285.

[102] C.F. Ijames and C.L. Wilkins, Anal. Chem., 62 (1990) 1295.

[103] E.R. Williams, K.D. Henry, F.W. McLafferty, J. Shabanowitz and D.F. Hunt, J. Am. Soc. Mass Spectrom., 1 (1990) 413.

[104] R.A. Zubarev, N.L. Kelleher and F.W. McLafferty, J. Am. Chem. Soc., 120 (1998) 3265.

[105] B.A. Budnik, K.F. Haselmann and R.A. Zubarev, Chem. Phys. Lett., 342 (2001) 299.

[106] A.G. Marshall, T.-C.L. Wang and T.L. Ricca, J. Am. Chem. Soc., 107 (1985) 7893.

[107] Y. Wang, S.D.-H. Shi, C.L. Hendrickson and A.G. Marshall, Int. J. Mass Spectrom., 198 (2000) 113.

[108] K. Biemann and S.A. Martin, Mass Spectrom. Rev., 6 (1987) 1.

[109] F.W. McLafferty, Tandem Mass Spectrometry, Wiley, New York, 1983.

[110] S.A. McLuckey, J. Am. Soc. Mass Spectrom., 3 (1992) 599.

[111] L. Sleno and D.A. Volmer, J. Mass Spectrom., 39 (2004) 1091.

[112] S. Guan, A.G. Marshall and M.C. Wahl, Anal. Chem., 66 (1994) 1363.

[113] R.M.A. Heeren, A.J. Kleinnijenhuis, L.A. McDonnell and T.H. Mize, Anal. Bioanal. Chem., 378 (2004) 1048.

[114] A.J.R. Heck, L.J. de Koning, F.A. Pinske and N.M.M. Nibbering, Rapid Commun. Mass Spectrom., 5 (1991) 406.

[115] E. Mirgorodskaya, P.B. O'Connor and C.E. Costello, J. Am. Soc. Mass Spectrom., 13 (2002) 318.

[116] S.A. McLuckey and D.E. Goeringer, J. Mass Spectrom., 35 (1997) 461.

[117] K. Hakansson, M.J. Chalmers, J.P. Quinn, M.A. McFarland, C.L. Hendrickson and A.G. Marshall, Anal. Chem., 75 (2003) 3256.

[118] K. Hakansson, H.J. Cooper, M.R. Emmett, C.E. Costello, A.G. Marshall and C.L. Nilsson, Anal. Chem., 73 (2001) 4530.

[119] K. Hakansson, M.R. Emmett, A.G. Marshall, P. Davidsson and C.L. Nilsson, J. Proteome Res., 2 (2003) 581.

[120] J.T. Adamson and K. Hakansson, J. Proteome Res., 5 (2006) 493.

[121] R.C. Dunbar, Mass Spectrom. Rev., 23 (2004) 127.

[122] Y. Ge, D.M. Horn and F.W. McLafferty, Int. J. Mass Spectrom., 210/211 (2001) 203.

[123] E.N. Kitova, D.R. Bundle and J.S. Klassen, J. Am. Chem. Soc., 124 (2002) 5902.

[124] J. Laskin and J.H. Futrell, Mass Spectrom. Rev., 22 (2003) 158.

[125] J. Laskin and J.H. Futrell, Mass Spectrom. Rev., 24 (2005) 135.

[126] H.J. Cooper, K. Hakansson and A.G. Marshall, Mass Spectrom. Rev., 24 (2005) 201.

[127] R.A. Zubarev, Mass Spectrom. Rev., 22 (2003) 57.

[128] R.A. Zubarev, Curr. Opin. Biotechnol., 15 (2004) 12.

[129] R.A. Zubarev, D.M. Horn, E.K. Fridriksson, N.L. Kelleher, N.A. Kruger, M.A. Lewis, B.K. Carpenter and F.W. McLafferty, Anal. Chem., 72 (2000) 563.

[130] K. Vekey, Mass Spectrom. Rev., 14 (1995) 195.

[131] F. Turecek, J. Am. Chem. Soc., 125 (2003) 5954.

[132] J.W. Jones, T. Sasaki, D.R. Goodlett and F. Turecek, J. Am. Soc. Mass Spectrom., 17 (2006) 432.

[133] J.M. Hogan, S.J. Pitteri, P.A. Chrisman and S.A. McLuckey, J. Proteome Res., 4 (2005) 628.

[134] J.E.P. Syka, J.J. Coon, M.J. Schroeder, J. Shabanowitz and D.F. Hunt, Proc. Natl. Acad. Sci. U.S.A., 101 (2004) 9528.

[135] D.M. Horn, Y. Ge and F.W. McLafferty, Anal. Chem., 72 (2000) 4778.

[136] J.H. Bowie, C.S. Brinkworth and S. Dua, Mass Spectrom. Rev., 21 (2002) 87.

[137] J. Jai-Nhuknan and C.J. Cassady, Anal. Chem., 70 (1998) 5122.

[138] F. Kjeldsen, O.A. Silivra, I.A. Ivonin, K.F. Haselmann, M. Gorshkov and R.A. Zubarev, Chem. Eur. J., 11 (2005) 1803.

[139] H.K. Kweon and K. Hakansson, 53rd ASMS Conference on Mass Spectrometry and Allied Topics: San Antonio, TX, 2005.

[140] J. Yang, J. Mo, J.T. Adamson and K. Hakansson, Anal. Chem., 77 (2005) 1876.

[141] J.T. Adamson and K. Hakansson, 53rd ASMS Conference on Mass Spectrometry and Allied Topics: San Antonio, Texas, 2005.

[142] J. Zaia, Mass Spectrom. Rev., 23 (2004) 161.

[143] M.T. Cancilla, S.G. Penn, J.A. Carroll and C.B. Lebrilla, J. Am. Chem. Soc., 118 (1996) 6736.

[144] G.E. Hofmeister, Z. Zhou and J.A. Leary, J. Am. Chem. Soc., 113 (1991) 5964.

[145] R.H. Holm, P. Kennepohl and E.I. Solomon, Chem. Rev., 96 (1996) 2239.

[146] L.M. Teesch and J. Adams, J. Am. Chem. Soc., 112 (1990) 4110.

[147] A.J.R. Heck and R.H.H. van den Heuvel, Mass Spectrom. Rev., 23 (2004) 368.

[148] J.A. Loo, Mass Spectrom. Rev., 16 (1997) 1.

[149] P. Juhasz and K. Biemann, Proc. Natl. Acad. Sci. U.S.A., 91 (1994) 4333.

[150] P. Juhasz and K. Biemann, Carbohydr. Res., 270 (1995) 131.

[151] P. Lecchi and L.K. Pannell, J. Am. Soc. Mass Spectrom., 6 (1995) 972.

[152] M.O. Glocker, S.H.J. Bauer, J. Kast, J. Volz and M. Przybylski, J. Mass Spectrom., 31 (1996) 1221.

[153] M. Moniatte, F.G. van der Goot, J.T. Buckley, F. Pattus and A. van Dorsselaer, FEBS Lett., 384 (1996) 269.

[154] B. Rosinke, K. Strupat, F. Hillenkamp, J. Rosenbusch, N. Dencher, U. Kruger and H.-J. Galla, J. Mass Spectrom., 30 (1995) 1462.

[155] W. Ens, K.G. Standing and I.V. Chernushevich, New methods for the study of bio-molecular complexes (Vol. 510), Kluwer Academic Publishers, Dordrecht, 1998.

[156] W.D. Lehmann, J. Wei, C.-W. Hung, H.-J. Gabius, D. Kirsch, B. Spengler and D. Kubler, Rapid Commun. Mass Spectrom., 20 (2006) 2404.

[157] E. Carafoli, Proc. Natl. Acad. Sci. U.S.A., 99 (2002) 1115.

Lectins: Analytical Technologies
C.L. Nilsson (Editor)
© 2007 Elsevier B.V. All rights reserved.

Chapter 15

Catalytically Inactive Endoglycosidases as Microbial Diagnostic Reagents: Chitinases and Lysozymes as Fungal and Bacterial Capture/Label Agents

Roger A. Laine[a,b,c], Jennifer W.-C. Lo[a,c] and Betty C.-R. Zhu[a]

[a]*Department of Biological Sciences, Louisiana State University, Baton Rouge, LA 70803, USA*
[b]*Department of Chemistry, Louisiana State University and A&M College, Baton Rouge, LA 70803, USA*
[c]*Anomeric, Inc., 755 Delgado Dr., Baton Rouge, LA 70808, USA*

1. Introduction

The use of enzyme-binding sites as carbohydrate recognition and detection systems was pioneered by Mega and Hase (1994), where they used a dimeric form of catalytically disabled lysozyme to generate a neolectin [1]. Parsons catalytically disabled lysozyme chemically in 1969, producing a carbohydrate-binding protein with no enzymatic activity [2]. Jack Kirsch and his group, in 1989, also generated a catalytically disabled lysozyme by mutating hen egg white lysozyme (HEWL) at E35Q or D52N [3] and showed that it bound to peptidoglycan with the same avidity as the active enzyme. These experiments were designed to elucidate the mechanism of the enzymes, which was finally clarified by Wither's group [4]. These mutated enzymes were not, however, considered in their time to be useful for detection or diagnosis of bacterial infections. A chance observation that we had in the laboratory led to the development of a fungal diagnostic when we found that *Vibrio parahemolyticus* chitinase, 94 kDa, [5] adhered to a particulate chitin affinity column with such avidity that it could not be removed by 4 M salts. The off-rate was not measurable. This evidence for avid binding induced us to try its efficacy for diagnostic staining of chitinous fungi. In 1996, we were issued a US patent for the use of a chitinase for diagnosis of fungal infections [6], using a cloned version of the chitinase from *V. parahemolyticus*, the gene and gene product for which we were issued a patent in 1994 [5]. In this we were successful, and several further patents were issued on the subject [6–10]. The low turnover number of the enzyme rendered the enzymatic activity null under laboratory fungal staining conditions, enabling the practical use of the binding specificity without degradation of the chitin target. A low turnover number can render the enzyme "catalytically inactive" in the time frame of the observation event. Also, regarding specificity, this *V. parahemolyticus* enzyme does not bind peptidoglycan and there appear to be no cross-reacting structures in mammalian tissues.

Therefore we employ endoglycosidases as fungal or bacterial detection systems in a "catalytically inactive mode", that is, utilizing only their binding specificity. We also isolated a 134 kDa chitin-binding lectin from *V. parahemolyticus*, which appeared to bind 9-mer oligosaccharides or larger, optimally [11, 12], but exhibited no enzyme activity. This lectin also has a potential as a fungal diagnostic [12]. Later, we fused the green–fluorescent protein (GFP) to the chitin-binding domain from *Bacillus circulans*, to generate a potential ready-made diagnostic agent [13]. The chitinase version of the stain was commercialized by Anomeric, Inc., Baton Rouge, LA, and used in human diagnostics by the Mayo Clinic in Rochester, MN to determine that >90% of chronic sinusitis is actually eosinophilic fungal rhinosinusitis (EFRS) [14, 15], has a fungal etiology, and can be treated by intranasal amphotericin B [16].

We followed up this example of a cell wall binding protein for fungi by developing a novel bacterial detection system [7–10] based on catalytically inactivated HEWL E35Q, which had been mutated in 1989 by the laboratory of Jack Kirsch [3]. Fluorescent labeling of E35Q provided a flow cytometric assay that could count all types of bacteria by cell wall staining [7–10]. As another example of an endoglycosidase used in diagnostics, Pelkonen et al. generated an enzymatically inactive polysialidase from a phage mutant, which could be used as a neolectin-like diagnostic agent [17]. This was later employed either as using the whole phage as a detection system [18], or using an engineered neolectin as a GFP fusion protein [19] (reviewed in this volume, Chapter 16).

There are significant advantages of cell wall polymer degradative enzymes as targeted diagnostic agents:

- High specificity for the target polymer
- High avidity
- Easily labeled with signal molecules (current experience)
- Much more economical than antibodies
- Produced fermentatively from bacteria or yeast
- High stability, long shelf life.
- No nonspecific binding in human tissues
- Low binding to cellulose linters (which stain with calcofluor white, the cotton whitener).

Market need for rapid and automatable microbial diagnostics: The current large volume *manual* clinical tests represent today's need for automation development – automation can include fluorescence imaging techniques, image-labeled pixel counting, and flow cytometry.

- Fungal analysis (now performed by KOH, silver stain, or long-term culture).
- Bacterial analysis (primarily urinary tract infection screening, currently the colony forming unit assay (CFU), 24 h).

- Antibiotic sensitivity (40–72 h, colony isolated from CFU, broth culture) (some automation exists today for minimum inhibitory concentration (MCI) (Dade, Becton-Dickinson, Biomeiurex) but no significant acceleration in time).

Catalytically inactive enzymes constitute a new class of capture/label reagents useful in the following types of assays:

- Slide-based stains, fluorescence
- ELISA (enzyme-linked immunosorbentassay)
- Dipsticks
- Latex agglutination assays
- Skin disclosure stains (skin, nail fungi, onchyomycosis)
- Flow cytometry (for bacteria or yeast forms of fungi)
- Surface plasmon resonance

1. Chitin and Fungal Staining with Chitinase

A bacterial chitinase is unlikely to bind or degrade peptidoglycan, since the product would attack the producing bacterial cell wall. Binding is therefore specific for:

Filamentous fungi (*Aspergillus*, for example, a common cause of sinus infection), yeast forms of fungi, *Candida sp.*, fungal spores (*Pneumocystis* does not stain).

For example, HEWL binds to both chitin and peptidoglycan. The chitinase reagent was therefore specific for chitinous fungi while HEWL binds both murein and chitin.

The *V. parahemolyticus* chitinase was cloned and expressed in *Escherichia coli* [2] and conjugated with fluorescein *N*-hydroxysuccinimide to form a fluorescent diagnostic agent with a stable covalent linkage. This reagent was found to be stable in solution at room temperature for at least one year, and indefinitely when lyophilized. The binding was stable to 1–2 M salt. The chitinase–fluorescein reagent was used to show the first evidence for the presence of chitin in bony fish cuticle [20]. Staining is performed either on thin sections or dried sample smears by first, treatment with acetone, followed by periodate oxidation (chitin has no vicinal hydroxyl groups) to degrade mannans and other carbohydrates and help expose chitin. The fluorescein-labeled or other fluorophore-labeled protein is directly applied to the slide for a few minutes and washed off with phosphate-buffered saline or water, covered with a glass cover slip using a drop of glycerol, and directly viewed with an excitation wavelength suitable for the attached dye. A product like Slow-Fade® can be added to prevent free-radical bleaching.

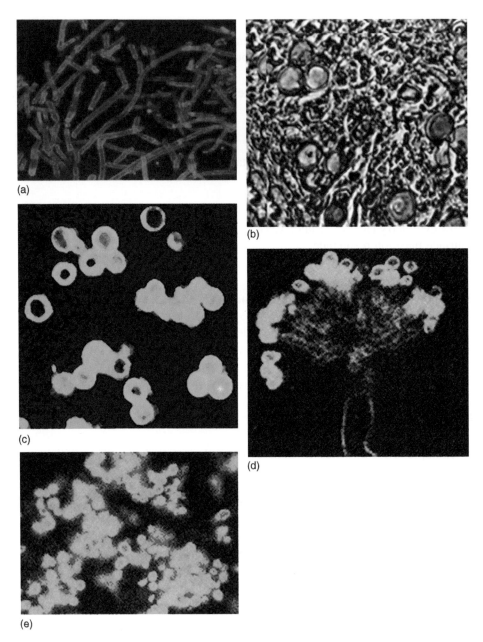

Figure 1. (a) Detection of *Alternaria* in culture using fluorexein-labeled chitinase. Adapted from Ref. [14]. (b) GMS Silver Stain (Sigma Fungi-tissue-trol®); (c) Chitinase stain: (Fungalase®) (Sigma Fungi-tissue-trol®); (d) *Aspergillus* conidia from corn; (e) Bone marrow *Histoplasmosis* from HIV patient (clustered in macrophages); (f) *Aspergillus sp.* from HIV patient; (g) Zygomycosis (HIV patient); (h) *Coccidioides sp.* from HIV patient; (i) *Phaeohyphomycosis* from HIV patient; (j) *Candida albicans* × *400*; (k) *Candida albicans* × *200*.

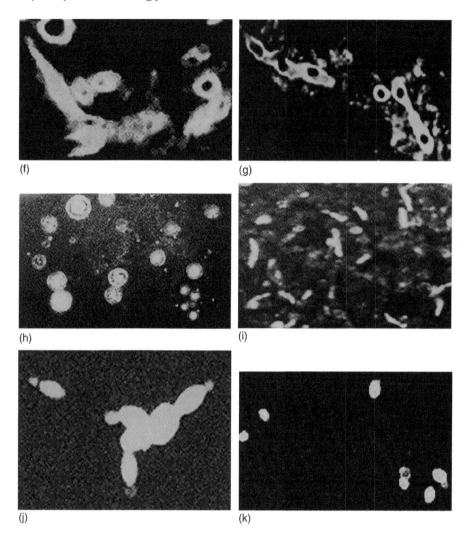

(f) (g)

(h) (i)

(j) (k)

Figure 1. (Continued)

Examples of use of the chitinase-based stain:

Staining of a common laboratory strain of *Alternaria* showed strong staining of the cell walls and septae as depicted in Fig. 1a.

We stained a number of thin sections containing several species of fungal hyphae and yeast forms with fluorescein-labeled *V. parahemolyticus* chitinase (Fungalase®). See, for examples, Figs. 1b–k.

Blastomyces in human skin:

In Fig. 2, the purified chitinase was gold labeled by Ponikau and his group, and electron micrographs showed extensive gold particle binding only in the fungal cell wall [14]. (Fig. 3)

When microtome slices are made of fixed sinus EFRS mucin samples, as shown in Fig. 4, it is clear that only the fungal cell walls in transverse and angled view are stained with no cross-reactivity of other components.

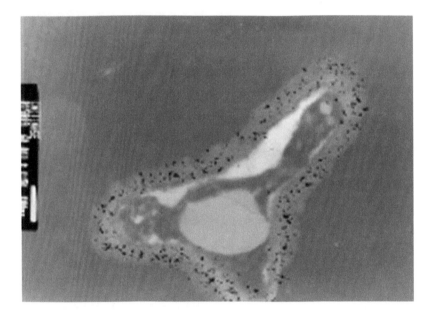

Figure 2. Gold-labeled *Vibrio parahemolyticus* chitinase was used to label cell wall of an *Aspergillus sp.*, and thin sections were examined by transmission electron microscopy. The location of gold particles indicates that the reagent can penetrate into the chitin cell wall structure and stain throughout the wall. Adapted from Ref. [14].

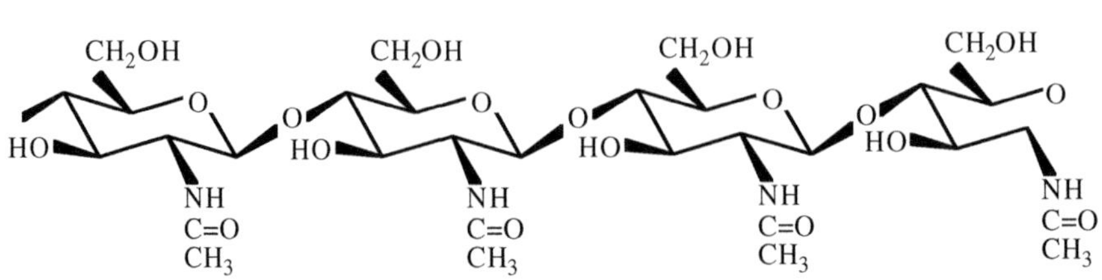

Figure 3. Chitin fungal cell wall polymer targeted by chitinases $(GlcNAc\beta1\text{->}4)_n$.

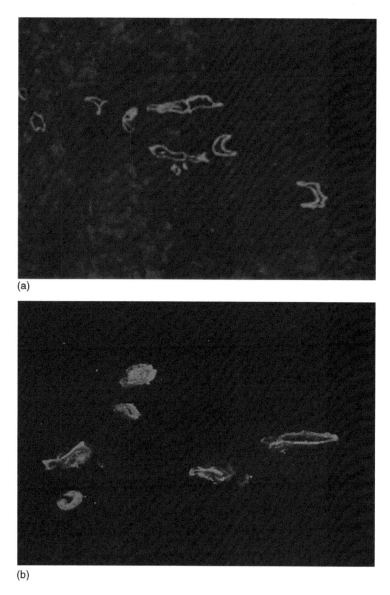

(a)

(b)

Figure 4. (a) Fungi in mucin sample from eosinophilic fungal rhinosinusitis (EFRS). Adapted from Ref. [14]. (b) EFRS of lacrimal sac. Fungal infection was found in the lacrimal sac by staining with fluorescein-labeled *V. parahemolyticus* chitinase (Fungalase®). Adapted from Ref. [15]. The photo (b) was featured on the cover of the cited journal [15].

2. Use of E35Q HEWL for Bacterial Diagnostics

One important concept we have, which is not in the mainstream of current diagnostic protocol is the following: For treatment, for example, of urinary tract infections, for therapeutic decisions the physician needs to know (1) whether there is an infection or inflammation (2) whether in positive samples the infective agent is a bacteria or a yeast, and (3) which antibiotic will be effective. In our view, the physician does not need to know the species of bacteria for initial, effective therapeutics, but only evidence for an effective antibiotic. Epidemiology can be done by conventional methods later.

These answers are in current clinical urology practice provided by streaking the urine sample on blood–agar plates (an estimated 50 million petri dishes per year in the USA) and making a 24-h assay of CFU, where >100,000/mL is generally considered an infection. After the colonies have been counted, further antibiotic sensitivity tests involve either a broth culture of one or more of the colonies, tested for bacterial growth in the presence or absence of antibiotics, or a bacterial lawn assay using filter paper disks with various concentrations of antibiotics (Kirby-Bauer) to determine minimum inhibitory concentration (MIC). The latter studies take an additional 30–72 h.

We consider that the identification of the bacterial species and strain is not necessary for rapid therapeutics, just the antibiotic efficacy is needed. Epidemiology can be done by a number of methods including differential growth medium, antibodies, etc. Therefore, if a rapid counting method can be developed, all of the downstream assays can be accelerated. For example, we have developed an E35Q HEWL 5 min flow cytometric assay that is capable of showing >90% positive and negative predictive value with 24 h petri dishes, including live/dead ratios.

The very fact that only 5 min is needed to count the bacteria can be used to accelerate the antibiotic sensitivity test, or any other growth medium tests. By growing positive cultures in seven different antibiotics in three concentrations (21 samples), with two growth and one killed control (24 samples), four patient antibiotic sensitivity tests can be conducted from one 96 well plate, using flow cytometry or imaging to count the bacteria. The sample times are about 1 min for the flow cytometer, so that one patient's samples can be read in about 30 min. With imaging techniques now available, simultaneous counting of pixels in several samples could be accomplished at once.

Often, for urinalysis, if a patient arrives in the clinic with a "burning sensation", broad-spectrum antibiotics are prescribed. Once started, the patient is obligated to take a full course of antibiotic *before the physician knows that the problem is a bacterial infection*. This causes a large number of unnecessary antibiotic treatments, which can have side effects and, over time, contribute to antibiotic resistant organisms.

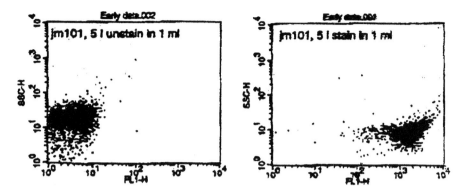

Figure 5. Left: Laboratory *Escherichia coli* strain JM101 unstained, analyzed by a modified FaxCount flow cytometer (Becton-Dickinson), showing light scatter on the ordinate and fluorescence intensity on the abscissa. Right: The sample has been stained with fluorescein E35Q. The light scatter indicates a size range of 1–2 microns, showing single bacteria and budding doubles. Also, the fluorescence intensity is about 300 times intensified. The bacteria could be directly counted by the event counting software, and specific windows can be chosen. For accurate counting in a given volume, fluorescent beads of a size not interfering with the assay can be added. Some flow cytometers can measure volumes accurately, such as Partek, Inc., Instruments (Germany) and the Optoflow Microcyte (Optoflow, Norway).

In a new paradigm which could be made possible by a "5 min screening", patients could be asked to contribute a urine sample in the morning. By afternoon the physician would know if the infection were bacteria or yeast, and which antibiotic to prescribe, if any. However, there is a great inertia against change in the medical/clinical system. This kind of novel bacterial detection scheme could provide the first large acceleration in microbial patient diagnostics in decades.

Some examples of flow cytometric assays using E35Q HEWL–fluorescence reagents are shown in Fig. 5.

HEWL also binds to chitin, therefore yeast can be measured in the same assay, using size difference to distinguish among bacteria and yeast, such as with urine screening as shown in Fig. 6.

Obviously, if bacteria can be counted in a 5 min total assay, any assay that can take advantage of the accelerated counting is useful. We have conducted a test of 600 clinical urine samples from urology clinics and have generated data that indicates a better than 90% positive and negative predictive value when compared with petri dishes (unpublished). Using a two-tube system where one tube is stained without an acetone treatment (only dead cells stain intensely within 2 min), and the second tube treated with acetone (live and dead cells both stain intensely). Subtracting the untreated from the solvent treated tubes gives a live/dead ratio. In many cases our system would count many more bacteria in the combined live/dead sample than

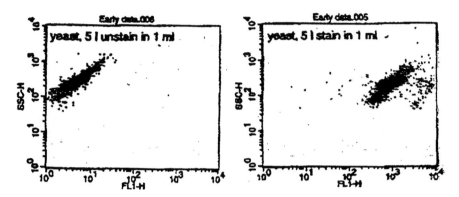

Figure 6. Left: unlabeled *Candida albicans*; Right: labeled with fluorescein-E35Q hen egg white (HEW). The light scatter indicates a size range of 7–14 microns, which indicates mainly single yeast and budding doubles. The fluorescence intensity is 300-fold increased over background. The yeast area of analysis is clearly different from the *E. coli* area shown in Fig. 5 and in these graphs.

shown on culture. In one case we found 72 million bacteria per mL that were dead bacteria being shed into the urine while the petri dish method and our live/dead method showed only 15,000/mL live bacteria. The petri dish alone would have missed the massive dead cell content. The patient was taking an antibiotic.

For an antibiotic sensitivity test, positive urine samples [>100,000/mL] are incubated in growth medium with seven antibiotics and a control and grown for 3 h. The slowest organism doubled in 3 h. *E. coli* (80% of the positive samples) grows eightfold in 3 h. Then the tubes are counted, and sensitivity to an antibiotic is easily depicted.

In Fig. 7, a *Streptococcus sp.* and a *Pseudomonad sp.* were incubated with 50μg/mL of Ampicillin for 3 h with a growth control. The flow cytometer used was a Sysmex (Kobe, Japan) experimental instrument with somewhat lower size resolution than the FaxCount. The *Streptococcus* was susceptible at this concentration but the *Pseudomonad* was resistant as shown in the right panel. The cluster of signals showing larger size (light scatter on the ordinate) is the *Streptococcus* which occurs in chains. An MIC could be found by using three or more concentrations.

In Fig. 7, only one concentration of one antibiotic is shown to illustrate the potential of the assay system. In the left panel it is clear that two populations exist, the smaller size being the *Pseudomonad* and the higher sized cluster (*Streptococci* showing various sized chains). In the right panel, Fig. 7, 50 μg/mL of Ampicillin had destroyed the *Streptococci*, while the *Pseudomonad* was obviously resistant to the antibiotic. This also shows the value of the assay in mixed infections. Now let us consider what this diagram would look like if there were a mixed infection of *E. coli* and *Enterococci*, for example. The populations would show up nearly in the

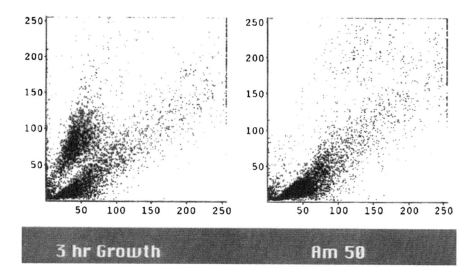

Figure 7. Flow cytometry (experimental Sysmex flow cytometer, Sysmex Inc., Kobe, Japan) of a mixed urinary tract infection of *Streptococci* and *Pseudomonas*. Light scatter is plotted shown on the ordinate and fluorescence intensity on the abscissa. Left panel: 3-h growth from a positive urine sample in nutrient medium; Right panel: Same sample with 50 μg/mL of Ampicillin added.

same area on the flow cytometric diagram, however, only an antibiotic that would be effective for both organisms would show a disappearance of the signals in the treated sample.

The practical use of carbohydrate-binding enzymes, used in a noncatalytic, binding-only mode (as "neolectins"), and targeting cell wall polymers in microbes may provide new practical methods for automation of medical and food science microbiology assays.

References

[1] T. Mega and S. Hase, Biochim. Biophys. Acta, 1200 (1994) 331.
[2] S.M. Parsons and M.A. Raftery, Biochemistry, 8 (1969) 4199.
[3] B.A. Malcolm, S. Rosenberg, M.J. Corey, J.S. Allen, A. de Baetselier and J.F. Kirsch, Proc. Natl. Acad. Sci. USA, 86 (1989) 133.
[4] D.J. Vocadlo, G.J. Davies, R.A. Laine and S.G. Withers, Nature, 412 (2001) 835.
[5] R.A. Laine, J.M. Jaynes and C.Y. Ou, United States Patent #5,352,607 (2004).
[6] R.A. Laine and J.W.C. Lo, United States Patent #5,587,292 (1996).
[7] R.A. Laine and J.W.C. Lo, United States Patent #5,935,804 (1999).
[8] R.A. Laine and J.W.C. Lo, United States Patent #6,090,573 (2000).
[9] R.A. Laine and J.W.C. Lo, United States Patent #6,159,719 (2000).

[10] R.A. Laine and J.W.C. Lo, United States Patent #6,184,027 (2001).

[11] O.S. Gildemeister, B.C.R. Zhu and R.A. Laine, Glycoconj. J., 11 (1994) 518.

[12] R.A. Laine, United States Patent #5,914,239 (1999).

[13] M. Hardt and R.A. Laine, Arch. Biochem. Biophys., 426 (2004) 286.

[14] M.J. Taylor, J.U. Ponikau, D.A. Sherris, E.B. Kern, T.A. Gaffey, G. Kephart and H. Kita, Otolaryngol. Head Neck Surg., 127 (2002) 377.

[15] M.L. Facer, J.U. Ponikau and D.A. Sherris, Laryngoscope, 113 (2003) 210.

[16] J.U. Ponikau, D.A. Sherris, H. Kita and E. B. Kern, J. Allergy Clin. Immunol., 110 (2002) 862.

[17] S. Pelkonen, J. Aalto and J. Finne, J. Bacteriol., 174 (1992) 7757.

[18] J. Aalto, S. Pelkonen, H. Kalimo and J. Finne, Glycoconj. J., 18 (2001) 751.

[19] A. Jokilammi, P. Ollikka, M. Korja, E. Jakobsson, V. Loimaranta, S. Haataja, H. Hirvonen and J. Finne, J. Immunol. Methods, 295 (2004) 149.

[20] G.P. Wagner, J.W.C. Lo and R.A. Laine, Experientia, 49 (1993) 317.

Lectins: Analytical Technologies
C.L. Nilsson (Editor)
© 2007 Elsevier B.V. All rights reserved.

Chapter 16

Generation of Lectins from Enzymes: Use of Inactive Endosialidase for Polysialic Acid Detection

Anne Jokilammi, Miikka Korja, Elina Jakobsson and Jukka Finne

Department of Medical Biochemistry and Molecular Biology, University of Turku, Kiinamyllynkatu 10, FI-20520 Turku, Finland

Lectins are defined as carbohydrate-binding proteins that have no catalytic activity. Glycosidases and glycosyltransferases are carbohydrate-binding molecules, which are not defined as lectins. They represent a wide spectrum of specific carbohydrate-binding molecules, which could be converted to lectins, if their enzymatic activity is inactivated without changing their binding properties.

Enzyme–substrate interactions are comparable to lectin–carbohydrate or antigen–antibody interactions in terms of specificity. The K_D values of lectins and enzymes are comparable (generally from millimolar to micromolar). Despite the enormous potential of specific interactions, enzyme–substrate interactions have generally not been exploited as a source for the construction of specific molecular probes, such as lectins. In the present paper we describe a catalytically inactive bacteriophage-derived endosialidase that can be used as a probe with lectin-like properties in the specific detection of its substrate, polysialic acid (polySia).

1. Polysialic Acid

Polysialic acid (polySia) [1] is a developmentally regulated carbohydrate polymer involved in neural cell differentiation [2, 3], organogenesis [4] and malignancies [5–8]. It is a linear homopolymer of *N*-acetylneuraminic acid units joined by α2–8 linkages [9]. The main carrier of polySia in eukaryotes is the neural cell adhesion molecule (NCAM), whose adhesion properties polySia modulates. PolySia is present in the embryonic but not the adult form of NCAM. PolySia has a pivotal role during normal neural development in cell motility and synaptic plasticity, while in the adult cerebrum polySia is involved in spatial learning and memory [2–4, 10, 11].

PolySia is also an oncofetal marker of a number of tumors [12–14]. As a poor immunogen, it disguises malignant tumor cells and enables their metastatic spread. In addition, polySia seems to favor the growth potential of tumor cells [8, 15]. The capsules of several pathogenic bacteria consist of host-mimicking polySia [16], which similarly to malignant tumor cells enables the escape of bacteria from immune defense [17].

PolySia represents an important developmental antigen, and its specific detection method could be utilized to study neural plasticity, various malignancies, and central nervous system infections. Inactivated endosialidase represents a new approach to detect polySia, which due to the poor immunogenicity has been a complicated target for the production of antibodies [18–20].

2. Endosialidase-Green Fluorescent Protein Fusion Construct

2.1. Construction of fusion protein

Phages infecting polySia-encapsulated *E. coli* K1 have been isolated from sewages [21]. Major part of these are lytic linear double-stranded DNA viruses, which differ in morphology. The anti-K1 phages have tailspike proteins, endosialidases, which specifically bind and cleave polySia [22].

We have previously isolated mutants of *E. coli* PK1A bacteriophages, which had lost their polySia-cleaving activity, while retaining their polySia-binding activity [22]. The mutant PK1A2 phage endosialidase was used for the subsequent construction of a GFP fusion protein with an N-terminal His-tag.

Direct fusion of the mutant PK1A2 endosialidase and GFP genes yielded a protein of the expected size, but in insoluble form when expressed in *E. coli*. To increase the amount of soluble protein, a part of the *Yersinia enterocolitica* adhesin (YadA) stalk [23] was used as a spacer between the endosialidase and GFP. In order to study influence of YadA-stalk lengths on the properties of the fusion protein construct, truncated forms of the stalk were made (Fig. 1A). Sodium dodecyl sulfate polyacrylamide gel electrophoresis of the pellets and supernatants of cell lysates (Fig. 1D and E) revealed that the longest truncated stalk (Fig. 1B) yielded the largest amounts of stable and soluble protein. In gel filtration the protein was eluted as a single peak (Fig. 1C) in a volume corresponding to the expected trimeric form of the fusion protein [24].

2.2. Binding affinity to polysialic acid

The binding of polySia to the fusion protein is dose dependent (Fig. 2A). The binding constant was determined by equilibrium binding and Scatchard plot analysis, and yielded a curved plot typical of bivalent interactions (Fig. 2B) [25, 26].

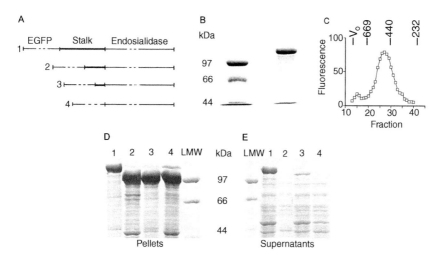

Figure 1. Schematic presentation of the fusion protein construct using a part of the YadA stalk as a spacer between the GFP and the endosialidase. (A) The spacer lengths are 399 (1), 183 (2), 84 (3), and 0 bp (4). (B) SDS-PAGE of the purified fusion protein. (C) gel filtration of the fusion protein. The void volume (V_0), the elution volumes, and molecular masses (kDa) of the reference proteins are indicated. (D) SDS-PAGE of the cell lysate pellets, and (E) supernatants expressing the protein constructs 1−4. Low molecular weight marker (LMW): 97, 66, and 44 kDa. Reprinted from Ref. [42], copyright (2004), with permission from Elsevier.

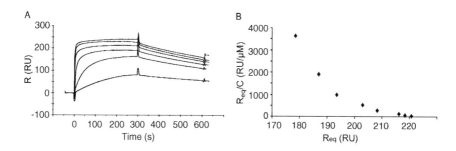

Figure 2. Binding of polySia (colominic acid) to the fusion protein. (A) Binding of different concentrations of polySia to the fusion protein. The concentrations from the bottom are 1.6 nM, 8.0 nM, 40 nM, 0.2 μM, 1 μM, and 25 μM. (B) Scatchard plot of the equilibrium binding. Reprinted from Ref. [42], copyright (2004), with permission from Elsevier.

The K_D values calculated separately for the two phases of the curve are 5.0×10^{-9} mol/L and 33×10^{-9} mol/L and the K_D (average) is 19×10^{-9} mol/L. These values were similar to the corresponding values 2.5×10^{-9} mol/L and 11×10^{-9} mol/L, and K_D (average) 7×10^{-9} mol/L observed for the binding of colominic acid to a class IgG monoclonal antibody to polySia, mAb735 [27].

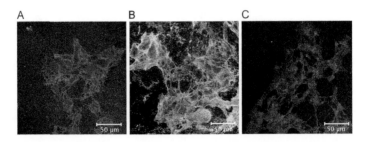

Figure 3. Staining of cells with catalytically active and inactive endosialidase fusion protein. Paraformaldehyde-fixed BHK-21 stained with (A) catalytically active fusion protein, (B) catalytically inactive fusion protein, and (C) catalytically inactive fusion protein in the presence of 100 µg/ml polySia (colominic acid). The scale bar is 50 µm.

2.3. Catalytic inactivation of endosialidase is essential for the detection of polysialic acid

Either the catalytically active form or the inactive form of endosialidase in the fusion protein was used to stain BHK-21 cells containing surface polySia. Only the fusion protein with the inactive form of the enzyme stains the cells (Fig. 3B). The fusion protein with active enzyme cleaves off the polySia chains from the cell surface (Fig. 3A). The staining was inhibited by the addition of free polySia (Fig. 3C).

3. Applications of the Fusion Protein

3.1. Detection of bacterial polysialic acid

The fusion protein with catalytically inactive endosialidase is bound specifically to *E. coli* which contains the K1 [28–30] polySia capsule, whereas no binding is observed to bacteria of the unrelated K2 capsular type (Figs. 4A and B). PolySia is also a capsular polysaccharide of the bacterial pathogens *Neisseria meningitidis* group B [31, 32], *Mannheimia (Pasteurella) haemolytica* A2 [31, 33], and *Moraxella nonliquefaciens* [31, 32].

The presence and types of polySia capsules in bacterial strains are summarized in Table 1.

The fusion protein with inactive endosialidase strongly stains the bacteria containing an α2–8 polySia capsule; *N. meningitidis* group B (Fig. 4G), *M. (Pasteurella) haemolytica* A2 (Figs. 4C and D), and *M. nonliquefaciens* (Fig. 4I), and staining is not observed with the closely related strains not containing a polySia capsule (Fig. 4E, F, H, and J). Preincubation of the fusion protein with free polySia (colominic acid) abolished binding in all cases (not shown).

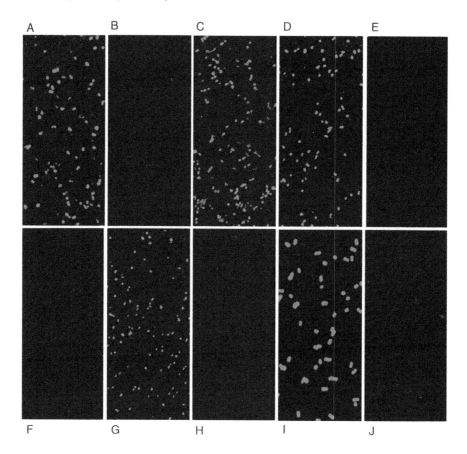

Figure 4. Detection of polySia in pathogenic bacteria by fluorescence microscopy. Bacterial strains with an α2–8 polySia capsule (+) and closely related strains with a capsule not containing polySia (−) were detected by fluorescence microscopy using the fusion protein. (A) *E. coli* K1 (+), (B) *E. coli* K2 (−), (C) *M. haemolytica* A2 (KU 201/83) (+), (D) *M. haemolytica* (KU 301/83) A2 (+), (E) *M. haemolytica* (−), (F) *N. meningitidis* group A (−), (G) *N. meningitidis* group B (+), (H) *N. meningiditis* group C (−); contains the isomeric α2–9 polySia capsule, (I) *M. nonliquefaciens* (+), and (J) *M. nonliquefaciens* (−). Reprinted from Ref. [42], copyright (2004), with permission from Elsevier.

3.2. Fluorometric detection of NCAM-bound and bacterial polysialic acid

The polySia-containing K1 bacteria stained with inactive endosialidase-containing fusion protein gave an intense signal in microtiter plate fluorometry, whereas the K2 bacteria with an unrelated capsule did not (Fig. 5A). The embryonic form of NCAM (contains polySia) was detected as a prominent fluorescence signal, whereas the adult, non-polysialylated form remained negative (Fig. 5B).

Table 1. Type of polysialic acid in bacterial strains.

Bacterial Strain	Capsule Type	Type of PolySia
E. coli IH3088	K1	α2–8
E. coli IH3083	K2	—
Neisseria meningitidis NCTC 10025	A	—
N. meningitidis NCTC 10026	B	α2–8
N. meningitidis NCTC 8554	C	α2–9
Mannheimia haemolytica KU 201/83	A2	α2–8
M. haemolytica KU 301/83	A2	α2–8
M. haemolytica KU 363/84		
Moraxella nonliquefaciens EF 10057		α2–8
M. nonliquefaciens KK 987/84		—

Note: —: No polySia.

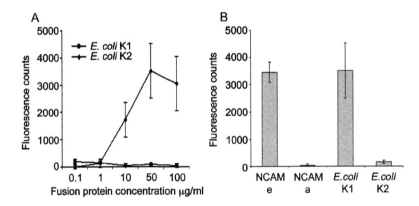

Figure 5. Fluorometric detection of polySia in microtiter plate assay. (A) Detection of *E. coli* K1 (polySia positive) and K2 (polySia negative) capsules by a fluorescence microtiter plate analyzer plotted as a function of the fusion protein concentration. (B) Detection of polySia-positive embryonic (e) and polySia-negative adult (a) neural cell adhesion molecule (NCAM), and *E. coli* K1 and K2 at a fusion protein concentration of 50 μg/ml. Reprinted from Ref. [42], copyright (2004), with permission from Elsevier.

3.3. Western analysis

In western blots of extracts of rat brain homogenates, the catalytically inactive fusion protein revealed a polySia-containing band corresponding by mobility to the carrier molecule of polySia, NCAM. As a control, the catalytically active form did not display binding (Fig. 6).

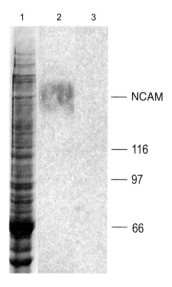

Figure 6. Western analysis using endosialidase fusion protein. Brain homogenates of newborn rats were subjected to sodium dodecyl sulfate polyacrylamide gel electrophoresis. (1) Gel proteins stained with Coomassie blue. (2) Blot stained for polySia with catalytically inactive fusion protein. (3) Blot stained with catalytically active fusion protein. Molecular masses (kDa) are shown on the right. Reprinted from Ref. [42], copyright (2004), with permission from Elsevier.

3.4. Detection of polysialic acid on human neuroblastoma sections

Neuroblastomas are malignancies of early childhood, and some of these highly malignant tumors are known to express polySia on their cell surface [6, 34]. The inactive form of the fusion protein specifically stained neuroblastoma cells in paraffin-embedded sections of neuroblastoma tumors (Fig. 7). The staining could be specifically blocked by the addition of free polySia (colominic acid), or pre-treatment of the sections with active endosialidase.

4. Comments

A majority of sequences in nucleotide and protein databases correspond to enzymes [35, 36], and almost 40,000 3-D structures (http://www.pdb.org/) have already been described. Furthermore, there are databases on the catalytic residues of the active sites of enzymes [37]. Thus, it should be possible to identify enzyme candidates for targeted amino acid substitutions that inactivate the catalytic activity of the enzyme, but preserve the binding activity. Alternatively, inactivated enzymes could be obtained through selection of natural mutants as described here.

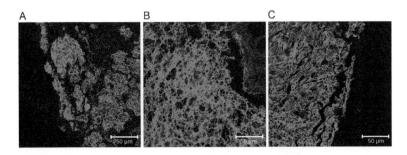

Figure 7. Fluorescence microscopy of paraffin-embedded tissue sections. Sections of human neuroblastomas (A, B, and C) stained with the polySia-binding fusion protein. The scale bar is 50 μm.

To date there are few examples of the conversion of a carbohydrate modifying enzyme to a lectin-like reagent. An inactive chitinase has been patented for identifying fungi by binding to fungal cell walls [38, 39] (Fungalase-F; Anomeric, Inc., Baton Rouge, LA). Chitin is a homopolymer of β(1,4)-linked *N*-acetylglucosamine, which is a component of the cell walls of most fungi, but is not found in mammals. The method has been used to detect fungal infections in human samples [40]. An inactivated lysozyme for proteoglycan and chitin detection has been made by the same company (http://membership.acs.org/S/SCHB/acs-NewOrleansTSA.html).

We have previously used whole mutant bacteriophages with catalytically inactive endosialidase for the detection of polySia [41]. The method involved the use of anti-phage antibody and secondary antibody conjugate. The present method represents a more convenient single-step approach for the detection of polySia.

5. Experimental Considerations

The materials and experimental methods are described in detail in the original paper [42]. Some considerations on the use of the fusion protein are given below:

(1) The yield of the expressed fusion protein is in the range of 10–20 mg/l of bacterial culture without optimization.
(2) The fusion protein can be stored in phosphate buffered saline (with or without Ca^{2+} and Mg^{2+}) in the presence of sodium azide at 4°C for at least six months without major loss in binding activity.
(3) As negative controls in stainings, the fusion protein can be coincubated with polySia (colominic acid) 100 μg/ml for 30 min; the specimens can be pretreated with active endosialidase (10 μg/ml, 30 min) or stained with fusion protein constructed from the active endosialidase.

(4) As positive control, a polySia expressing bacteria (Table 1) or eukaryotic cells, such as baby hamster kidney BHK-21 [C-13] (ATCC: CCL-10) or human neuroblastoma SHSY-5Y (ECACC: 94030304) cell line can be used. The cells can be fixed with 4% paraformaldehyde, acetone, or 0.25% glutaraldehyde.

References

[1] J. Finne, J. Biol. Chem., 257 (1982) 11966.

[2] J.Z. Kiss and G. Rougon, Curr. Opin. Neurobiol., 7 (1997) 640.

[3] E. Ong, J. Nakayama, K. Angata, L. Reyes, T. Katsuyama, Y. Arai and M. Fukuda, Glycobiology, 8 (1998) 415.

[4] L.M. Garcia-Segura, M. Duenas, S. Busiguina, F. Naftolin and J.A. Chowen, J. Steroid Biochem. Mol. Biol., 53 (1995) 293.

[5] M. Fukuda, Cancer Res., 56 (1996) 2237.

[6] H. Hildebrandt, C. Becker, S. Glüer, H. Rosner, R. Gerardy-Schahn and H. Rahmann, Cancer Res., 58 (1998) 779.

[7] F. Dall'Olio and M. Chiricolo, Glycoconj. J., 18 (2001) 841.

[8] R. Seidenfaden, A. Krauter, F. Schertzinger, R. Gerardy-Schahn and H. Hildebrandt, Mol. Cell. Biol., 23 (2003) 5908.

[9] J. Finne, U. Finne, H. Deagostini-Bazin and C. Goridis, Biochem. Biophys. Res. Commun., 112 (1983) 482.

[10] H. Cremer, R. Lange, A. Christoph, M. Plomann, G. Vopper, J. Roes, R. Brown, S. Baldwin, P. Kraemer and S. Scheff, Nature, 367 (1994) 455.

[11] U. Rutishauser, J. Cell. Biochem., 70 (1998) 304.

[12] M. Lipinski, M.R. Hirsch, H. Deagostini-Bazin, O. Yamada, T. Tursz and C. Goridis, Int. J. Cancer, 40 (1987) 81.

[13] J. Roth, C. Zuber, P. Wagner, D.J. Taatjes, C. Weisgerber, P.U. Heitz, C. Goridis and D. Bitter-Suermann, Proc. Natl. Acad. Sci. U.S.A., 85 (1988) 2999.

[14] M. Miettinen and W. Cupo, Hum. Pathol., 24 (1993) 62.

[15] C.E. Moolenaar, E.J. Muller, D.J. Schol, C.G. Figdor, E. Bock, D. Bitter-Suermann and R.J. Michalides, Cancer Res., 50 (1990) 1102.

[16] E. Vimr, S. Steenbergen and M. Cieslewicz, J. Ind. Microbiol., 15 (1995) 352.

[17] J. Finne, M. Leinonen and P.H. Mäkelä, Lancet, 2 (1983) 355.

[18] C. Dubois, A. Okandze, D. Figarella-Branger, C. Rampini and G. Rougon, J. Immunol. Methods, 181 (1995) 125.

[19] C. Sato, K. Kitajima, S. Inoue, T. Seki, F.A. Troy 2nd and Y. Inoue, J. Biol. Chem., 270 (1995) 18923.

[20] M. Frosch, I. Görgen, G.J. Boulnois, K.N. Timmis and D. Bitter-Suermann, Proc. Natl. Acad. Sci. U.S.A., 82 (1985) 1194.

[21] R.J. Gross, T. Cheasty and B. Rowe, J. Clin. Microbiol., 6 (1977) 548.

[22] S. Pelkonen, J. Aalto and J. Finne, J. Bacteriol., 174 (1992) 7757.

[23] E. Hoiczyk, A. Roggenkamp, M. Reichenbecher, A. Lupas and J. Heesemann, EMBO J., 19 (2000) 5989.

[24] M. Mühlenhoff, K. Stummeyer, M. Grove, M. Sauerborn and R. Gerardy-Schahn, J. Biol. Chem., 278 (2003) 12634.

[25] J.L. Pellequer and M.H. Van Regenmortel, Mol. Immunol., 30 (1993) 955.

[26] C.R. MacKenzie, T. Hirama, S.J. Deng, D.R. Bundle, S.A. Narang and N.M. Young, J. Biol. Chem., 271 (1996) 1527.

[27] J. Häyrinen, S. Haseley, P. Talaga, M. Mühlenhoff, J. Finne and J.F. Vliegenthart, Mol. Immunol., 39 (2002) 399.

[28] T.K. Korhonen, M.V. Valtonen, J. Parkkinen, V. Väisänen-Rhen, J. Finne, F. Ørskov, I. Ørskov, S.B. Svenson and P.H. Mäkelä, Infect. Immun., 48 (1985) 486.

[29] S. Pelkonen and J. Finne, FEMS Microbiol. Lett., 42 (1987) 53–57.

[30] S. Pelkonen, Curr. Microbiol., 21 (1990) 23.

[31] H.J. Jennings, E. Katzenellenbogen, C. Lugowski, F. Michon, R. Roy and D.L. Kasper, Pure Appl. Chem., 56 (1984) 893.

[32] S.J. Devi, R. Schneerson, W. Egan, W.F. Vann, J.B. Robbins and J. Shiloach, Infect. Immun., 59 (1991) 732.

[33] H.J. Adlam, J.M. Knights, A. Mugridge, J.M. Williams and J.C. Lindon, FEMS Microbiol. Lett., 42 (1987) 23.

[34] I.Y. Cheung, A. Vickers and N.K. Cheung, Int. J. Cancer, 119 (2006) 152.

[35] S.T. Cole, R. Brosch, J. Parkhill, T. Garnier, C. Churcher, D. Harris, S.V. Gordon, K. Eiglmeier, S. Gas, C.E. Barry 3rd, F. Tekaia, K. Badcock, D. Basham, D. Brown, T. Chillingworth, R. Connor, R. Davies, K. Devlin, T. Feltwell, S. Gentles, N. Hamlin, S. Holroyd, T. Hornsby, K. Jagels and B.G. Barrell, Nature, 393 (1998) 537.

[36] J. Kawai, A. Shinagawa, K. Shibata, M. Yoshino, M. Itoh, Y. Ishii, T. Arakawa, A. Hara, Y. Fukunishi, H. Konno, J. Adachi, S. Fukuda, K. Aizawa, M. Izawa, K. Nishi, H. Kiyosawa, S. Kondo, I. Yamanaka, T. Saito, Y. Okazaki, T. Gojobori, H. Bono, T. Kasukawa, R. Saito, K. Kadota, H. Matsuda, M. Ashburner, S. Batalov, T. Casavant, W. Fleischmann, T. Gaasterland, C. Gissi, B. King, H. Kochiwa, P. Kuehl, S. Lewis, Y. Matsuo, I. Nikaido, G. Pesole, J. Quackenbush, L.M. Schriml, F. Staubli, R. Suzuki, M. Tomita, L. Wagner, T. Washio, K. Sakai, T. Okido, M. Furuno, H. Aono, R. Baldarelli, G. Barsh, J. Blake, D. Boffelli, N. Bojunga, P. Carninci, M.F. de Bonaldo, M.J. Brownstein, C. Bult, C. Fletcher, M. Fujita, M. Gariboldi, S. Gustincich, D. Hill, M. Hofmann, D.A. Hume, M. Kamiya, N.H. Lee, P. Lyons, L. Marchionni, J. Mashima, J. Mazzarelli, P. Mombaerts, P. Nordone, B. Ring, M. Ringwald, I. Rodriguez, N. Sakamoto, H. Sasaki, K. Sato, C. Schonbach, T. Seya, Y. Shibata, K.F. Storch, H. Suzuki, K. Toyo-oka, K.H. Wang, C. Weitz, C. Whittaker, L. Wilming, A. Wynshaw-Boris, K. Yoshida, Y. Hasegawa, H. Kawaji, S. Kohtsuki, Y. Hayashizaki and RIKEN Genome Exploration Research Group Phase II Team and the FANTOM Consortium., Nature, 409 (2001) 685.

[37] C.T. Porter, G.J. Bartlett and J.M. Thornton, Nucleic Acids Res., 32 Database issue (2004) D129.

[38] R.A. Laine, C.Y. Ou and J.M. Jaynes, US Patent no. 5,532,607 (1994) 395.

[39] R.A. Laine and E.C.J. Lo, US Patent no. 5, 587,292 (1996).

[40] M.J. Taylor, J.U. Ponikau, D.A. Sherris, E.B. Kern, T.A. Gaffey, G. Kephart and H. Kita, Otolaryngol. Head Neck Surg., 127 (2002) 377.

[41] J. Aalto, S. Pelkonen, H. Kalimo and J. Finne, Glycoconj. J., 18 (2001) 751.

[42] A. Jokilammi, P. Ollikka, M. Korja, E. Jakobsson, V. Loimaranta, S. Haataja, H. Hirvonen and J. Finne, J. Immunol. Methods, 295 (2004) 149.

Lectins: Analytical Technologies
C.L. Nilsson (Editor)
© 2007 Elsevier B.V. All rights reserved.

Chapter 17

Probing Cell Surface Lectins with Neoglycoconjugates

Eugenia M. Rapoport, Elena I. Kovalenko, Ivan M. Belyanchikov and Nicolai V. Bovin

Shemyakin and Ovchinnikov Institute of Bioorganic Chemistry RAS, 117997, ul. Miklukho-Maklaya 16/10, Moscow, Russia

1. Introduction

Lectins are expressed on various cells and are involved in numerous processes related to vital functions of organisms [1–4]. C-lectins, galectins-1 and -3, for example, are the counter receptors for integrins, whereas galectin-1, -3, siglec-1, -2 – for proteins of immune cells. Lectins mediate cell–cell and cell–matrix adhesion, cell proliferation, and apoptosis [2, 5, 6]. Selectins, galectin-3, and galectin-9 are inflammation mediators or chemoattractants [7]. Lectins expressed on macrophages, dendritic cells, and NK cells are involved in antitumor, antimicrobial, and antiviral protection of the organism [6, 8–10].

Knowledge of carbohydrate-binding properties of cell surface lectins is important for understanding their functions. At the same time revealing cell lectins and studying their specificity are associated with a number of difficulties. Firstly, the same lectins can be exposed not only on cell membrane surface but also be localized inside the cell. Secondly, different lectins can recognize similar carbohydrate ligands in probe composition. Thirdly, lectins expressed on the cell surface can be masked with the glycoconjugates of the same cells (*cis*-interaction) [11, 12], and therefore not bind external ligands, including probes. Finally, lectin topography on the cell affects its binding with carbohydrate probes.

A particularity of carbohydrate–protein recognition is that both affinity and specificity are achieved due to polyvalent interaction, whereas affinity of single ligand to lectin can be very low or not registered at all [13–15]. The use of polyvalent probes allows increasing of the binding due to multipoint interaction [15–18].

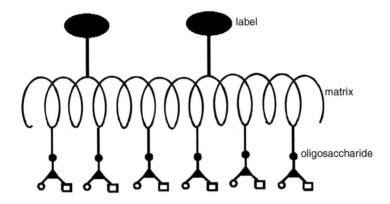

Figure 1. Design of PAA-based glycoconjugates.

2. Design of Neoglycoconjugates as Probes

Multivalency of binding requires a specific design of the probes; several copies of saccharide should be attached to a polymer carrier at an appropriate distance. The probe bears a direct label (e.g. fluorescein) or a tag for subsequent detection (e.g. biotin), see Fig. 1. Importantly, the probes for *cell studies* should correspond to enormously strict requirements in respect of nonspecific interaction with cell components and matrix. Obviously, label or tag, linker between saccharide (Glyc) and matrix and matrix itself must be maximally hydrophilic and neutral or weakly negatively charged. Another important necessary property of cell probes is the matrix flexibility as, in opposite case, due to the fixed (inappropriate) distance between ligands on matrix, the probe may lose its ability to multipoint binding. Polyacrylamide-based glycoconjugates used for a long time for the study of specificity of soluble and membrane-associated lectins fit very well these requirements [15–18].

The polyacrylamide carrier is a random coil, i.e. the distances between oligosaccharides are not fixed (Fig. 1). This allows the conjugate to adjust itself to several lectin molecules on a cell [16–18]; importantly, the molecule possesses a weak negative charge and lacks hydrophobic groups completely. Additionally, chemistry of polyacrylamide probes allows easy optimizing (varying) of sugar and label loading, as well as molecular weight (size of the probe) (see the details in earlier reviews [15, 16]).

3. Experimental Approaches

Biotinylated glycoprobes are convenient for lectin revealing based on CELISA (cell ELISA) technique, whereas fluorescein-labeled ones are appropriate for use in flow cytometry [16–18]. Both biotinylated and fluorescein-labeled glycoprobes

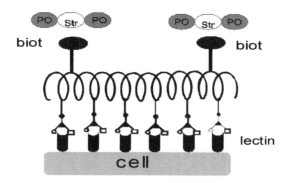

Figure 2. Scheme of solid-phase assay for the study of cell lectins.

can be used in cytological and histological studies. Sometimes, more intense staining is observed when the probe is attached to a fluorescent particle (see below). Glycoparticles prepared by immobilization of glycoconjugates on magnetic beads are convenient for isolation of a particular cell population, expressing carbohydrate-binding molecules [19, 20].

3.1. CELISA

Cells are cultivated in plates followed by the addition of a biotinylated glycoprobe; the degree of Glyc-PAA-biot binding to cells is determined with the help of Str-PO conjugate (Fig. 2) [18].

3.2. Flow cytometry

Flow cytometry has a number of advantages over CELISA, because it allows working with living cells and makes it possible to isolate a narrow population of lectin expressing cells from a heterogeneous population. In contrast to CELISA, the flow cytometry technique does not require continuous incubations and washings. The technique of cell preparation for the assay is simple: cells are incubated with Glyc-PAA-fluo followed by washing off nonreacted reagent by centrifugation. The degree of cell binding with glycoprobe is detected by the intensity of fluorescent signal [15, 18].

3.3. Cell separation

Magnetic separation methods are widely used for isolation of a variety of cell types. Magnetic particles with immobilized antibodies to various antigens have been employed for the rapid isolation of populations T-(CD4+, CD3+, CD8+) and B- (CD19+) of lymphocytes, NK cells, and monocytes. Similarly, immobilization of glycoconjugates on magnetic beads allows the isolation of cell populations expressing a particular carbohydrate-recognizing molecule [19, 20]. Glycosylated magnetic beads can be prepared by loading biotinylated probes onto streptavidin-coated magnetic beads. The glycoparticles are then incubated with a cell suspension and the subpopulation of interest is fished out by means of a magnetic device [20].

3.4. Microscopy (histology and cytology)

To study the role of lectins in inflammation processes and cell transformation, it is necessary to gather information about their expression on cell surface and localization in intracellular compartments. Labeled glycoprobes can provide such information. For microscopy studies, cells are fixed in mild conditions (95% ethanol or 2% paraformaldehyde), incubated with 3% BSA-PBS, in order to prevent nonspecific binding, followed by incubation with a glycoprobe. After washing, the bound Glyc-PAA-biot is revealed with the help of Str-PO conjugate [20–23].

There are three techniques for fluorescent microscopy: (1) direct observation of Glyc-PAA-fluo probes binding [21]; (2) two-step procedure based on Glyc-PAA-biot followed by Str-fluo staining [24]; and (3) binding of fluorescent glycoparticles (Fig. 3).

3.5. Confocal spectral imaging system

Laser scanning confocal fluorescence microscopy is useful for the localization of biologically active molecules, the study of their targets and mechanisms of action [25].

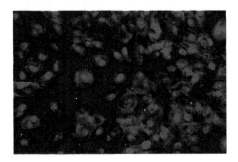 

Figure 3. Staining of normal (left) and activated (E-selectin positive, right) HUVECs by fluorescent beads bearing SiaLe^x. Courtesy of Dr. J. Magnani (GlycoTech, USA).

Using a confocal scheme of signal filtration, this technique allows measurement of the 3D distribution of fluorescent molecules in a cell. Detailed study of spectra obtained with the help of scanning confocal microspectrometer is the base of the method of confocal spectral imaging technique for identifying the distribution of biomolecules in a single cell [25]. The use of fluorescein-labeled glycoprobes allows for studies of carbohydrate-binding molecules [21]. The use of glycoprobes in confocal spectral imaging technique can demonstrate the lectin distribution in a single cell [21]. Thus, cells of lung carcinoma OAT75 and B-lymphoma Raji bound (GlcNAc)$_2$-PAA-fluo (structures of all abbreviated aligosaccharides are given in Table 1) though lectin accumulation was observed on cell membrane surface in case of lung carcinoma and inside cytoplasm in cell vesicles in case of B-lymphoma [21]. Several examples of glycoprobe application for revealing lectins on cells and the study of their specificity are given below.

Table 1. List of oligosaccharide abbreviations.

Abbreviation	Oligosaccharide Structure
LacNAc	Galβ1-4GlcNAc
LacdiNAc	GalNAcβ1-4GlcNAc
TF	Galβ1-3GalNAcα
Le^c	Galβ1-3GlcNAc
3′-O-Su-Le^c	*3HSO$_3$*-Galβ1-3GlcNAc
T$_{\beta\beta}$	Galβ1-3GalNAcβ
6′-O-Su-LacNAc	*6HSO$_3$* -Galβ1-4GlcNAc
H$_{di}$	Fucα1-2Gal
A$_{di}$	GalNAcα1-3Galβ
(GlcNAc)$_2$	GlcNAcβ1-4GlcNAc
Sia$_2$	Neu5Acα2-8Neu5Acα
Galα-3′LacNAc	Galα1-3Galβ1-4GlcNAc
A$_{tri}$	GalNAcα1-3(Fucα1-2)Galβ
B$_{tri}$	Galα1-3(Fucα1-2)Galβ
H (type1)	Fucα1-2Galβ1-3GlcNAc
H (type2)	Fucα1-2Galβ1-4GlcNAc
Le^a	Galβ1-3(Fucα1-4)GlcNAc
3′-O-SuLe^a	*6HSO$_3$*-Galβ1-3(Fucα1-4)GlcNAc
3′SiaLacNAc	Neu5Acα2-3Galβ1-4GlcNAc
6′SiaLac	Neu5Acα2-6Galβ1-4Glc
LNnT	Galβ1-4GlcNAcβ1-3Galβ1-4Glc
aGM1	Galβ1-3GalNAcβ1-4Galβ1-4Glc
SiaLe^a	Neu5Acα2-3Galβ1-3(Fucα1-4)GlcNAc
6′-O-Su-SiaLe^x	Neu5Acα2-3(*6HSO$_3$*)Galβ1-4(Fucα1-3)GlcNAc
(LN)$_3$	Galβ1-4GlcNAcβ1-3Galβ1-4GlcNAcβ1-3Galβ1-4GlcNAc

4. Examples of Application

4.1. CELISA

Biotinylated glycoprobes (Glyc-PAA-biot) were used for the study of the lectin pattern of MDCK cells. The aim of the study was to reveal cell lectins as additional receptors that can participate in interactions with influenza virus, as an alternative to the common recognition performed by influenza virus hemagglutinin and cell sialooligosaccharides. MDCK cells were selected because this cell line can be infected with influenza virus. Binding of LacNAc and LNnT to cells has been observed (Fig. 4), i.e. molecules recognizing a lactosamine fragment were revealed on the surface of MDCK cells. No binding with other probes, e.g. 6'SiaLac, was observed, which indicates the specificity of the binding. Thus, the presence of a β-galactoside binding lectin, probably a galectin, on MDCK cells is evidenced. Galectin-3, whose affinity to LacNAc and LNnT has been demonstrated earlier [26, 27], is expressed on the surface of MDCK cells. It is possible that galectin-3 interacts with multiple galactose-terminated saccharide chains on influenza virus glycoproteins, thus facilitating virus-to-cell binding [28].

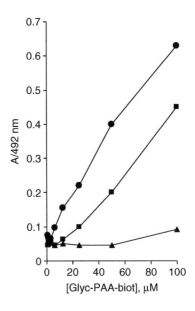

Figure 4. Binding of LacNAc-PAA-biot (•), and LNnT-PAA-biot (■) to MDCK cells. 6'SiaLac-PAA-biot (▲) serves as negative control. Trypsinized cells were coated on ELISA plates and incubated with probes followed by Str-PO.

4.2. Flow cytometry

4.2.1. Siglecs

The specificity of siglecs-1, -5, -7, and -9 transfected into CHO cells was studied by means of the fluorescein-labeled glycoconjugates Glyc-PAA-fluo [29]. In particular, 3'SiaLacNAc but not 6'SiaLac bound CHO-siglec-1, CHO-siglec-5, and CHO-siglec-9 (Fig. 5, data for cells CHO-siglec-1 are shown).

Disaccharide Neu5Acα2-8Neu5Ac in composition of Sia_2-PAA-fluo displayed affinity to siglec-7; its binding with siglec-5 was significantly lower. This probe did not bind the cells CHO-siglec-1 (data not shown) or CHO-siglec-9 (Fig. 6). These data correlate well with the specificity of the corresponding soluble siglecs [30–33].

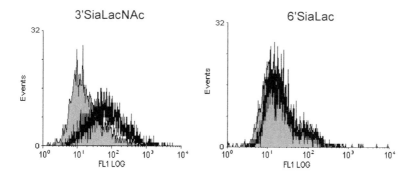

Figure 5. Probing CHO-siglec-1 with Glyc-PAA-fluo. Cells were pretreated with neuraminidase to unmask sialic acid binding sites of siglec. Grey histograms show background fluorescence of CHO-WT cells incubated with Glyc-PAA-fluo (negative control). The logs of fluorescence intensities are plotted against cell numbers. CHO-WT means mock-transfected cells.

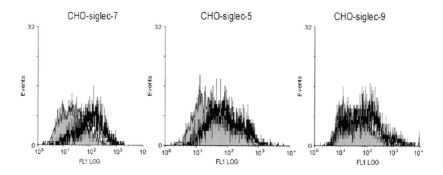

Figure 6. Probing the CHO-siglec cells with Sia_2-PAA-fluo. Cells were pretreated with neuraminidase to unmask sialic acid binding sites of siglec. Grey histograms show background fluorescence of CHO-WT cells incubated with the same probe (negative control). The logs of fluorescence intensities are plotted against cell numbers.

4.2.2. Galectins

Galectins are a family of β-galactoside binding lectins, homologous by amino acid sequence of the carbohydrate-binding site [2]. Studies of cellular galectins are complicated by the following factors: (1) galectins do not have GPI anchor, being anchored on the cell via interaction of CRD with cell glycoconjugates and (2) galectins do not have signal N-terminal sequence, making difficult the preparation of transfected cell lines expressing lectins on the surface [34]. Taking these into consideration, we artificially loaded galectins onto Raji cells, which normally do not express galectins [35]. Exogenous galectin-1 or -3 was loaded onto these cells by incubation with the soluble protein. Galectin binding with cells was verified using the corresponding antibodies. Galectin-1 in cell composition bound LNnT, whereas Galα1-3′LacNAc was the highest affinity probe for galectin-3 (Fig. 7A and B).

4.2.3. Glycoprofiling of T-lymphocytes

Glycoprobes can be used for cell "glycophenotyping." The aim of this study is to reveal the population of cells that expresses lectins. Data on glycophenotyping of T-lymphocytes are given as the example. T-lymphocyte populations CD4+ and CD8+ include two functionally different subpopulations, namely naïve cells (mature lymphocytes that have not had any contact with antigens) and memory cells (formed along with effector cells in the process of immune response after antigen recognition). These cells can be distinguished by expression of isoforms of leukocyte-associated antigen CD45. Naïve cells express a "full" variant of CD45 molecule, CD45RA, whereas memory T-cells express a truncated version, CD45R0. It was demonstrated using glycoprobes that naïve cells of T-helpers (CD4+ CD45RA+) and cytotoxic T-lymphocytes (CD8+ CD45RA+) bound

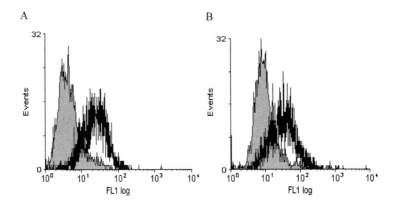

Figure 7. Probing the galectin-1- (A) and galectin-3- (B) -loaded Raji cells with Glyc-PAA-fluo, flow cytometry analysis. Grey histograms show background fluorescence of galectin-free cells incubated with the same probe (negative control). The logs of fluorescence intensity are plotted against cell number. Unpublished data.

3′-O-Su-Le^a-PAA-fluo, Fig. 8D. Interestingly, the highest expression levels of a lectin that recognizes this glycoprobe has been observed in a T-helper cell population (CD4+ CD45R0+), probably L-selectin (Fig. 8E) [36]. These data are in agreement with the results obtained earlier [37]. In combination with other receptors of T-helper naïve cells, L-selectin seems to mediate the ability of these cells to migrate in secondary lymphoid organs, where T-helper cells stimulate activity of effector cells or antibody production [37].

4.2.4. Monocytes, macrophages, and tumor cells

The aim of lectin profiling of *monocytes* was to reveal the receptor involved in monocyte adhesion to vascular epithelium of recipient during xenotransplantation [38]. Rejection of transplanted tissues takes place as a result of recognition of

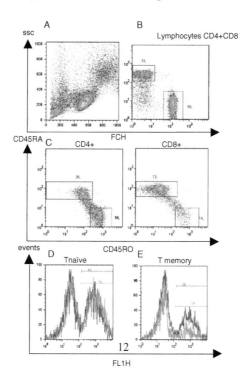

Figure 8. 3′-O-SuLe^a-PAA-fluo binding to T-lymphocytes from human blood, flow cytometry analysis. (A) Forward scatter versus side scatter profile of isolated leukocytes. The R1 window corresponds to lymphocytes. (B) The fluorescence dot plots show double labeling of cells in R1 window with anti-CD4-phycoerythrin-cyanin-7 and antiCD8-allophycocyanin. (C) The fluorescence dot show labeling of CD4+ and CD8+ cells with antiCD45RA-phycoerythrin-cyanin-5 and antiCD45R0-phycoerythrin. (D and E) Histograms of naïve (CD45RA+) and memory (CD45R0) cells bound 3′-O-SuLe^a-PAA, cells positive for CD4+ (red) and CD8 (green) are shown. The logs of fluorescence intensity are plotted against cell number. Values in the quadrants and histograms show the percentage of gated cells.

alien histocompatibility tissue antigens on transplant cells by recipient's immune system. In the case of xenotransplantation, the immune barrier is usually unbreakable: the xenograft becomes rapidly rejected upon response action of the recipient's immune system. The role of monocytes in this process is production of antiinflammatory cytokines and increased expression of adhesion molecules causing monocyte adhesion to transplant endothelium [39, 40]. A C-type lectin capable of recognizing the xenoantigen Galα1-3′LacNAc has been detected on monocytes with the help of glycoprobes [38]. It was demonstrated that the lectin is not involved in intercellular adhesion, but instead stimulates the expression of β2 integrins on monocytes. Binding with ICAM-1 and ICAM-2 integrins induces monocyte adhesion to vascular endothelium, which leads to xenograft rejection.

Glycoprobes were also used for the study of lectin pattern of macrophages. The aim of this work was to identify lectins taking part in phagocytosis of apoptotic bodies by macrophages. Elimination of apoptotic bodies is one of the important functions of macrophages. A decrease in this function is related to the inhibition of cytotoxic activity of macrophages and the loss of antigen-presenting properties. In combination with other factors, this leads to suppression of activity of cytotoxic lymphocytes and irreversible proliferation of tumor cells [41]. Interaction of glycoprobes with THP-1 cells (macrophage origin) has been studied. These cells bound β-galactoside containing probes, in particular LacNAc, asialoGM1, LacdiNAc. In parallel, it was demonstrated that THP-1 cells bound antibodies to galectins-1, -2, and -4 but not antibodies to galectins -3 and -7, i.e. galectins-1, -2, and -4 seemed to be expressed on THP-1 cells after activation. Phagocytosis of apoptotic bodies by THP-1 cells was then studied. The engulfment of apoptotic bodies by THP-1 cells was specifically inhibited by aGM1-PAA [42] and antibodies to galectin-1 or galectin-4. It seems that macrophage galectin-1 and galectin-4, which bind aGM1-PAA [43, 44], take part in lectin-mediated phagocytosis via binding with Galβ1-3GalNAcβ fragment of glycans on apoptotic bodies.

We have also studied the binding of tumor cells with β-galactoside probes (Glyc-PAA-fluo) in order to determine whether galactoside-binding affinity of the cells could be a general marker of malignancy. Binding of the probe aGM1 with whole cells that originated from spontaneous mammary carcinoma was one order of magnitude higher than that of LacdiNAc, Galα1-3′LacNAc, and LacNAc; whereas $T_{\beta\beta}$, TF, and LNnT did not bind to the cells (Fig. 9) [45].

Lectins localized in cytoplasm were detected only after membrane permeabilization. For instance, accumulation of carbohydrate-binding molecules, which recognized aGM1, has been observed in composition of mammary carcinoma cells but not the colonies of the same cells metastasizing to the liver and lungs. After permeabilization with saponin the cells bound aGM1, i.e. accumulation of carbohydrate-binding molecules takes place in cytoplasm of metastazing cells [45].

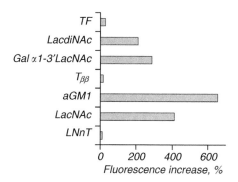

Figure 9. Probing the tumor cells with Glyc-PAA-fluo, flow cytometry analysis. Fluorescence increase was calculated as $[(F_i/F_0) \times 100] - 100\%$, where F_i, F_0 are fluorescence intensities of tumor cells after and before incubation with Glyc-PAA-fluo, respectively.

4.3. Cell separation

Immobilization of carbohydrate ligands on a magnetic carrier allows sorting/enrichment of cell subpopulations differing by lectin patterns. Following enrichment on magnetic beads, subpopulations of cells can be analyzed microscopically or by flow cytometry methods, because magnetic beads do not interfere with the performance of the assay [19, 20]. Glycobeads were obtained by loading of Glyc-PAA-biot onto streptavidin-coated magnetic beads. Populations of L-selectin-enriched cells were isolated from lymphoma cells MOLT4 and CCRF-CEM using glycobeads with immobilized 3'-O-SuLe[a] (Fig. 10). Using beads with variable carbohydrate ligands, it was possible to isolate subpopulations binding Le[c], H (type 1), and A_{di} from lung carcinoma cells [20].

4.4. Cytology and histology studies

Glycoprobes are useful in cytological and histological studies. Glycoprobes in combination with microscopy demonstrates the lectin pattern of cellular compartments [17, 22–25, 46, 47]. For example, samples of normal skin, oral mucosa, and basal cell carcinoma of epidermis were probed with A_{tri}-PAA-biot [24, 46, 47]. The aim was the study of lectin pattern of Langerhans cells in normal and malignant tissues. Langerhans cells are members of the myeloid-type family of dendritic, professional antigen-presenting cells. These cells are present predominantly in epithelia and express distinct markers such as CD1a and the mannose-specific lectin langerin [48]. The majority of epidermal Langerhans cells and Langerhans cells located in hair follicles bound the probe A_{tri}, whereas carcinoma cells were completely negative [47]. Using antigalectin-3 antibodies as an additional probe, it was shown that

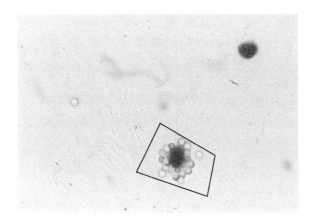

Figure 10. Selection of acute lymphoblastic leukemia cells CCRF-CEM by 3'-O-SuLe[a] modified magnetic beads. Hematoxylin stained rosetted cell and acridine orange fluorescent stained cells were gated. Courtesy of Dr. P. Rye (Norway).

galectin-3 was localized in the cytoplasm of Langerhans cells [24, 46, 47]. A possible receptor of dendritic cells, galectin-3 mediates intercellular adhesion by activation, in concert with other receptors, the process of antigen presentation to T-helpers and cytotoxic lymphocytes [46]. Employing the same double labeling methodology, galectin-3 was revealed in the cytoplasm of alveolar macrophages [24, 46].

Histology study of colon carcinoma cells was performed with glycoprobes. The aim of this study was to define differences between tumor and corresponding normal cells in lectin profiles. Accumulation of lectin capable of recognizing H_{di} was observed in the cytoplasm and nuclei of carcinoma epithelial cells of gastric mucosa, whereas normal tissue cells did not bind the probe [23].

5. Discussion

As a rule, lectins are revealed on cells with the help of the corresponding antibodies, thus demonstrating the presence of lectins on the surface of cell membrane. One of the most intriguing, though least studied, function of lectins is mediation of intercellular adhesion, for which at least one CRD must be free, i.e. unengaged during lectin anchoring on cell. Antibodies are not a suitable probe for the study of generic lectin function, i.e. its ability to bind glycans of the same membrane or another cell to be recognized. Problems of that kind can be solved with the assistance of glycoprobes discussed in this review. Due to their multivalent design, such probes possess the required affinity to allow quantitative measurements of their binding with cells. A set of Glyc-PAA-fluo probes differing only by the nature of the carbohydrate ligand allows for studies of cell lectin specificity. Moreover, glycoprobes allow the study of lectins, to which antibodies have not been obtained yet, and can even reveal novel lectins by the analysis of their carbohydrate specificity pattern. Detection of an αGal-binding C-type lectin on monocytes [38] is one example. Another example of this kind, from the field of virology, is the discovery of a specific 6'-O-Su-LacNAc-binding ability of influenza viruses [49] a feature not yet assigned to any particular protein.

The affinity of a glycoprobe to a cellular lectin is usually lower than that of the corresponding antibodies; therefore, about two orders of magnitude higher concentration of glycoprobes than antibodies is necessary for studies of cellular lectins. However, it should be noted that due to the special design of the probes (see above) a high working concentration (100 µg/ml) does not lead to nonspecific adhesion of glycoprobes on cell surfaces. The average fluorescence intensity is somewhat lower than that of the cells stained with antibodies, but the percentage of bound cells is practically the same. Thus, 25–30% of the total cell population was stained both by antibodies to E-selectin and SiaLe^x probe during the revealing of E-selectin on the surface of the corresponding gene transfected cells [22].

Key matters are the proof of binding specificity and correct selection of negative control. Inhibition with carbohydrate ligands (i), direct binding with a set of several probes (both similar and very different by glycan) (ii), and cell treatment with suitable glycosidases (iii) are used for the proof of the binding specificity. For example, binding of H$_{di}$-biot with rat intestine carcinoma cell (see Section 4.4) was inhibited with free disaccharide Fucα1-2Gal and related oligosaccharides H (type 1), and H (type 2) but not with A$_{tri}$ and B$_{tri}$ [23]. Another example is binding of 3'-O-SuLe^a-PAA-fluo with macrophages, which has been inhibited with monomeric 3'-O-SuLe^a or its polyacrylamide conjugate (Fig. 11) but not with SiaLe^a [50].

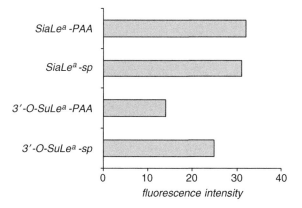

Figure 11. Inhibition of the 3'-O-SuLe^a-PAA-fluo binding to macrophages by free saccharides (20 mM) and Glyc-PAA (100 µM). Fluorescence intensity is given in arbitrary units [50].

A combination of Glyc-PAA-label with other probes gives additional information about features of lectin accommodation in glycocalix. This can be exemplified by the binding of LacNAc-PAA-fluo with galectin-1, which increased after cell degalactosylation (Fig. 12). Thus, though LacNAc is a specific ligand for galectin-1, the unknown yet cell surface glycan is more specific, even multimeric LacNAc cannot displace the natural ligand from the complex with galectin.

Selection of negative control(s) during the work with glycoprobes is not a trivial matter; it requires a separate approach in each particular case. In our experience, evidence that carbohydrate-free conjugate PAA-fluo is poor negative control, whereas PAA-biot is very poor control, this being illustrated with the example given below (Fig. 13).

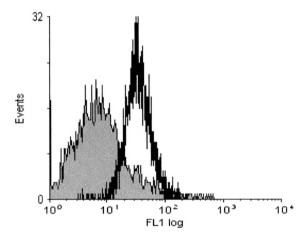

Figure 12. Probing the galectin-1-loaded Raji cells before (filled histogram) and after degalactosylation (unfilled histogram) with LacNAc-PAA-fluo, flow cytometry analysis. The logs of fluorescence intensity are plotted against cell number. Unpublished data.

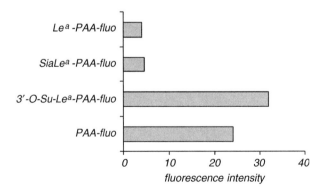

Figure 13. Probing the macrophages, flow cytometry analysis. Fluorescence intensity is given in arbitrary units [50].

At a glance, it looks paradoxical that the carbohydrate-free conjugate PAA-fluo interacts with cells closely to 3'-O-SuLe^a, whereas Le^a and SiaLe^a display low interaction. We explain high level of the binding of nonglycosylated probes by hydrophobicity of the label. Though fluorescein content in the conjugate is low (1 mol% or 3–4 residues per chain), the molecule *as a whole* acquires some hydrophobicity resulting in background detectable in flow cytometry. Attachment of any carbohydrate residue to PAA-fluo quenches this hydrophobicity; as a result nonspecific binding becomes decreased to a practically acceptable value. Therefore, it becomes obvious that such monosaccharides as Fuc or GlcNAc in composition of Glyc-PAA-fluo probes are much worse than less hydrophobic monosaccharides and that basically oligosaccharides are more preferable as negative controls. Experience of the use of various saccharides also evidences that glucopyranoside and open form of glucose (attached to polymer as amino alditol) produce high background values too, possibly due to interactions with a glucose transporter or scavenger receptors. Thus, there is no negative control, which is universal for all cell types, and most probably such a negative control will never exist. So, in practice, we take several glycoprobes as the negative control, which correspond to cell type, expected specificity of the studied lectin, and to the structure of the "main" probe (3'-O-SuLe^a in the last example).

Comparison of a number of molecules Glyc-PAA-label made especially for selection of the least "noisy" probes demonstrated the radioactive label to be the best followed by fluorescein one, whereas digoxigenin and biotin ones were notably inferior to the first two labels [22]. Nevertheless, it is possible to use the biotinylated probes, but it is better to limit oneself with individual homogenous cells as high background risk becomes increased in case of heterogenous cell populations. Higher noise of biotinylated probes can be explained not only by hydrophobicity of biot residue but by the fact that standard PAA-biot conjugates contain 5 mol% of biot, i.e. five times more than the number of fluo residues in analogous probes [17].

In parallel, it is desirable to use suitable cells that do not express the lectin under study as the negative control. For example, mock-transfected cells were taken as controls during the study of siglecs on siglec-transfected cells (see Section 4.2.1).

Returning to *antibody vs. glycoprobe* selection, it is obvious that a combination of both tools provides complementary information. Thus, the combined use of antibodies and probes in inhibitory test of apoptotic bodies phagocytosis by THP-1 cells allowed identifying a galectin on THP-1 cells and revealing of carbohydrate ligand, which bound galectin, on apoptotic bodies. Due to this combination it was demonstrated that phagocytosis was mediated by galectin-1 or galectin-4 that recognized Galβ1-3GalNAcβ-fragment of the engulfed apoptotic bodies (see Section 4.2.4).

So, what unique tasks can be solved with the help of corresponding glycoprobes besides a simple lectin revealing on cell or profiling of cell carbohydrate-binding properties?

(1) *Specificity of lectin in composition of cell membrane.* Does the specificity of soluble lectin fully coincide with lectin specificity on cell membranes? It was demonstrated in the example of galectins that though specificity patterns towards wide range of saccharides are close, there are some differences in them. Probes $(LN)_3$ and LNnT displayed high affinity to cell galectin-1 and its soluble form [43, 51]. At the same time, cellular galectin-1 did not bind the probe 3'-O-Su-Le^c in contrast to the soluble form, which demonstrated the best binding with this probe [43]. Galα1-3'LacNAc is the most potent ligand for galectin-3 [27, 43].

(2) *Cis-interaction.* Cell lectins can bind glycoconjugates exposed on the same membrane, thus hindering (generally – regulating) their binding with glycoconjugates of another cell [11, 12]. Thus, cellular siglecs are masked with *cis*-ligands and capable of binding with exogenous glycoconjugates or glycans of another cell only after total desialylation of cell surface. However, if an external glycan displays notably higher affinity, it can displace the *cis*-ligand. Indeed, this effect was proved at the example of cellular siglec-2, the binding of which to probe was observed without preliminary desialylation when the probe contained an "improved," higher affinity, ligand than the cell *cis*-glycan [52]. This explains how a lectin characterized by broad specificity can nevertheless provide narrow specificity of intercellular recognition: *cis*-binding abolishes all interactions lower than some affinity threshold but does not hinder highly specific recognition.

(3) *Role of multivalency.* In the above example *cis*-interaction was surmounted due to increased affinity of the external glycan, i.e. the latter had a different structure. The next example demonstrates that the same result can be achieved by increased multivalency of glycoprobe. Thus, siglec-8, an eosinophil receptor, is masked with glycoproteins of the same cells and binds to the standard conjugate 6'-O-Su-SiaLe^x (30 kDa) only after cellular desialylation [53]. However, its interaction with an analogous conjugate having considerably higher MW, 1,000 kDa, which has a greater potential of multilocular interaction, was observed without neuraminidase treatment [54].

(4) *Study of tumor cell lectin pattern.* Changes in exposed carbohydrates and lectins are well known to occur in cancerous cells. Increased expression of lectins, in particular galectins, seems to be the reason of formation of tumor cell resistance to innate and specific antitumor immunity (immunological tolerance) [55]. For example, galectin-1 and -3 induce apoptosis of cytotoxic lymphocytes via binding to T-cell antigens [55, 56], whereas the interaction of these lectins with cell matrix proteins leads to cell aggregates resistant to proapoptotic agents [56]. Galectins interfere with cell–matrix and cell–cell adhesion, thereby promoting uncontrollable cell proliferation and tumor invasiveness [55–57]. Therefore, it appears that lectins are good targets for cancer therapy. In order to investigate this, we have studied the galactoside-binding potency of breast carcinoma cells. Asialo-GM1 bound most strongly to cells, whereas the level of the binding of

other β-galactoside-containing probes was notably lower (see Section 4.2.4). The identity of the lectin revealed with this probe remains unknown, but we believe it could have been galectin-1 or galectin-4. According to the literature, these lectins are overexpressed on breast cancer cells [58, 59]; however, the known carbohydrate binding pattern of these galectins differs from binding profile determined with the help of Glyc-PAA-fluo. Particularly, the binding of LNnT (its potent ligand [43]) with tumor cells was not observed probably due to masking of galectins-1 and -4 by tumor associated glycoconjugates.

In summary, glycoprobes allow not only the discovery of lectins on cells but also studies of lectins' functional activities on cell surfaces, including interaction with *cis*- and *trans*-ligands. Probes of the type Glyc-PAA-label are convenient for CELISA, flow cytometry, and microscopy. At the same time, the accumulated experience of cell studies with the help of probes, which have been initially designed for the work with individual proteins, has resulted in valuable experience and general strategies for the use of these tools as cell probes. One strategy is to employ the probes having principally different MWs. A second strategy is to use less "noisy" labels, i.e. more hydrophilic ones and that produce higher signal than fluorescein.

Acknowledgments

This work was supported in part by the grant of Russian Foundation for Basic Research 04-04-49689 and 07-04-00969, NIH grant 5 U54 GM062116-05, and by the Russian Academy of Sciences Program "Molecular and Cell Biology."

References

[1] H.J. Gabius, Eur. J. Biochem., 243 (1997) 543.
[2] F.-T. Liu, Int. Arch. Allergy Immunol., 136 (2005) 385.
[3] P.R. Crocker, Curr. Opin. Pharmacol., 5 (2005) 431.
[4] A.N. Zelensky and J.E. Gready, FEBS J., 272 (2005) 6179.
[5] J. Almkvist and A. Karlsson, Glycoconj. J., 19 (2004) 575.
[6] S. Gordon, Cell, 111 (2002) 927.
[7] F.T. Liu, Expert. Opin. Ther. Targets, 6 (2002) 461.
[8] S. Seki, Y. Habu, T. Kawamura, K. Takeda, H. Dobashi, T. Ohkawa and H. Hiraide, Immunol. Rev., 174 (2000) 35.
[9] S.A. Linehan, L. Martinez-Pomares and S. Gordon, Microbes Infect., 2 (2000) 279.
[10] K. Lock, J. Zhang, J. Lu, S.H. Lee and P.R. Crocker, Immunobiology, 209 (2004) 199.
[11] P.R. Crocker and A. Varki, Immunology, 103 (2001) 137.
[12] P.R. Crocker, Trends Glycosci. Glycotechnol., 16 (2005) 357.

[13] P.R. Crocker and S. Kelm, Methods for studying the cellular binding properties of lectin-like receptors. In: L.A. Herzenberg, D.M. Weir and C. Blackwell (Eds.), Weir's handbook of experimental immunology), Blackwell Sciences, Cambridge, 1996, pp. 166.1–166.11.

[14] C.F. Brewer, M.C. Miceli and L.G. Baum, Curr. Opin. Struct. Biol., 12 (2002) 616.

[15] N.V. Bovin, Biochem. Soc. Symp., 69 (2002) 143.

[16] N.V. Bovin, Glycoconj. J., 15 (1998) 431.

[17] N.V. Bovin, E. Yu Korchagina, T.V. Zemlyanukhina, N.E. Byramova, O.E. Galanina, A.E. Zemlyakov, A.E. Ivanov, V.P. Zubov and L.V. Mochalova, Glycoconj. J., 10 (1993) 142.

[18] N.N. Bovin, Neoglycoconjugates as probes in glycobiology. In: M.P. Schneider (Ed.), Chemical probes in biology), Kluwer Academic Publishers, Netherlands, 2003, pp. 207–223.

[19] H.-L. Sun, W. Cui, C. Haller and E.L. Chaikof, Chembiochem, 5 (2004) 1593.

[20] Ph.D. Rye and N.V. Bovin, Glycobiology, 7 (1997) 179.

[21] O.E. Galanina, A.V. Feofanov, A.B. Tuzikov, E.M. Rapoport, P.R. Crocker, A.I. Grishin, M. Egret-Charlier, P. Vigny, J. LePendu and N.V. Bovin, Acta A Mol. Biomol. Spectrosc., 57 (2001) 2285.

[22] O.E. Galanina, A.B. Tuzikov, E.M. Rapoport, J. LePendu and N.V. Bovin, Anal. Biochem., 265 (1998) 282.

[23] O. Galanina, F. Hallouin, C. Coupille, N. Bovin and J. LePendu, Int. J. Cancer, 76 (1998) 136.

[24] K. Smetana, J. Homolka, V. Fronkova, Z. Holikova, N.V. Bovin, S. Andre, D.C. Rijken, Fu-T. Liu and H.-J. Gabius, Simultaneous detection of lectins by immuno- and glycocytochemistry. In: A. Kotyk (Ed.), Fluorescence microscopy and fluorescent probes, Espero Publishing, Prague, 1999, pp. 235–242.

[25] A.V. Feofanov, A.I. Grishin, I.A. Kudelina, L.A. Shitova, T.A. Karmakova, R.I. Iakubovskaia, M. Egret-Charlier and P. Vigny, Rus. J. Bioorg. Chem., 25 (1999) 793.

[26] Q. Bao and R.C. Hughes, J. Cell Sci., 108 (1995) 2791.

[27] J. Hirabayashi, T. Hashidate, Y. Arata, N. Nishi, T. Nakamura, M. Hirashima, T. Urashima, T. Oka., M. Futai, W.E.G. Muller, F.Yagi and K.-I. Kasai, Biochem. Biophys. Acta, 1572 (2002) 232.

[28] S.J. Stray, R.D. Cummings and G.M. Air, Glycobiology, 10 (2000) 649.

[29] E.M. Rapoport, G.V. Pazynina, M.A. Sablina., P.R. Crocker and N.V. Bovin, Biochemistry (Moscow), 71 (2006) 496.

[30] A.L. Cornish, S. Freeman, G.Forbes, J. Ni, M. Zhang, M. Cepeda, R. Gentz, M. Augustus, K.C. Carter and P.R. Crocker, Blood, 92 (1998) 2123.

[31] G. Nicoll, J. Ni, D. Liu, P. Klenerman, J. Munday, S. Dubock, M.-G. Mattei and P.R. Crocker, J. Biol. Chem., 274 (1999) 34089.

[32] T. Angata and A. Varki, Glycobiology, 10 (2000) 431.

[33] O. Blixt, B.E. Collins, I.M. van den Nieuwenhof, P.R. Crocker and J.C. Paulson, J. Biol. Chem., 278 (2003) 31007.

[34] W. Nickel, Eur. J. Biochem., 270 (2003) 2109.

[35] H. Lahm, S. Andre, A. Hoeflich, H. Kaltner, H.-Ch. Sieber, B. Sordat., C.-W. von der Lieth, C. Wolf and H.-J. Gabius, Glycoconj. J., 20 (2004) 227.

[36] N. Ikeda, H. Eguchi, S. Nishihara, H. Narimatsu, R. Kannagi, T. Irimura, M. Okta, H. Matsuda, N. Taniguchi and K. Honke, J. Biol. Chem., 276 (2001) 38588.

[37] B. Moser and P. Loetscher, Nat. Immunol., 2 (2001) 123.

[38] M.D. Peterson, R. Jin, S. Hyduk, P. Duchesneau, M.I. Cybulsky and T.K. Waddell, J. Immunol., 15 (2005) 174, 8072.

[39] M.L. Blakely, W.J. van der Werf, M.C. Berndt, A.P. Dalmasso, F.H. Bach and W.W. Hancock, Transplantation, 58 (1994) 1059.

[40] M. Geczy, K.M. Stuhlmeier, D.J. Goodman, C. Ferran and F.H. Bach, Transplantation, 63 (1997) 421.

[41] M. Jakobisiak, W. Lasek and J. Golab, Immunol. Lett., 90 (2003) 103.

[42] E. Rapoport, S. Khaidukov, O. Baidina, V. Bojenko, E. Moiseeva, G. Pasynina, U. Karsten, N. Nifant'ev, J. LePendu and N. Bovin, Eur. J. Cell Biol., 82 (2003) 295.

[43] http://www.functionalglycomics.org.

[44] H.C. Siebert, K. Born, S. Andre, M. Frank, H. Kaltner, C.W. von der Lieth, A.J. Heck, J. Jimenez-Barbero, J. Kopitz and H.J. Gabius, Chem. Eur. J., 12 (2005) 388.

[45] E.V. Moiseeva, E.M. Rapoport, N.V. Bovin, A.I. Miroshnikov, A.V. Chaadaeva., M.S. Krasilshschikova, V.K. Bojenko, C. Bijleveld, J.E. van Dijk and W.D. Otter, Breast Cancer Res. Treat., 91 (2005) 227.

[46] V. Fronkova, Z. Holikova, F.T. Liu, J. Homolja, D.C. Rijken, S. Andre, N.V. Bovin, K. Smetana Jr. and H.J. Gabius, Folia Biol. (Praha), 45 (1999) 157.

[47] J. Plzak, Z. Holikova, B. Dvorankova, K. Smetana Jr., J. Betka, J. Hercogova, S. Saeland, N.V. Bovin and H.J. Gabius, Histochem. J., 34 (2002) 247.

[48] Z. Plzakova, M. Chovanec, K. Jr. Smetana, J. Plzak, J. Stork and S. Saeland, Folia Biol. (Praha), 50 (2004) 71.

[49] E.M. Rapoport, L.V. Mochalova, H.-J. Gabius, J. Romanova and N.V. Bovin, Glycoconj. J., 23 (2006) 115.

[50] E. Rapoport, L. Bretaudeau and J. LePendu, Unpublished data.

[51] S.R. Stowell, M. Dias-Baruffi, L. Penttila, O. Renkonen, A.K. Nyame and R.D. Cummings, Glycobiology, 14 (2004) 157.

[52] S. Han, B.E. Collins, P. Bengtson and J.C. Paulson, Nat. Chem. Biol., 1 (2005) 93.

[53] H. Tateno, P.R. Crocker and J.C. Paulson, Glycobiology, 15 (2005) 1125.

[54] H. Tateno, H. Li, N. Bovin, P.R. Crocker and J.C. Paulson, submitted for publication, (2006).

[55] Fu-T. Liu and G.A. Rabinovich, Nat. Rev. Cancer, 5 (2005) 29.

[56] S. Nakahara, N. Oka and A. Raz, Apoptosis, 10 (2005) 267.

[57] G.A. Rabinovich, Br. J. Cancer, 92 (2005) 1188.

[58] F. van den Brule, S. Califice and V. Castronovo, Glycoconj. J., 19 (2004) 537.

[59] M. Huflejt and H. Leffler, Glycoconj. J., 20 (2004) 247.

417

Appendix 1

Table of Lectin Affinities

Carol L. Nilsson

National High Magnetic Field Laboratory, Tallahassee, FL 32310, USA

Table 1. Lectins described in this volume and selected other lectins used in analytical settings, and their known carbohydrate specificities.

Organism	Lectin Name	Specificity	Reference
Abrus precatorius (jequirity bean)	Abrin	Galα(1,4)Gal, Gal	[1]
Aegopodium podagraria (ground elder)	APP	GalNAc>Lac≫Gal	[2]
Agaricus bisporus (mushroom)	Agaricus bisporus lectin Agaricus bisporus agglutinin	Galβ(1,3)GalNAc	[3]
Agrocybe cylindracea	Fungal galectin	Neu5Acα(2,3)Lac, LacNAc	[4]
Aleuria aurantia (orange peel mushroom)	Fucose-specific lectin	α-Fuc	[5]
Allomyrina dichotoma (Japanese beetle)	Allo A	β-Gal	[6]
Anguilla anguilla (eel)	Eel lectin (fucolectin)	Fuc (blood group fucosylated antigens)	[7]

Table 1. Continued

Organism	Lectin Name	Specificity	Reference
Arachis hypogaea (peanut)	PNA	Galβ(1,3)GalNAc>Lac>β-Gal≫GalNAc	[8]
Artocarpus integrifolia (jackfruit)	Jacalin	Gal, Galβ(1,3)GalNAc	[9]
Bauhinia purpurea alba (camel's foot tree)	BPA	GalNAc	[10]
Boletopsis leucomelas	BLL	Agalacto-*N*-linked glycans	[11]
Bryonia dioica (white bryony)	BDA	GalNAc>Lac>Melibiose	[12]
Canavalia ensiformis (horse bean)	Concanavalin A (con A)	3,6-di-*O*-(α-Man)αMan> Manα(1,3)Man>methyl-α-Man> Man, Glc	[13], Ch. 4
Cancer antennarius (California crab)	CCA	9-*O*-acetyl-Neu5Ac, 4-*O*-acetyl-Neu5Ac	[14]
Caragana arborescens (pea tree)	CAA	GalNAc	[15, 16]
Castanea crenata (Japanese chestnut)	CCA	High mannose>Man, Glc	[17],Ch. 10
Colchicum autumnale (Autumn crocus)	CA	Lac>GalNAc>Gal	[18]
Coprinus cinerea (inky cap fungus)	CGL-I, CGL-II	Galβ(1,3)GalNAc	[19]
Cratylia mollis	Mannose/glucose-specific lectin	Glc/Man	[20]
Crotalus atrox (rattlesnake)	RSL	Gal	[21]
Cycas revoluta (Japanese cycad)	CRLL	High mannose structures>Man, Glc	[22], Ch. 10

Cytisus scoparius (Scotch broom)	CSA	GalNac≫Lac>Gal	[23]
Datura stramonium (thorn apple)	TAA	GlcNAcβ(1,4)GlcNAcβ(1,4)GlcNAc> GlcNAcβ(1,4)GlcNAc>GlcNAc	[24]
Dioclea grandiflora	DGL	3,6-di-*O*-(αMan)αMan> Manα(1,3)Man>methyl-αMan> Man, Glc	[25], Ch. 4
	Anti-H(O) lectin II CSA-II	Gal, Lac	[26]
Dolichos biflorus (horse gram)	DBA	α-GalNAc	[27]
Erythrina corallodendron	ECL	LacNAc>Lac>GalNAc>Gal	[28]
Erythrina cristagalli	ECA	LacNAc>Lac>GalNAc>Gal	[29]
Escherichia coli	CTB	GM1 ganglioside	[30]
	LTB	Blood group A pentasaccharide	[30]
	FimH adhesin	Man	[31]
	GafD adhesin	GlcNAc	[32]
	PapG adhesin	Galα(1,4)Galβ	[33]
Euonymus europaeus (spindle tree)	EEA	Galα(1,3)-[Fucα(1,2)]Galβ(1,3/4)GlcNAc, Fucα(1,2)Galβ(1,3/4)GlcNAc, Galα(1,3)Galβ(1,3/4)GlcNAc	[34]
Galanthus nivalis (snowdrop)	GNA	Manα(1,3)Man	[35]
Geodia cydonium	GCA	Galβ(1,4)GlcNAc>Galβ(1,3) GlcNAc≫Gal	[36], Ch. 7
Glechoma hederacea (ivy)	GHA, glehedin	GalNAcα-1-Ser/Thr≫GalNAc≫Gal	[37]
Glycine max (soybean)	SBA	GalNAcα(1,3)Gal>GalNAc, Gal	[38]
Griffonia simplicifolia	GS-I-A4	Galα(1,3)Gal>Galα(1,4)Gal>GalNAcα (1,3)Gal>α-Gal, α-GalNAc	[39], Ch. 7

Table 1. Continued

Organism	Lectin Name	Specificity	Reference
	GS-I-B4	Galα(1,3)Gal, α-Gal	[39]
	GS-II	GlcNAc	[40]
	GS-IV	Difucosylated structures	[41]
Homarus americanus (California lobster)	Lag-1	Neu5Ac	[42]
	Lag-2	GalNAc	[43]
Hypnea cervicornis (Brazilian red alga)	HCA	GalNAc	[44]
Iris x hollandica (Dutch iris)	A-disaccharide-binding lectin	GalNAcα(1,3)Gal>GalNAcα(1,6)Gal	[45]
Laburnum alpinum (scotch laburnum)	LAA-I	di-*N*-acetylchitobiose	[46]
	LAA-II	Lac	[46]
Laetiporus sulphureus	LSL	LacNAc	[47]
Lens culinaris (lentil)	LcH-A, LcH-B	methyl-α-Man, Man, Glc	[48]
Lotus tetragonolobus (asparagus pea)	Lotus lectin	Fucα(1,6)GlcNAc, Galβ(1,4[Fucα(1,3)] GlcNAc, αFuc	[49, 50]
Lycopersicon esculentum (tomato)	LEA	GlcNAcβ(1,4)GlcNAc	[51]
Maackia amurensis	MAL, MAM	Neu5Acα(2,3)Galβ(1,4)GlcNAc/Glc	[52]
Maclura pomifera (osage orange)	MPA	Galα(1,6)Glc, Galβ(1,3)GalNAc>α-Gal	[53]
Morus nigra (black mulberry)	Morniga G	Galβ(1,3)GalNAc-Ser/Thr> GalNAcα-1-Ser/Thr> GalNAcβ(1,3)Gal>GalNAcβ(1,3)Gal ≫Gal, GalNAc	[54]
Narcissus pseudonarcissus (daffodil)	NPA	α(1,6)Man	[55]

Phaseolus lunatus (lima bean)	LBA	GalNAcα(1,3)[Fucα1,2]Gal> GalNAcα(1,2) Gal	[56]
Phaseolus vulgaris (red kidney bean)	PHA-E	Poly Galβ(1,4)GlcNAc> H>A,B	[57]
	PHA-L	Multivalent Galβ(1,4)GlcNAc> Galβ(1,4)GlcNAc	Ch. 7
Phytolacca americana (pokeweed)	PWM	Oligomers of poly-*N*-acetyllactosamine, GlcNAc	[58]
Pisum sativum (garden pea)	PSA	Man dendrimers>methylα-Man>Man	[59]
Pseudomonas aeruginosa	PA-IL	Gal	[60]
	PA-IL	Le^a>Fuc	[61]
Psophocarpus tetragonolobus (winged bean)	WBA-II	GalNAc, Galβ(1,3)GalNAc	[62]
Rhizopus stolonifer	Alpha-(1-6)-linked fucose-specific lectin	α-(1,6)Fuc	[64]
	Agglutinin isolectin 3 WGA3	GlcNAc, Neu5Ac	[65]
Ricinus communis (castor bean)	RCA-I	Galβ(1,4)GlcNAc> Galβ(1,3) GlcNAc>β-Gal	[63], Ch. 7
	RCA-II	GalNAc, β-Gal	[63]
Sambucus nigra (elderberry)	SNA	Neu5Acα(2,6)Gal, Neu5Acα(2,6)GalNAc	[66]
Solanum tuberosum (potato)	STA	[GlcNAcβ(1,4)]₃GlcNAc> [GlcNAcβ(1,4)]₂GlcNAc> GlcNAcβ(1,4)GlcNAc	[67]
Trichosanthes kirilowii (China gourd)	TKA	Lac>Gal	[68]

Table 1. Continued

Organism	Lectin Name	Specificity	Reference
Trifolium repens (white clover)	RTA	2-deoxy-Glc	[69]
Triticum vulgare (wheat)	Wheat germ agglutinin (WGA)	GlcNAcβ(1,4)GlcNAcβ(1,4)GlcNAc> GlcNAcβ(1,4)GlcNAc>GlcNAc≫ Neu5Ac≫GalNAc	[70]
Ulex europeus (Furze)	UEA-I, UEA-II	Fuc	[71]
Urtica dioica (Stinging nettle)	UDA	GlcNAc	[72]
Vicia faba (fava or broad bean)	VFA, favin	Man>Glc>GlcNac	[73]
Vicia graminea	VGA, anti-N lectin	O-linked Galβ(1,3)GalNAc	[74, 75]
Vicia villosa (hairy vetch)	VVA	GalNAc, GalNAc-α-1-Ser/Thr	[76]
Viscum album (European mistletoe)	ML-I, VAA-I	Galα(1,4)Gal, Gal, Neu5Ac	[77, 78]
	ML-II, VAA-II	Gal, GalNAc	[78]
	ML-II, VAA-III	GalNAc	[78]
	Chitin-binding lectin	Galβ(1,4)GlcNAc, GlcNAc	[79]
Wisteria floribunda	WFA	GalNAc≫ Galβ(1,4)GlcNAc>Gal	[80]
Xenopus laevis (clawed frog)	Cortical granule lectin, CGL	Terminally sulfated, fucosylated oligosaccharides	Ch. 13
Xerocomus spadicius	Inulin-specific lectin	Inulin	[81]

Carbohydrate abbreviations: A: blood group A antigen; B: blood group B antigen; Chitobiose: dimer of β(1,4) linked GlcNAc; Fuc: fucose; Gal: galactose; GalNAc: N-acetyl galactosamine; Glc: glucose; GlcNAc: N-acetyl glucosamine; H: blood group O antigen; Lac: lactose; LacNAc: lactosamine; Le^a: Lewis a antigen; Man: mannose; Melibiose: Galα1-6Glc; Neu5Ac: sialic acid.

References

[1] S. Olsnes, E. Saltvedt and A. Pihl, J. Biol. Chem., 249 (1974) 803–810.

[2] W.J. Peumans, M. Nsima-Lubaki, B. Peeters and W.F. Broekaert, Planta, 164 (1985) 75–82.

[3] C.A. Presant and S. Kornfeld, J. Biol. Chem., 247 (1972) 6937–6945.

[4] M. Ban, H.J. Yoon, E. Demirkan, S. Utsumi, B. Mikami and F. Yagi, J. Mol. Biol., 351 (2005) 695–706.

[5] F. Fukumori, N. Takeuchi, T. Hagiwara, K. Ito, N. Kochibe, A. Kobata and Y. Nagata, FEBS Lett., 250 (1989) 153–156.

[6] K. Umetsu, S. Kosaka and T. Suzuki, J. Biochem. (Tokyo), 95 (1984) 239–245.

[7] A.M. Wu, J.H. Wu, T. Singh, J.H. Liu and A. Herp, Life Sci. 75 (2004) 1085–1103.

[8] R. Banerjee, K. Das, R. Ravishankar, K. Suguna, A. Surolia and M. Vijayan, J. Mol. Biol., 259 (1996) 281–296.

[9] K. Hagiwara, D. Collet-Cassart, K. Kobayashi and J.P. Vaerman, Mol. Immunol., 25 (1988) 69–83.

[10] N.M. Young, D.C. Watson and R.E. Williams, FEBS Lett., 182 (1985) 403–406.

[11] Y. Koyama, T. Suzuki, S. Odani, S. Nakamura, J. Kominami, J. Hirabayashi and M. Isemura, Biosci. Biotechnol. Biochem., 70 (2006) 542–545.

[12] W.J. Peumans, M. Nsima-Lubaki, A.R. Carlier and E. Van Driessche, Planta, 160 (1984) 222–228.

[13] J.W. Becker, G.N. Reeke Jr., J.L. Wang, B.A. Cunningham and G.M. Edelman, J. Biol. Chem., 250 (1975) 1513–1524.

[14] M.H. Ravindranath, H.H. Higa, E.L. Cooper and J.C. Paulson, J. Biol. Chem., 260 (1985) 8850–8856.

[15] R. Bloch, J. Jenkins, J. Roth and M.M. Burger, J. Biol. Chem., 251 (1976) 5929–5935.

[16] O. Makela, Ann. Med. Exp. Biol. Fenn., 35 (1957) 1–133.

[17] K. Nomura, S. Nakamura, M. Fujitake and T. Nakanishi, Biochem. Biophys. Res. Commun., 276 (2000) 23–28.

[18] W.J. Peumans, A.K. Allen and B.P. Cammue, Plant Physiol., 82 (1986) 1036–1039.

[19] R.P. Boulianne, Y. Liu, M. Aebi, B.C. Lu and U. Kuees, Microbiology, 146 (2000) 1841–1853.

[20] M.T.S. Correia and L.C.B.B. Coelho, Appl. Biochem. Biotechnol., 55 (1995) 261–273.

[21] J. Hirabayashi, T. Kusunoki and K. Kasai, J. Biol. Chem., 266 (1991) 2320–2326.

[22] F. Yagi, T. Iwaya, T. Haraguchi and I.J. Goldstein, Eur. J. Biochem., 269 (2002) 4335–4341.

[23] N.M. Young, D.C. Watson and R.E. Williams, Biochem. J., 222 (1984) 41–48.

[24] J.F. Crowley, I.J. Goldstein, J. Arnarp and J. Lonngren, Arch. Biochem. Biophys., 231 (1984) 524–533.

[25] D. Gupta, S. Oscarson, T.S. Raju, P. Stanley, E.J. Toone and C.F. Brewer, Eur. J. Biochem., 242 (1996) 320–326.

[26] Y. Konami, K. Yamamoto and T. Osawa, Biol. Chem. Hoppe-Seyler, 372 (1991) 103–111.

[27] M.E. Etzler and E.A. Kabat, Biochemistry, 9 (1970) 869–877.

[28] A. Surolia, N. Sharon and F.P. Schwarz, J. Biol. Chem. 271 (1996) 17697–17703.

[29] J.L. Iglesias, H. Lis and N. Sharon, Eur. J. Biochem., 123 (1982) 247–252.

[30] S. Teneberg, T.R. Hirst, J. Angstrom and K.A. Karlsson, Glycoconj. J., 11 (1994) 533.

[31] K.A. Krogfelt, H. Bergmans and P. Klemm, Infect. Immun. 58 (1990) 1995–1998.

[32] S. Saarela, S. Taira, E.L. Nurmiaho-Lassila, A. Makkonen and M. Rhen, J. Bacteriol., 177 (1995) 1477–1484.

[33] B. Lund, F. Lindberg, B.-I. Marklund and S. Normark, Proc. Natl. Acad. Sci. U.S.A., 84 (1987) 5898–5902.

[34] J. Petryniak, T.K. Huard, G.D. Nordblom and I.J. Goldstein, Arch. Biochem. Biophys., 244 (1986) 57–66.

[35] N. Shibuya, I.J. Goldstein, E.J. van Damme and W.J. Peumans, J. Biol. Chem., 263 (1988) 728–734.

[36] W.E. Muller, J. Conrad, C. Schroder, R.K. Zahn, B. Kurelec, K. Dreesbach and G. Uhlenbruck, Eur. J. Biochem., 133 (1983) 263–267.

[37] T. Singh, J.H. Wu, W.J. Peumans, P. Rouge, E.J. van Damme, R.A. Alvarez, O. Blixt and A.M. Wu, Biochem. J., 393 (2006) 331–341.

[38] V.S. Rao, K. Lam and P.K. Qasba, J. Biomol. Struct. Dyn., 15 (1998) 853–860.

[39] J. Lescar, R. Loris, E. Mitchell, C. Gautier, V. Chazalet, V. Cox, L. Wyns, S. Perez, C. Breton and A. Imberty, J. Biol. Chem., 277 (2002) 6608–6614.

[40] K. Zhu, J.E. Huesing, R.E. Shade, R.A. Bressan, P.M. Hasegawa and L.L. Murdock, Plant Physiol., 110 (1996) 195–202.

[41] M. Vandonselaar, L.T. Delbaere, U. Spohr and R.U. Lemieux, J. Biol. Chem., 262 (1987) 10848–10849.

[42] J.L. Hall and D.T. Rowlands, Biochemistry, 13 (1974) 821–827.

[43] J.L. Hall and D.T. Rowlands, Biochemistry, 13 (1974) 828–832.

[44] C.S. Nagano, H. Debray, K.S. Nascimento, V.P.T. Pinto, B.S. Cavada, S. Saker-Sampaio, W.R.L. Farias, A.H. Sampaio and J.J. Calvete, Protein Sci., 14 (2005) 2167–2176.

[45] H. Mo, E.J.M. van Damme, W.J. Peumans and I.J. Goldstein, J. Biol. Chem., 269 (1994) 7666–7673.

[46] Y. Konami, K. Yamamoto, T. Tsuji, I. Matsumoto and T. Osawa, Hoppe-Seylers Z Physiol. Chem., 364 (1983) 397–405.

[47] J.M. Mancheno, H. Tateno, I.J. Goldstein, M. Martinez-Ripoll and J.A. Hermoso, J. Biol. Chem., 280 (2005) 17251–17259.

[48] N.M. Young, M.A. Leon, T. Takahashi, I.K. Howard and H.J. Sage, J. Biol. Chem., 246 (1971) 1596–1601.

[49] M.E. Pereira and E.A. Kabat, Biochemistry, 13 (1974) 3184–3192.

[50] J.P. Susz and G. Dawson, J. Neurochem., 32 (1979) 1009–1013.

[51] M.S. Nachbar, J.D. Oppenheim and J.O. Thomas, J. Biol. Chem., 255 (1980) 2056–2061.

[52] R.N. Knibbs, I.J. Goldstein, R.M. Ratcliffe and N. Shibuya, J. Biol. Chem., 266 (1991) 83–88.

[53] J.N. Bausch and R.D. Poretz, Biochemistry, 16 (1977) 5790–5794.

[54] T. Singh, J.H. Wu, W.J. Peumans, P. Rouge, E.J.M. Van Damme and A.M. Wu, Mol. Immunol., 44 (2007) 1451–1462.

[55] H. Kaku, E.J. van Damme, W.J. Peumans and I.J. Goldstein, Arch. Biochem. Biophys., 279 (1990) 298–304.

[56] D.D. Roberts and I.J. Goldstein, J. Biol. Chem., 259 (1984) 903–908.

[57] R. Kornfeld and S. Kornfeld, J. Biol. Chem., 245 (1970) 2536–2545.

[58] T. Irimura and G.L. Nicolson, Carbohydr. Res., 120 (1983) 187–195.

[59] D. Page and R. Roy, Bioconj. Chem., 8 (1997) 714–723.

[60] G. Cioci, E.P. Mitchell, C. Gautier, M. Wimmerova, D. Sudakevitz, S. Perez, N. Gilboa-Garber and A. Imberty, FEBS Lett., 555 (2003) 297.

[61] S. Perret, C. Sabin, C. Dumon, M. Pokorna, C. Gautier, O. Galanina, S. Ilia, N. Bovin, M. Nicaise, M. Desmadril, N. Gilboa-Garber, M. Wimmerova, E.P. Mitchell and A. Imberty, Biochem. J., 389 (2005) 325.

[62] S.R. Patanjali, S.U. Sajjan and A. Surolia, Biochem. J., 252 (1988) 625–631.

[63] D. Ganguly and C. Mukhopadhyay, Biopolymers, 83 (2006) 83–94.

[64] Y. Oda, T. Senaha, Y. Matsuno, K. Nakajima, R. Naka, M. Kinoshita, E. Honda, I. Furuta and K. Kakehi, J. Biol. Chem., 278 (2003) 32439–32447.

[65] K. Harata, H. Nagahora and Y. Jigami, Acta Crystallogr. D, 51 (1995) 1013–1019.

[66] N. Shibuya, I.J. Goldstein, W.F. Broekaert, M. Nsima-Lubaki, B. Peeters and W.J. Peumans, J. Biol. Chem., 262 (1987) 1596–1601.

[67] A.K. Allen and A. Neuberger, Biochem. J., 135 (1973) 307–314.

[68] H.W. Yeung, T.B. Ng, D.M. Wong, C.M. Wong and W.W. Li, Int. J. Pept. Protein Res., 27 (1986) 208–220.

[69] F.B. Dazzo, W.E. Yanke and W.J. Brill, Biochim. Biophys. Acta, 539 (1978) 276–286.

[70] A.K. Allen, A. Neuberger and N. Sharon, Biochem. J., 131 (1973) 155–162.

[71] Y. Konami, K. Yamamoto and T. Osawa, Biol. Chem. Hoppe-Seyler, 372 (1991) 95–102.

[72] M.P. Chapot, W.J. Peumans and A.D. Strosberg, FEBS Lett., 195 (1986) 231–234.

[73] A.K. Allen, N.N. Desai and A. Neuberger, Biochem. J., 155 (1976) 127–135.

[74] D. Blanchard, A. Asseraf, M.J. Prigent, J.J. Moulds, D. Chandanayingyong and J.P. Cartron, Hoppe-Seylers Z Physiol. Chem., 365 (1984) 469–478.

[75] M. Duk, E. Lisowska, M. Kordowicz and K. Wasniowska, Eur. J. Biochem., 123 (1982) 105–112.

[76] S.E. Tollefsen and R. Kornfeld, J. Biol. Chem., 258 (1983) 5165–5171.

[77] A.M. Wu, S.C. Song, P.Y. Hwang, J.H. Wu and U. Pfuller, Biochem. Biophys. Res. Commun., 214 (1995) 396–402.

[78] S. Olsnes, F. Stirpe, K. Sandvig and A. Pihl, J. Biol. Chem., 257 (1982) 13263–13270.

[79] W. Voelter, R. Wacker, M. Franz, T. Maier and S. Stoeva, J. Prakt. Chem., 342 (2000) 812–818.

[80] T. Kurokawa, M. Tsuda and Y. Sugino, J. Biol. Chem., 251 (1976) 5686–5693.

[81] Q. Liu, H. Wang and T.B. Ng, Peptides, 25 (2004) 7–10.

Lectins: Analytical Technologies
C.L. Nilsson (Editor)

Appendix 2

List of Abbreviations

Carol L. Nilsson

National High Magnetic Field Laboratory, Tallahassee, FL 32310, USA

A, blood group A determinant, GalNAcα1 − 3Gal
A_{di}, GalNAcα1-3Galβ
A_{tri}, GalNAcα1-3(Fucα1-2)Galβ
$[A]_0$, initial concentration of analyte
Å, angstrom, 10^{-10} m
2-AA, 2-aminobenzoic acid
AAA, *Anguilla anguilla* agglutinin
2-AB, 2-aminobenzamide
AAL, lectin of *Aleuria aurantia*
ABA, *Agaricus bisporus* agglutinin (ABL)
ABIN, 2,2′-azobisisobutyronitrile
ABL, lectin of *Agaricus bisporus* (ABA)
ACG, lectin of *Agrocybe cylindricea*
ACL, *Amaranthus caudatus* lectin
ACN, acetonitrile
ADP, adenosine diphosphate
AGC, automatic gain control
aGM1, Galβ1-3GalNAcβ1-4Galβ1-4Glc
AGP, α-acid glycoprotein
A_h, GalNAcα1-2 (LFucα1-3)Gal
AMAC, aminoacridone
AOX, alcohol oxidase
ASOR, asialoorosomucoid
B, human blood group B determinant, Galα1 − 3Gal
BabA, blood-group antigen binding adhesin A
biot, biotin
BLL, *Boletopsis leucomelas* lectin
BMCD, biomolecular crystallization databases
BPA, lectin of *Bauhinia purpurea alba*

BSA, bovine serum albumin
B_t, effective ligand content expressed in moles
B_{tri}, Galα1-3(Fucα1-2) Galβ
C, chitin disaccharide, GlcNAcβ1-4GlcNAc, (GlcNAcβ1-4)$_n$, repeat unit of GlcNAcβ1-4
CAD, collision-induced dissociation/decomposition
CE, capillary electrophoresis
ΔC_p, change in heat capacity
CCA, *Castanea crenata* agglutinin
CCUG, culture collection of Göteborg University (Sweden)
CEC, capillary electrochromatography
Chitobiose, GlcNAcβ1-4GlcNAc
CID, collision-induced dissociation
CFU, colony forming unit
CGL, cortical granule lectin from *Xenopus laevis*
CGL2, lectin of *Coprineus cinerea*
CHO, chinese hamster ovary
CMP, cytidine monophosphate
Con A, concanavalin A
COSY, correlation spectroscopy
CRD, carbohydrate recognition domain
CRINEPT, cross relaxation-enhanced polarization transfer
CRLL, *Cycas revoluta* leaf lectin
CTB, cholera toxin subunit
2,4-D, 2,4-dichlorophenoxyacetic acid
Da, Dalton
DADA, *N,N'*-diaminooxyacetic acid amide of 1,3-diaminopropane
DBA, *Dolichos biflorus* agglutinin
DC-SIGN, a C-type lectin present on dendritic cells
1D-GE, one-dimensional gel electrophoresis
2-DE, two-dimensional gel electrophoresis (2D-GE)
DGL, lectin of *Dioclea grandiflora*
DIGE, differential in-gel electrophoresis
2D-LPE, two dimensional liquid phase electrophoresis
DMF, dimethylformamide
DNA, deoxyribonucleotide
DOSY, diffusion ordered spectroscopy
DSA, *Datura stramonium* agglutinin
DSC, *N,N'*-disuccinimidyl carbonate
ECD, electron capture dissociation
EDC, *N*-ethyl-*N'*-(dimethylaminopropyl)carbodiimide
EDD, electron detachment dissociation
EDMA, ethylene dimethacrylate

EDS, electronic density server
EFRS, eosinophilic fungal rhinosinusitis
EHEC, enterohemorrhagic *E. coli*
ELISA, enzyme-linked immunosorbent assay
Endo, endoglycosidase
EOF, electroosmotic flow
ESI-MS, electrospray ionization mass spectrometry
ETD, electron transfer dissociation
ETEC, enterotoxic *E. coli*
F_s, Forssman antigen, GalNAcα1$-$3GalNAc
FAB, fast atom bombardment
FAC, frontal affinity chromatography
Fc, calculated structure factor amplitude
FGF, fibroblast growth factor
FimH, pili-related lectin from uropathogenic *E. coli*
FITC, fluorescein isothiocyanate
FLAG, octapeptide tag (DYKDDDDK) used to purify overexpressed proteins
fluo, fluorescein residue
Fo, observed structure factor amplitude
Fru, fructose
FT-ICR, Fourier transform ion cyclotron resonance
Fuc, D- or L-fucopyranose (fucose)
Fve, lectin of *Flammulina velutipes*
ΔG, change in free energy
GafD, pili-related lectin from enterotoxic *E. coli*
Gal, D-galactopyranose (galactose)
GalNAc, 2-acetamido-2-deoxy-D-galactopyranose (*N*-acetyl galactosamine)
GalT3, α3-galactosyltransferase
GalT4, β4-galactosyltransferase
GFP, green fluorescent protein
Glc, D-glucopyranose (glucose)
GlcNAc, 2-acetamido-2-deoxy-D-glucopyranose (*N*-acetyl glucosamine)
Glyc, carbohydrate residue
Glyc-PAA, polyacrylamide glycoconjugate
GM1, a monosialylated ganglioside
GMA, glycidyl methacrylate
GNT3, β3-*N*-acetylglucosaminyltransferase
GSA, lectin of *Griffonia* (*Bandeiraea*) *simplicifolis*
GSL, *Griffonia simplicifolia* lectin
GST, glutathione S-transferase
ΔH, change in enthalpy
H (type 1), Fucα1-2Galβ1-3GlcNAc
H (type 2), Fucα1-2Galβ1-4GlcNAc

HDA, hexadecylaniline
H_{di}, Fucα1-2Gal
HEWL, hen eggwhite lysozyme
Hex, hexose
HexNAc, *N*-acetylated hexose
HIV, human immunodeficiency virus
HLf, human lactoferrin
HMO, human milk oligosaccharide
HP, hinge-region peptides
HPA, snail (*Helix pomatia*) agglutinin
HPLAC, high-performance lectin affinity chromatography
HPLC, high-performance liquid chromatography
HSQC, heteronuclear single quantum correlation
Hz, hertz
IEC, ion exchange chromatography
IEF, isoelectric focusing
IMAC, immobilized metal affinity chromatography
IRMPD, infrared multiphoton dissociation
ITC, isothermal titration calorimetry
K_a, association constant
K_d, dissociation constant
LAC, lectin affinity chromatography
Lac, Galβ1-4Glc
LacdiNAc, GalNAcβ1-4GlcNAc
LacNAc, Galβ1-4GlcNAc
LBA, lima bean agglutinin
LC, liquid chromatography
LCA, lectin of *Lens culinaris*
LC-MS/MS, liquid chromatography-tandem mass spectrometry
Le^a, Lewis a saccharide, Galβ1-3(Fucα1-4)GlcNAc
Le^c, Lewis c saccharide, Galβ1-3GlcNAc
LN, Galβ1-4GlcNAc
(LN)$_3$, Galβ1-4GlcNAcβ1-3Galβ1-4GlcNAcβ1-3Galβ1-4GlcNAc
LNDFH-1, a branched oligosaccharide from human milk
LNnT, Galβ1-4GlcNAcβ1-3Galβ1-4Glc
LPA, *Limulus polyphemus* agglutinin
LSL, lectin of *Laetiporus sulphureus*
LTA, *Lotus tetragonolobus* agglutinin
LTB, subunit of heat-labile toxin from enterotoxic *E. coli*
MAA, *Maackia amuriensis* agglutinin
MAD, multiple anomalous dispersion
MAETA, [2-(methacryloyloxy)ethyl]trimethylammonium chloride

MALDI-TOF MS, matrix-assisted laser desorption ionization mass spectrometry - time of flight mass spectrometry

Man, D-mannopyranose (mannose)

Melibiose, Galα1-6Glc

MF, mating factor

MIC, minimum inhibitory concentration

MIR, multiple isomorphous replacement

ML-I, mistletoe lectin I

MPD, methylpentane diol

MR, molecular replacement

mRNA, messenger RNA

MS, mass spectrometry

MSI, matrix suppression of ionization

MTX, methotrexate

MUDPIT, multidimensional protein identification technology

MUP, major urinary protein

MVDA, multivariate data analysis

MW, molecular weight

m/z, mass-to-charge ratio

NCAM, neural cell adhesion molecule

NCS, non-crystallographic symmetry

NeuAc, *N*-acetylneuraminic acid (sialic acid)

Neu5Ac, *N*-acetylneuraminic acid (sialic acid)

Neu5Gc, *N*-glycolylneuraminic acid (sialic acid)

NeuNAc, *N*-acetylneuraminic acid (sialic acid)

NHS, *N*-hydroxysuccinimide

NMR, nuclear magnetic resonance

NOE, nuclear Overhauser effect

NOESY, nuclear Overhauser effect spectroscopy

OMP, outer membrane protein

PA, pyridylaminated

PA-IL, galactose-specific lectin of *Pseudomonas aeruginosa*

PA-IIL, fucose-specific lectin of *Pseudomonas aeruginosa*

PapG, pili-related lectin (adhesin) from uropathogenic *E. coli*

PBS, phosphate-buffered saline

PCA, principal component analysis

PDB, Protein Data Bank

PE, phosphatidylethanolamine

PEEK, polyetheretherketone (polyketone)

PEG, polyethylene glycol

PHA, phytohemoagglutinin, lectin of *Phaseolus vulgaris*

PmB, polymixin B

PNA, peanut agglutinin
PNGaseF, protein *N*-glycosidase F
pNP, *p*-nitrophenyl
PO, peroxidase
polySia, polysialic acid
PSA, *Pisum sativum* agglutinin
PTM, post-translational modification
PVDF, polyvinylidine difluoride
PVL, lectin of *Psathyrella velutina*
q, charge
Q-oa-TOF, quadrupole orthogonal acceleration time-of-flight
Raffinose, Galα1-6Glcβ1-2Fru
RCA, lectin of *Ricinus communis*
RDC, residual dipolar coupling
R-factor, reliability factor or residual
ROESY, rotational frame nuclear Overhauser effect spectroscopy
RPLC, reversed-phase liquid chromatography (RPC)
RT, room temperature
RU, resonance unit
SA, Neu5Ac
SabA, sialic acid-binding adhesin A
SAM, self-assembled monolayer
SARS-CoV, severe acute respiratory syndrome corona virus
SAT3/6, α3/6sialyltransferase
SBA, soybean agglutinin
scFv, single chain antibody variable domains
Sda antigen, Neu5Acα2-3[GalNAcβ1-4]Galβ1-4GlcNAc
SDS–PAGE, sodium dodecylsulfate polyacrylamide electrophoresis
Se-Cys, seleneocysteine
Se-Met, selenomethionine
Sia$_2$, Neu5Acα2-8Neu5Acα
SiaLe^a, Neu5Acα2-3Galβ1-3(Fucα1-4)GlcNAc
SID, surface-induced dissociation
SLAC, serial lectin affinity chromatography
SNA, *Sambucus nigra* agglutinin
SORI, sustained off-resonance irradiation
SPR, surface plasmon resonance
SSA, *Salvia sclarea* lectin
Stachyose, Galα1-6Galα1-6Glcβ1-2Fru
STD, saturation transfer distance
Str, streptavidin
T, tesla
TF, Galβ1−3GalNAcα (TF)

TAA, thorn apple (*Datura stramonium*) agglutinin
T_b, Galβ1−3GalNAcβ ($T_{ββ}$)
TFA, trifluoroacetic acid
THGP, Tamm-Horsfall glycoprotein
TLC, thin layer chromatography
Tn, GalNAcα1→Ser/Thr
TOCSY, total correlation spectroscopy
TRIM, trimethylopropane trimethylacrylate
TRIS, tris(hydroxymethyl)aminomethane
*t*RNA, transfer RNA
TR-NMR, transferred NMR
UDP, uridine diphosphate
UEA, *Ulex europaeus* agglutinin
UV, ultraviolet
VNO, vomeronasal organ
$V − V_0$, elution volume
VVL, *Vicia villosa* lectin
WFA, *Wisteria floribunda* agglutinin
WGA, wheat germ agglutinin
XCL, lectin of *Xerocomus chrysenteron*
YadA, *Yersinia enterocolitica* adhesin

Subject Index

Printed and bound by CPI Group (UK) Ltd, Croydon, CR0 4YY

08/05/2025

01864806-0005